Topics in Boundary Element Research

Edited by C. A. Brebbia

Volume 2:
Time-dependent and
Vibration Problems

With 140 Figures and 5 Tables

Springer-Verlag Berlin Heidelberg GmbH

Editor:
Dr. Carlos A. Brebbia

Wessex Institute of Technology
52 Henstead Road
Southampton SO1 2DD
England

Library of Congress Cataloging in Publication Data
(Revised for volume 2)

Main entry under title: Topics in boundary element research. Includes bibliographies and index.
Contents: v. 1. Basic principles and applications – v. 2. Time-dependent and vibration problems.
1. Boundary value problems. 2. Transients (Dynamics) 3. Vibration. I. Brebbia, C. A.
TA347.B69T67   1984   620'.001'51535   84-10644
ISBN 978-3-662-28142-0   ISBN 978-3-662-29651-6 (eBook)
DOI 10.1007/978-3-662-29651-6

© Springer-Verlag Berlin Heidelberg 1985
Originally published by Springer-Verlag Berlin Heidelberg New York Tokyo in 1985
Softcover reprint of the hardcover 1st edition 1985

2061/3020-5 4 3 2 1 0

Contributors

C.A. Brebbia	Wessex Institute of Technology and Computational Mechanics Centre, Southampton, England	(Chaps. 3, 7)
P.T.T. Esperanca	Federal University of Rio de Janeiro, Brazil	(Chap. 6)
S. Kobayashi	Kyoto University, Japan	(Chap. 1)
T. Kuroki	Fukuoka University, Japan	(Chap. 8)
W.J. Mansur	Federal University of Rio de Janeiro, Brazil	(Chap. 4)
D. Nardini	Gradjevinski Institute, Yugoslavia	(Chap. 7)
K. Onishi	Fukuoka University, Japan	(Chap. 8)
R. Shaw	State University of New York at Buffalo, U.S.A.	(Chap. 2)
P. Skerget	University of Maribor, Yugoslavia	(Chap. 3)
S.H. Sphaier	Federal University of Rio de Janeiro, Brazil	(Chap. 6)
M. Tanaka	Shinshu University, Japan	(Chap. 8)
W.L. Wendland	Technical University of Darmstadt, F.R.G.	(Chap. 9)
L. Wrobel	Federal University of Rio de Janeiro, Brazil	(Chap. 6)

Introduction to the Series
"Topics in Boundary Element Research"

The continuing interest in the application of Boundary Element Methods in engineering has generated a series of books and numerous scientific papers, not least those regularly presented at the International Conferences on Boundary Elements which have been held under my direction since 1978. Most recently a new journal, "Engineering Analysis", has been launched which concentrates on new developments in this important area. In spite of all this activity, the need exists for a serial publication in which the most recent advances in the method are documented in a more complete form than is usually the case in papers presented at conferences or scientific gatherings. This unfulfilled need prompted me to launch the present series.

Each volume in this series will comprise chapters describing new applications of the method. The emphases will be on contributions which are self-contained and explain a particular topic in sufficient detail for the analytical engineer or scientist to be able to understand the theory and in due course to write the relevant computer software. All chapters are written by scientists who are actively involved in Boundary Element research, the internationally best known names being balanced with those of new researchers who have recently made significant contributions in this area.

Another objective of the series is to report work for direct application by the practising engineer. Furthermore, I feel that it is important to include sections which discuss the modelling strategies and presentation of results as well as theoretical chapters. The relationship between Boundary Element analysis codes and computer aided design packages will be discussed in subsequent volumes to achieve the right perspective on the application of Boundary Elements. It is all too easy when dealing with these types of analytical techniques to forget that they exist within the framework of the final engineering product.

It is my intention that the series should be open to all those researchers who have made significant contributions to the advancement of the new method. In this regard I shall be happy to receive any suggestions that such members of the scientific community may wish to make, in an effort to produce a publication that is indispensable to all concerned with the advancement of Boundary Elements.

Carlos A. Brebbia
Editor

Preface

This series has been developed in response to the interest shown in boundary elements by scientists and engineers. Whilst Volume 1 was dedicated to basic principles and applications, this book is concerned with the state of the art in the solution of time-dependent problems. Since papers have recently been published on this important topic it is time to produce a work of a more permanent nature.

The volume begins with a chapter on the Fundamentals of Boundary Integral Equation Methods in Elastodynamics. After reviewing the basic equations of elastodynamics, the wave equation and dynamic reciprocal theorems are stated and the direct and indirect boundary element formulations are presented. Eigenvalue problems are discussed together with the case of the Fourier transformations. Several applications illustrate the effectiveness of the technique for engineering.

Chapter 2 examines some of the various boundary integral equation formulations available for elastodynamic problems. In particular the displacement-traction formulation is compared with the displacement-potential case. The special characteristics of the elastodynamics fundamental solutions are discussed in detail and a critical comparison with the elastostatics case is presented. While the chapter is not meant to be a complete review of the work in the field, the original presentation of the problem and the suggestions for further work make an important contribution to the development of the method.

Time-dependent non-linear potential problems are discussed in Chap. 3, where the cases of constant and temperature (or potential) dependent conductivity are solved. The solution of the latter problem using the Kirchhoff transform is illustrated by several representative examples stressing the applicability of the technique for solving practical engineering problems.

Chapter 4 is addressed to the solution of the transient scalar wave equation. The direct boundary element formulation is used throughout for two and three dimensional cases in conjunction with time and space dependent fundamental solutions. The chapter presents a series of new results concentrating on the solution of the problem without having to define internal cells.

The case of transient elastodynamics (Chap. 5) is solved also by using a time-dependent fundamental solution. The chapter is concerned with solving problems with domains extending to infinity as well as closed domains. Three and two dimensional integral representations are devised although most of the ensuing discussions concentrate on the latter case as this is where most of the mathematical complications occur. Several numerical examples are presented to illustrate the accuracy of the technique in comparison with other numerical methods. It is important to point out that the results were obtained integrating on time starting always from the original conditions rather than with the usual scheme which updates the initial conditions at the end of each time step. In this way it is usually possible to avoid carrying out the integrations and all terms need to be defined only on the boundary. This technique is obviously of great importance for domains extending to infinity.

The propagation of surface waves and their interaction with fixed or floating bodies is of great interest to engineers. Chapter 6 studies the problems, starting with the basic aspects of the general three-dimensional problem and its solution using boundary elements. Since general formulations are usually costly to run in a computer due to their complexity, simpler particular approaches applicable to more restricted geometric configurations are discussed in this chapter. A special section studies the propagation of waves which are no longer harmonic. Finally the complete non-linear problem is presented and possible solution schemes discussed in detail.

Chapter 7 introduces a novel and attractive approach for dynamic analysis in solid mechanics. The method permits the formulation of a mass matrix in function of the boundary nodes only. The approach is developed for two and three dimensional bodies and allows the natural frequencies of the system to be found as well as its transient response. For the former case the main advantage of this formulation is that the boundary integrals need to be computed only once as they are frequency independent. Hence the procedure is extremely economic for free vibrations when compared with previous numerical techniques. The elegance of the approach and the accuracy for the numerical results are demonstrated by some of the examples presented at the end of the chapter.

Laminar Viscous Flow and Convective Diffusion problems are discussed in Chap. 8. The authors applied the direct boundary element method using the governing equations for thermal problems based on Boussinesq approximation. New types of boundary conditions on vorticity are described and an upwind boundary element scheme is presented which increases the stability of the computational scheme. The formulation is validated by comparison of boundary element and other numerical results.

The last chapter (Chap. 9) discusses the asymptotic accuracy and convergence of boundary integral solutions using point collocation methods. Although the chapter concentrates on elliptical problems, it is important in practice as it provides the user with guidelines to assess the asymptotic convergence of the method for time harmonic boundary value problems.

The above contributions refer to work that in spite of having been completed recently can already be applied to solve practical engineering problems. The rapid development of the technique is witness to the considerable effort that is presently being put by the international scientific and engineering community into new applications of the method, thereby creating a new and practical tool for engineering analysis.

Carlos A. Brebbia

Contents

Chapter 1

Fundamentals of Boundary Integral Equation Methods in Elastodynamics

by S. Kobayashi

1.1 Introduction

The boundary integral equation method (BIEM) is one of the most effective techniques for elastodynamics, specifically suitable for exterior problems since it is able to manage the infinite domain directly. This is also advantageous over the conventional domain type methods such as the finite element method (FEM) and the finite difference method (FDM). Moreover, BIEM reduces the number of unknowns drastically compared with the domain type techniques.

The use of integral equation formulations in the analysis of initial-boundary value problems has a long history, perhaps surprisingly it traces back over one and a half centuries to Green's essay (1828) [1], in which he derived the so-called Green's formula and expressed a harmonic function by the aid of simple and double layer potentials with densities distributed over the boundary. This may be the origin of the indirect BIEM. Green's work was overlooked for a long time until Kelvin refound the formula and extended it (c. 1850).

In relation to elasticity, Betti [2] derived the celebrated reciprocal theorem in 1872 and Rayleigh generalized it about ten years later. Somigliana [3] obtained so-called Somigliana formula in 1885 and 1886, which is a basis of the direct BIEM. In the same period, Kirchhoff [4] derived an expression called Helmholtz-Kirchhoff integral formula by use of a retarded time potential, which is interpreted as a mathematical expression of Huygen's principle, Baker and Copson [5].

Fredholm [6], among others, studied the integral equation theory for elasticity. In the beginning two decades of this century, the integral expressions were further studied in order to prove the existence of solutions. At the same time potential theory was further developed, Kellogg [7]. Studies of singular integral equations were made a great progress by Tbilisi school of applied mathematicians of USSR. Their works are summarized in the excellent books, by Muskhelishivili [8], Mikhlin [9], and Kupradze [10, 11]. Kupradze extensively discussed integral equation method for elastodynamics by use of potentials.

Some details of classical works done in elastodynamics are contained in Love [12], Morse and Feshbach [13], Kupradze [10, 11], Eringen and Suhubi [14], Achenbach [15], Graff [16], Pao and Mow [17], Ewing, Jardetzky and Press [18], Wheeler and Sternberg [19], among others.

Although the basic integral equation formulations for wave propagation problems have been known for a long time, their application in obtaining numerical solutions of initial-boundary value problems is not very old. Some of the earliest such developments were made in the early 1960's, e.g. Friedman and Shaw [20], Chen and Schweikert [21], and Banaugh and Goldsmith [22] in acoustics or similarly anti-plane elastodynamics, and Banaugh and Goldsmith [23] in the steady-state in-plane elastic wave propagation.

In 1967, Rizzo [24] solved elastostatic problems by directly formulating the boundary integral equation using boundary displacements and tractions based on the Somigliana's formula. This pioneering work is the starting point of the direct boundary integral equation method.

In the field of elastodynamics, Cruse and Rizzo [25] and Cruse [26] derived BIE approach in conjunction with the Laplace transform in order to solve a half-plane wave propagation problem. A modified version of their approach was used later by Manolis and Beskos [27], Manolis [28] and Manolis and Beskos [29] to investigate the scattering of transient elastic waves by a circular opening and tunnel.

The steady-state solution and reconstitution of the transient response by the Fourier transform synthesis was done by Niwa et al. [30, 31, 32] and Kobayashi and Nishimura [33, 34].

Eigenvalue problems are also investigated by Tai and Shaw [35], De Mey [36], Augirre-Ramirez and Wong [37], and Hutchinson [38] in acoustics, Vivoli and Filippi [39] in plate vibration, and Niwa et al. [40, 41, 42] in elastodynamics and plate vibration.

A time domain formulations for elastic wave scattering problems were studied by Friedman and Shaw [20], Cole et al. [43], Mansur and Brebbia [44], for the anti-plane problems and by Niwa et al. [45], Mansur and Brebbia [46] for the general two-dimensional case.

More informations of BIEM in elastodynamics will be found in Shaw [47], Kleinman and Roach [48], Banerjee and Butterfield [49], Brebbia and Walker [50], Dominguez et al. [51], among others.

In this chapter, we restrict ourselves to fundamentals of BIEM for elastodynamics. In Sect. 1.2 elastodynamic problems are briefly defined and wave equations and dynamic reciprocal theorem are stated for the later use. In Sect. 1.3 BIEs in time-space domain are formulated directly by the use of the reciprocal theorem. In Sect. 1.4 the direct and indirect BIEs are formulated in a integral transformed domain. As an extension of the BIE formulation in the Fourier transformed domain, integral equations for the inhomogeneous domain are also derived in Sect. 1.5.

Eigenfrequency problems are discussed in Sect. 1.6. In Sect. 1.7 several inherent problems of BIEM in elastodynamics are discussed. In Sect. 1.8 versatility and effectiveness of BIEM are demonstrated by typical application examples.

1.2 Elastodynamic Problems

1.2.1 Basic Equations

1) Field Equations

In this section, the linear theory of elastodynamics is briefly summarized. For details, readers are advised to refer to Eringen and Suhubi [14], Achenbach [15], Graff [16], Wheeler and Sternberg [19], or Dominguez et al. [51], among others.

In the two- and three-dimensional Euclidian space $\mathbb{R}^n$ $(n = 2, 3)$, basic equations of linear elastodynamics are given as follows:

$$\nabla \cdot \boldsymbol{\sigma} + \varrho \boldsymbol{b} = \varrho \ddot{\boldsymbol{u}} \qquad \text{(Equation of motion)}, \tag{1.1}$$

$$\boldsymbol{\sigma} = \boldsymbol{c} : \boldsymbol{\varepsilon} \qquad \text{(Constitutive equation)}, \tag{1.2}$$

$$\boldsymbol{\varepsilon} = \tfrac{1}{2}(\nabla \boldsymbol{u} + \boldsymbol{u}\nabla) \qquad \text{(Kinematical relation)}, \tag{1.3}$$

specifically, for isotropic solids the constitutive equation is given by

$$\boldsymbol{\sigma} = \lambda \operatorname{tr} \boldsymbol{\varepsilon} \, \boldsymbol{1} + 2\mu \, \boldsymbol{\varepsilon} \equiv \lambda \, \nabla \cdot \boldsymbol{u} \, \boldsymbol{1} + \mu (\nabla \boldsymbol{u} + \boldsymbol{u}\nabla), \tag{1.4}$$

where notations are used; $\boldsymbol{\sigma}$, $\boldsymbol{\varepsilon}$ and $\boldsymbol{c}$ for stress, strain and elasticity tensors, respectively; $\boldsymbol{u}$, $\boldsymbol{b}$ and $\ddot{\boldsymbol{u}}$ for displacement, body force and acceleration vectors; ϱ for mass density, λ and μ for Lame's constants; ∇ and $\boldsymbol{1}$ for the del (nabla) operator and the unit tensor; $\operatorname{tr} \boldsymbol{\varepsilon}$ for the trace of $\boldsymbol{\varepsilon}$; and super dot indicates time differentiation.

In the following, the indicial notation refering to the rectilinear Cartesian coordinate system $(0; i_1 \, i_2 \, i_3)$ will be used interchangeably with direct notation and summation covention rule is implied over the repeated indices. In indicial notation, for example, $\boldsymbol{u} = u_k \, i_k$, $\nabla(\ ) = (\ )_{,k} \, i_k = i_k \, \partial(\ )/\partial x_k$, and $\nabla \cdot \boldsymbol{\sigma}, \boldsymbol{c} : \boldsymbol{\varepsilon}$ and $\nabla \times \boldsymbol{u}$ imply $\sigma_{ij,i}$, $c_{ijkl} \varepsilon_{kl}$ and $e_{ijk} \, \partial_j \, u_k$, where comma indicates space differentiation, and e_{ijk} denotes permutation symbol, i.e. $+1$ (-1) if ijk represent an even (odd) permutation of 1, 2, 3 and 0 otherwise.

By substituting Eqs. (1.3) and (1.2) or (1.4) into Eq. (1.1), we obtain the celebrated field equation of Navier-Cauchy,

$$\nabla \cdot (\boldsymbol{c} : \nabla \boldsymbol{u}) + \varrho \boldsymbol{b} = \varrho \ddot{\boldsymbol{u}}, \tag{1.5}$$

and for isotropic solids

$$(\lambda + \mu) \, \nabla\nabla \cdot \boldsymbol{u} + \mu \nabla^2 \boldsymbol{u} + \varrho \boldsymbol{b} = \varrho \ddot{\boldsymbol{u}}, \tag{1.6}$$

or introducing an operator E

$$E \, \boldsymbol{u} + \varrho \boldsymbol{b} = \varrho \ddot{\boldsymbol{u}},$$
$$E \equiv (\lambda + \mu) \, \nabla\nabla \cdot + \mu \nabla^2 = (\lambda + 2\mu) \, \nabla\nabla \cdot - \mu \nabla \times \nabla \times, \tag{1.7}$$

where

$$\nabla^2 \boldsymbol{u} = \nabla\nabla \cdot \boldsymbol{u} - \nabla \times \nabla \times \boldsymbol{u}, \tag{1.8}$$

is used. The traction vector for isotropic solid is defined by

$$t = \boldsymbol{n} \boldsymbol{\sigma} \equiv \overset{n}{T} \boldsymbol{u} = \lambda \, \boldsymbol{n} \, \nabla \cdot \boldsymbol{u} + \mu \boldsymbol{n} \cdot (\nabla \boldsymbol{u} + \boldsymbol{u}\nabla)$$
$$= \lambda \, \boldsymbol{n} \nabla \cdot \boldsymbol{u} + 2\mu (\boldsymbol{n} \cdot \nabla) \, \boldsymbol{1} \boldsymbol{u} + \mu \boldsymbol{n} \times \nabla \times \boldsymbol{u}. \tag{1.9}$$

2) Wave Equations

Taking the divergence and curl of Eq. (1.6) and making use of Eq. (1.8), we have

$$c_L^2 \nabla^2 I_e + \nabla \cdot \boldsymbol{b} = \ddot{I}_e, \tag{1.10}$$

$$c_T^2 \nabla^2 \boldsymbol{w} + \nabla \times \boldsymbol{b} = \ddot{\boldsymbol{w}}, \tag{1.11}$$

where I_e and $\boldsymbol{w}$ are the dilatation and rotation defined by $\qquad$ (1.12)

$$I_e = \nabla \cdot \boldsymbol{u}, \quad \boldsymbol{w} = \nabla \times \boldsymbol{u}$$

and

$$c_L = \sqrt{\frac{\lambda + 2\mu}{\varrho}}, \quad c_T = \sqrt{\frac{\mu}{\varrho}} \tag{1.13}$$

are phase speeds of dilatational (longitudinal, irrotational) and distorsional (transverse, equivoluminal, shear) waves, respectively.

If we make use of the Stokes-Helmholtz decomposition of the displacement vector

$$\boldsymbol{u} = \boldsymbol{u}^L + \boldsymbol{u}^T, \quad \boldsymbol{u}^L = \nabla \phi, \quad \boldsymbol{u}^T = \nabla \times \boldsymbol{\psi} \tag{1.14}$$

and body force vector

$$\boldsymbol{b} = \nabla f + \nabla \times \boldsymbol{F}, \tag{1.15}$$

the Navier-Cauchy Eq. (1.6) becomes

$$\nabla[(\lambda + 2\mu) \nabla^2 \phi + \varrho f - \varrho \ddot{\phi}] + \nabla \times [\mu \nabla^2 \boldsymbol{\psi} + \varrho \boldsymbol{F} - \varrho \ddot{\boldsymbol{\psi}}] = \boldsymbol{0}. \tag{1.16}$$

Equation (1.16) is satisfied if ϕ and $\boldsymbol{\psi}$ are solutions of

$$c_L^2 \nabla^2 \phi + f = \ddot{\phi}, \quad c_T^2 \nabla^2 \boldsymbol{\psi} + \boldsymbol{F} = \ddot{\boldsymbol{\psi}}, \tag{1.17}$$

where ϕ and $\boldsymbol{\psi}$ are called Lamé's potentials. Such solutions are proved complete in the sense that every solution of Navier-Cauchy equation admits representation (1.14), which is obtained from Eq. (1.16), Sternberg and Gurtin [52].

It is noted that Eq. (1.16) is obviously satisfied by $\boldsymbol{u}^L$ and $\boldsymbol{u}^T$ obtained from

$$c_L^2 \nabla^2 \boldsymbol{u}^L + \nabla f = \ddot{\boldsymbol{u}}^L, \quad c_T^2 \nabla^2 \boldsymbol{u}^T + \nabla \times \boldsymbol{F} = \ddot{\boldsymbol{u}}^T. \tag{1.18}$$

Since

$$\nabla \times \boldsymbol{u}^L = \boldsymbol{0}, \quad \nabla \cdot \boldsymbol{u}^T = 0,$$

$\boldsymbol{u}^L$ and $\boldsymbol{u}^T$ represent the irrotational and equivoluminal parts of displacement field $\boldsymbol{u}$.

3) Two-Dimensional Case

In the two-dimensional case, in which the displacement vector and body force vector are assumed as functions of (x_1, x_2) and t only, i.e.

$$\boldsymbol{u}(\boldsymbol{x}, t) = u_\alpha(x_1, x_2, t) \, \boldsymbol{i}_\alpha, \quad \boldsymbol{b}(\boldsymbol{x}, t) = b_\alpha(x_1, x_2, t) \, \boldsymbol{i}_\alpha, \quad (\alpha = 1, 2, 3).$$

Navier-Cauchy equation takes the following form:

$$(\lambda + \mu) u_{\beta, \beta\alpha} + \mu u_{\alpha, \beta\beta} + \varrho b_\alpha = \varrho \ddot{u}_\alpha, \quad (\alpha, \beta = 1, 2) \tag{1.19}$$

$$\mu u_{3, \alpha\alpha} + \varrho b_3 = \varrho \ddot{u}_3. \tag{1.20}$$

In a similar way, the stress-displacement relations are expressed as

$$\left.\begin{array}{l} \sigma_{\alpha\beta} = \lambda u_{\gamma,\gamma}\,\delta_{\alpha\beta} + \mu\,(u_{\alpha,\beta} + u_{\beta,\alpha}), \\ \sigma_{33} = \lambda u_{\alpha,\alpha}, \end{array}\right\} \quad (\alpha,\beta\,\gamma = 1,2) \tag{1.21}$$

$$\sigma_{\alpha3} = \mu u_{3,\alpha}, \tag{1.22}$$

where $\delta_{\alpha\beta}$ denotes Kronecker's delta.

From these Equations, we immediately realize that the plane motion $u_{\alpha}(x,t)$ and anti-plane motion $u_3(x,t)$ are uncoupled. Therefore, such two-dimensional motions can be analysed independently, provided that the boundary and initial conditions are given similarly.

Using the Lamé's potentials $\phi(x_1,x_2,t)$ and $\psi(x_1,x_2,t)$ and decomposing displacement and body force as

$$u(x_1,x_2,t) = \nabla\phi + \nabla\times\psi\,i_3 + u_3\,i_3,$$

$$b(x_1,x_2,t) = \nabla f + \nabla\times F\,i_3 + F_3\,i_3,$$

Eqs. (1.19) and (1.20) can be satisfied by ϕ, ψ and u_3,

$$c_L^2\,\nabla^2\phi + f = \ddot{\phi}, \quad c_T^2\,\nabla^2\psi + F = \ddot{\psi}, \tag{1.23}$$

$$c_T^2\,\nabla^2 u_3 + F_3 = \ddot{u}_3. \tag{1.2.4}$$

Displacements in plane problems can be determined from ϕ and ψ, the solutions of the two scalar wave equations, whereas in anti-plane problems only one scalar equation is needed to solve.

1.2.2 Elastodynamic Problems

1) Statement of the Problem

Let D be an interior or exterior domain in an Euclidian space $\mathbb{R}^n$ $(n = 2,3)$ bounded by ∂D and D_c be complement of D. The unit outward normal vector n is defined at a position x on ∂D. Also let T^+ be $[0, \infty^+)$ in time.

The elastodynamic initial-boundary value problems for isotropic solids may be stated as follows:

"To find a solution $u(x,t)$ of Eq. (1.6) or (1.7) in $(x,t) \in D \times T^+$ subject to the initial conditions

$$u(x,0) = u_0(x), \quad \dot{u}(x,0) = v_0(x) \quad \text{in } x \in D \tag{1.25}$$

and boundary conditions

$$u(x,t) = \bar{u}(x,t) \quad \text{on } \partial D_1 \times T^+,$$

$$t(x,t) = n\sigma = \overset{n}{T}u(x,t) \equiv \lambda\,n\,\nabla\cdot u + \mu\,n\cdot(\nabla u + u\nabla) \tag{1.26}$$

$$= \bar{t}(x,t) \quad \text{on } \partial D_2 \times T^+, \quad \partial D_1 \cup \partial D_2 = \partial D$$

where $u_0(x)$, $v_0(x)$, $\bar{u}(x,t)$, and $\bar{t}(x,t)$ denote the prescribed initial displacement, initial velocity, boundary displacement, and boundary traction."

In two-dimensional case, Eqs. (1.6) or (1.7) must be replaced by Eq. (1.19) and (1.20).

More specifically, problems are classified to the first (displacement), the second (traction) and the third (mixed) boundary value problems, according to boundary conditions prescribed on $\partial D_1 = \partial D$, $\partial D_2 = \partial D$, and $\partial D_1 \cup \partial D_2 = \partial D$, respectively.

2) Radiation Conditions

For the exterior problems, physical reasons demand that there exists no wave propagating from infinity back into the medium. The conditions for it are given as follows, Sommerfeld [53];

$$\frac{\partial u^L}{\partial r} + \frac{1}{c_L} \frac{\partial u^L}{\partial t} = o\left(r^{-\frac{n-1}{2}}\right), \quad u^L = o\left(r^{-\frac{n-3}{2}}\right),$$

$$\frac{\partial u^T}{\partial r} + \frac{1}{c_T} \frac{\partial u^T}{\partial t} = o\left(r^{-\frac{n-1}{2}}\right), \quad u^T = o\left(r^{-\frac{n-3}{2}}\right) \tag{1.27}$$

as $r = |x| \to \infty$. In the above expression, $f(x) = o[g(x)]$ implies that $f(x)/g(x)$ vanishes when x approaches to a certain limit, that is "o" means that $f(x)$ is of higher order than $g(x)$.

The above conditions are often called "radiation conditions". Roughly speaking, displacement field due to a source in an unbounded domain must be of the type of function $f(r - ct)$ but not $f(r + ct)$ as $r \to \infty$.

The radiation conditions are crucial in determining the unique solution in unbounded regions, particularly in time-harmonic problems.

1.2.3 Reciprocal Theorem

1) Elastodynamic State

We here define an elastodynamic state on $D \times T$, $T = (-\infty, \infty)$. The ordered pairs on $\mathscr{S} = [u, t]$ is called an elastodynamic state corresponding to a body force ϱb, mass density ϱ, longitudinal and transverse wave speeds c_L and c_T, provided that

(i) $u \in C^{2,2}(D \times T)$, $u \in C^{1,1}(\partial D \times T)$, $t \in C^{0,0}(D \times T)$,
   $b \in C^{0,0}(D \times T)$, $\varrho > 0$, $c_L > (2/\sqrt{3}) c_T > 0$;

(ii) u, t, b, ϱ, c_L, and c_T satisfy Eqs. (1.1) and (1.4) with Lamé's constants replaced by the use of Eq. (1.13)

$$\sigma_{ij,j} + \varrho b_i = \varrho \ddot{u}_i,$$

$$\sigma_{ij} = \varrho(c_L^2 - 2 c_T^2) u_{k,k} \delta_{ij} + \varrho c_T^2 (u_{i,j} + u_{j,i}). \tag{1.14'}$$

We denote the class of all elastodynamic state by $\mathscr{S}$ and write

$$\mathscr{S} \in \mathscr{E}(b, \varrho, c_L, c_T; D \times T).$$

If $T^- = (-\infty, 0)$ and $u = 0$ on $D \times T^-$, we call it the elastodynamic state with a quiescent past.

Uniqueness of the elastodynamic state on $D \times T^+$ is shown for appropriate initial and boundary conditions, Eringen and Suhubi [14], and Wheeler and Sternberg [19].

2) Dynamic Reciprocal Theorem

The dynamic reciprocal theorem, which is as extension of the classical reciprocal theorem of Betti-Rayleigh in elastostatics, is expressed as follows, Love [12], and Wheeler and Sternberg [19].

For two distinct elastodynamic states

$$\mathscr{S} = [u, t] \in \mathscr{E}(b, \varrho, c_L, c_T; D \times T^+),$$

$$\mathscr{S}' = [u', t'] \in \mathscr{E}'(b', \varrho, c_L, c_T; D \times T^+)$$

defined on the same regular region D (in the sense of Kellog [7]) with initial conditions

$$u(x, 0) = u_0(x), \quad \dot{u}(x, 0) = v_0(x),$$

$$u'(x, 0) = u_0'(x), \quad \dot{u}'(x, 0) = v_0'(x) \quad \text{in } D,$$

we have

$$\int_{\partial D} t * u' \, ds + \int_D \varrho \{b * u' + v_0 \cdot u' + u_0 \cdot \dot{u}'\} \, dv$$

$$= \int_{\partial D} t' * u \, ds + \int_D \varrho \{b' * u + v_0' \cdot u + u_0' \cdot \dot{u}\} \, dv \qquad (1.28)$$

where t and t' are tractions on ∂D_2 of states $\mathscr{S}$ and $\mathscr{S}'$ respectively, and $\phi * \psi$ is convolution integral defined by

$$[\phi * \psi](x, t) = \begin{cases} \displaystyle\int_0^t \phi(x, t - \tau) \, \psi(x, \tau) \, d\tau & (x, t) \in D \times T^+ \\ 0 & (x, t) \in D \times T^- \end{cases} \qquad (1.29)$$

3) Green's Identity

Replacing ϱb and $\varrho b'$ in the reciprocal theorem by using Eq. (1.7), i.e.

$$\varrho b = \varrho \ddot{u} - E u, \quad \varrho b' = \varrho \ddot{u}' - E u'$$

and also t, t' by $\overset{n}{T} u$ and $\overset{n}{T} u'$, we have

$$\int_D \{(E u - \varrho \ddot{u}) * u' - (E u' - \varrho \ddot{u}') * u\} \, dv$$

$$= \int_{\partial D} \{\overset{n}{T} u * u' - \overset{n}{T} u' * u\} \, ds + \int_D \varrho \{v_0 \cdot u' + u_0 \cdot \dot{u}' - v_0' \cdot u - u_0' \cdot \dot{u}\} \, dv. \qquad (1.30)$$

This may be an extension of Green's identity.

If the field is independent of time, the well-known Green's identity for self-adjoint operators recovers, i.e.

$$\int_D (E u \cdot u' - E u' \cdot u) \, ds = \int_{\partial D} (\overset{n}{T} u \cdot u' - \overset{n}{T} u' \cdot u) \, dv.$$

This identity is easily derived by integration by parts.

1.3 BIE Formulations in Time-Space Domain

1.3.1 Integral Representation and Fundamental Solutions

1) Integral Representation

In the reciprocal relation (1.28), choosing one state $\mathscr{S}$ as the actual and the another state $\mathscr{S}''$ for the state corresponding to a unit impulse at y in the direction of x_k-axis in the infinite region, i.e. for fixed k,

$$\varrho b' = \delta(t)\,\delta(x - y)\,i_k, \quad \varrho b'_i = \delta(t)\,\delta(x - y)\,\delta_{ik}$$

and zero initial conditions, i.e.

$$u'_0 = v'_0 = 0,$$

we immediately have the following representation;

$$\int_{\partial D} [t_i * u'_i - t'_i * u_i]\, ds(x) + \int_D \varrho b_i * u'_i\, dv(x) + \int_D \varrho(v_{0i}\, u'_i + u_{0i}\, \dot{u}'_i)\, dv(x)$$

$$= \begin{cases} u_k(y, t), & y \in D \\ 0, & y \in D_c, \end{cases} \tag{1.31}$$

where the property of Dirac's delta function is used, i.e.

$$\int_D \delta(t)\,\delta(x - y)\,\delta_{ik} * u_i\, dv(x)$$

$$= \int_D \int_0^t \delta(t - \tau)\,\delta(x - y)\,\delta_{ik}\, u_i(x, \tau)\, d\tau\, dv(x) = \begin{cases} u_k(y, t), & y \in D \\ 0, & y \in D_c. \end{cases}$$

If we here introduce new tensor functions, so-called the fundamental solutions, U and T defined as for fixed k at y

$$u'_i(x) = U_{ik}(x, t; y, 0) \quad \text{in } D \tag{1.32}$$

$$\sigma'_{ij}(x) = T_{ijk}(x, t; y, 0) = \{\lambda\, U_{lk,l}\,\delta_{ij} + \mu(U_{ik,j} + U_{jk,i})\}$$

$$= \{\varrho(c_L^2 - 2c_T^2)\, U_{lk,l}\,\delta_{ij} + \varrho\, c_T^2(U_{ik,j} + U_{jk,i})\} \tag{1.33}$$

and

$$t'_i = \sigma'_{ij}\, n_j = T_{ijk}\, n_j = n_j\{\lambda\, U_{lk,l}\,\delta_{ij} + \mu(U_{ik,j} + U_{jk,i})\} \equiv T_{ik} \quad \text{on } \partial D, \tag{1.34}$$

then Eq. (1.31) is reduced to

$$\varepsilon(y)\, u_k(y, t) = \int_{\partial D} [U_{ik} * t_i - T_{ik} * u_i]\, ds(x)$$

$$+ \int_D \varrho U_{ik} * b_i\, dv(x) + \int_D \varrho(v_{0i}\, U_{ik} + u_{0i}\, \dot{U}_{ik})\, dv(x) \tag{1.35}$$

or simply using $T = \overset{n}{T} U$,

$$\varepsilon(y)\, u(y, t) = \int_{\partial D} [U * t - T * u]\, ds(x)$$

$$+ \int_D \varrho U * b\, dv(x) + \int_D \varrho(v_0 \cdot U + u_0 \cdot \dot{U})\, dv(x) \tag{1.35}'$$

where

$$\varepsilon(y) = \begin{cases} 1, & y \in D \\ 0, & y \in D_c \end{cases}.$$

This expression is often called the integral representation, which is an extension of the Helmholtz-Kirchhoff formula for wave equation.

2) Fundamental Solutions

The fundamental solution or sometimes called free space Green's function is defined as a solution of the equation

$$(\lambda + \mu) U_{jk,ij} + \mu U_{ik,jj} - \varrho \ddot{U}_{ik} = - \delta(t - \tau) \delta(x - y) \delta_{ik} \qquad (1.36)$$

or

$$EU - \varrho \ddot{U} = - \delta I.$$

The meaning of U_{ik} is apparent. It gives the displacement component in the x_i-direction at the point x (field point) at t due to a concentrated unit impulsive force acting at the point y (source point) in the x_k-direction and at $t = 0$.
The explicit expression of U_{ij} is given for $\tau = 0$ as follows;

$$U_{ij}(x, t; y, 0) = \frac{1}{4\pi\varrho} \left\{ \frac{t}{r^2} \left(\frac{3 r_i r_j}{r^3} - \frac{\delta_{ij}}{r} \right) \left[H\left(t - \frac{r}{c_L}\right) - H\left(t - \frac{r}{c_T}\right) \right] \right.$$

$$+ \frac{r_i r_j}{r^3} \left[\frac{1}{c_L^2} \delta\left(t - \frac{r}{c_L}\right) - \frac{1}{c_T^2} \delta\left(t - \frac{r}{c_T}\right) \right] + \left. \frac{\delta_{ij}}{r c_T^2} \delta\left(t - \frac{r}{c_T}\right) \right\}$$

$$= \frac{t}{4\pi\varrho} \left\{ \left(\frac{1}{r}\right)_{,ij} \left[H\left(t - \frac{r}{c_L}\right) - H\left(t - \frac{r}{c_T}\right) \right] \right.$$

$$- r_{,i} \left(\frac{1}{r}\right)_{,j} \left[\frac{1}{c_L} \delta\left(t - \frac{r}{c_L}\right) - \frac{1}{c_T} \delta\left(t - \frac{r}{c_T}\right) \right] + \left. \frac{\delta_{ij}}{r^2 c_T} \delta\left(t - \frac{r}{c_T}\right) \right\}$$

$$= \frac{t}{4\pi\varrho} \left\{ \left(\frac{1}{r}\right)_{,ij} \left[H\left(t - \frac{r}{c_L}\right) - H\left(t - \frac{r}{c_T}\right) \right] \right. \qquad (1.37)$$

$$- r_{,i} \left(\frac{1}{r}\right)_{,j} \left[\frac{1}{c_L} \delta\left(t - \frac{r}{c_L}\right) - \frac{1}{c_T} \delta\left(t - \frac{r}{c_T}\right) \right] + \left. \frac{\delta_{ij}}{r^2 c_T} \delta\left(t - \frac{r}{c_T}\right) \right\}$$

where

$$r = |x - y|, \quad ()_{,i} = \frac{\partial()}{\partial x_i} = -\frac{\partial()}{\partial y_i} = -()_{,i}, \quad ()_{,ij} = \frac{\partial^2()}{\partial x_i \partial x_j} = \frac{\partial^2()}{\partial y_i \partial y_j} = ()_{,ij}.$$

In the expression (1.37), terms containing c_L and c_T obviously correspond to the longitudinal and shear waves, respectively.
The tensor-valued Green's function depends on t and τ only via the combination $t - \tau$ as seen in Eq. (1.36). Hence

$$U_{ik}(x, t; y, \tau) = U_{ik}(x, t - \tau; y, 0) = U_{ik}(x, -\tau; y, -t) \qquad (1.38)$$

which is a reciprocal relation for source and observer times, as being proved by the use of the reciprocal theorem and the "causality" requirement that U and $\partial U/\partial t = 0$ everywhere for $t < \tau$.

Green's function also satisfies reciprocity relations and translation identity for time and space, i.e.

$$U_{ik}(y_2, t + \tau_2; y_1, \tau_1) = U_{ki}(y_1, t - \tau_1; y_2, -\tau_2)$$

$$U_{ik}(x, t; y; \tau) = U_{ik}(x - y, t; 0, \tau)$$

$$U_{ik}(x, t; y, \tau) = U_{ik}(x, t + t_0; y, \tau + t_0). \tag{1.39}$$

The stress tensor σ_{ij} is obtained from Eq. (1.33), for fixed k. T_{ijk} is given as follows:

$$T_{ijk}(x, t; y, 0) = \frac{1}{4\pi} \left\{ -6c_T^2 \left[5 \frac{r_i r_j r_k}{r^5} - \frac{\delta_{ij} r_k + \delta_{ik} r_j + \delta_{jk} r_i}{r^3} \right] \right.$$

$$\cdot \frac{t}{r^2} \left[H\left(t - \frac{r}{c_L}\right) - H\left(t - \frac{r}{c_T}\right) \right]$$

$$+ 2 \left[6 \frac{r_i r_j r_k}{r^5} - \frac{\delta_{ij} r_k + \delta_{ik} r_j + \delta_{jk} r_i}{r^3} \right] \left[\delta\left(t - \frac{r}{c_T}\right) - \frac{c_T^2}{c_L^2} \delta\left(t - \frac{r}{c_L}\right) \right]$$

$$+ 2 \frac{r_i r_j r_k}{r^4 c_T} \left[\dot{\delta}\left(t - \frac{r}{c_T}\right) - \frac{c_T^3}{c_L^3} \dot{\delta}\left(t - \frac{r}{c_L}\right) \right]$$

$$- \frac{r_k \delta_{ij}}{r^3} \left(1 - 2\frac{c_T^2}{c_L^2}\right) \left[\delta\left(t - \frac{r}{c_L}\right) + \frac{r}{c_L} \dot{\delta}\left(t - \frac{r}{c_L}\right) \right]$$

$$- \frac{\delta_{ik} r_j + \delta_{jk} r_i}{r^3} \left[\delta\left(t - \frac{r}{c_T}\right) + \frac{r}{c_T} \dot{\delta}\left(t - \frac{r}{c_T}\right) \right] \right\}. \tag{1.40}$$

The convolution integrals in Eq. (1.35) can be easily evaluated, when they are necessary.

Displacement arising from a given initial displacement and velocity, that is, the last integral of Eq. (1.35), can be given by a more explicit form, Love [12].

Differentiating Eq. (1.35) with respect to y_l and substituting into (1.4), we have stress components

$$\sigma_{kl}(y, t) = \int_{\partial D} [\bar{T}_{ikl} * u_i - \bar{U}_{ikl} * t_i] ds(x) - \varrho \int_D \bar{U}_{ikl} * b_i \, dv(x)$$

$$- \varrho \int_{\partial D} [v_{0i} \bar{U}_{ikl} + u_{0i} \dot{\bar{U}}_{ikl}] \, dv(x) \tag{1.41}$$

where $\bar{U}_{ikl}$ and $\bar{T}_{ikl}$ are the displacement in D and on ∂D and the traction on ∂D of the elastodynamic state

$$\bar{\mathcal{T}}_{kl} = \varrho(c_L^2 - 2c_T^2) \mathring{\mathcal{S}}_{mm} \delta_{kl} + \varrho c_T^2 (\mathring{\mathcal{S}}_{kl} + \mathring{\mathcal{S}}_{lk}). \tag{1.42}$$

We have, of course,

$$\{\bar{U}_{ikl}, \bar{T}_{ikl}\} = \varrho(c_L^2 - 2c_T^2)\{U_{imm}, T_{imm}\}\,\delta_{kl} + \varrho c_T^2\{U_{ikl} + U_{ilk}, T_{ikl} + T_{ilk}\},$$

$$U_{ikl}(x, t; y, 0) = \frac{\partial U_{ik}}{\partial x_l} = -\frac{\partial U_{ik}}{\partial y_l},$$

$$T_{ikl} = T_{ijkl}\,n_j = \frac{\partial T_{ijk}}{\partial x_l}\,n_j = -\frac{\partial T_{ijk}}{\partial y_l}\,n_j.$$

1.3.2 Two-Dimensional Representation

We consider an elastodynamic state defined on $D \times T^+$, where D is a regular plane region bounded by a closed curve C. With prescribed initial conditions

$$u(x, 0) = u_0(x), \quad \dot{u}(x, 0) = v_0(x)$$

for every $(x_\alpha, t) \in D \times T^+$ $(\alpha = 1, 2)$, the integral representation (1.35) is reduced to

$$\varepsilon(y)\,u_k(y, t) = \int_C [V_{ik} * t_i - W_{ik} * u_i]\,dc(x)$$

$$+ \varrho \int_D V_{ik} * b_i\,ds(x) + \varrho \int_D [v_{0i} V_{ik} + u_{0i} \dot{V}_{ik}]\,ds(x), \qquad (1.43)$$

where the functional operators are defined by integrating over the x_3-coordinate

$$V_{ik}(x_\alpha, t; y_\beta, 0) \equiv \int_{-\infty}^{\infty} U_{ik}(x, t; y_\beta, 0)\,dx_3 \quad \text{in } D \cup C,$$

$$W_{ik}(x_\alpha, t; y_\beta, 0) \equiv \int_{-\infty}^{\infty} T_{ik}(x, t; y_\beta, 0)\,dx_3 \quad \text{on } C. \qquad (1.44)$$

These operators may be expressed explicitly by taking $r^2 + x_3^2$, $r^2 = (x_\alpha - y_\alpha)(x_\alpha - y_\alpha)$ in place of r^2 in $\mathbf{R}^3$, as follows;

$$V_{3\alpha} = 0,$$

$$V_{33} = \frac{H\left(t - \dfrac{r}{c_T}\right)}{2\pi\varrho c_T^2 \sqrt{t^2 - \dfrac{r^2}{c_T^2}}}, \qquad (1.45)$$

$$V_{\alpha\beta} = \frac{1}{2\pi\varrho} \left\{ \left(\frac{\left[2t^2 - \dfrac{r^2}{c_L^2}\right] H\left(t - \dfrac{r}{c_L}\right)}{\sqrt{t^2 - \dfrac{r^2}{c_L^2}}} - \frac{\left[2t^2 - \dfrac{r^2}{c_T^2}\right] H\left(t - \dfrac{r}{c_T}\right)}{\sqrt{t^2 - \dfrac{r^2}{c_T^2}}} \right) \frac{r_\alpha r_\beta}{r^4} \right.$$

$$\left. - \left[H\left(t - \dfrac{r}{c_L}\right)\sqrt{t^2 - \dfrac{r^2}{c_L^2}} - H\left(t - \dfrac{r}{c_T}\right)\sqrt{t^2 - \dfrac{r^2}{c_T^2}} \right] \frac{\delta_{\alpha\beta}}{r^2} + \frac{H\left(t - \dfrac{r}{c_T}\right)}{c_T^2 \sqrt{t^2 - \dfrac{r^2}{c_T^2}}}\,\delta_{\alpha\beta} \right\}. \qquad (1.46)$$

$W_{\alpha\beta}$ can be obtained similarly.

Using these fundamental solutions, we have

$$\varepsilon(y)\, u_\alpha(y, t) = \int_C [V_{\beta\alpha} * t_\beta - W_{\beta\alpha} * u_\beta]\, dc(x)$$

$$+ \varrho \int_D V_{\beta\alpha} * b_\beta\, ds(x) + \varrho \int_D [v_{0\beta} V_{\beta\alpha} + u_{0\beta} \dot{V}_{\beta\alpha}]\, ds(x), \qquad (1.47)$$

$$\varepsilon(y)\, u_3(y, t) = \int_C [V_{33} * t_3 - W_{33} * u_3]\, dc(x)$$

$$+ \varrho \int_D V_{33} * b_3\, ds(x) + \varrho \int_D [v_{03} V_{33} + u_{03} \dot{V}_{33}]\, ds(x) \qquad (1.48)$$

where

$$x = (x_1, x_2), \quad y = (y_1, y_2) \quad \text{and} \quad \varepsilon(y) = \begin{cases} 1, & y \in D \\ 0, & y \in D_c \end{cases}.$$

1.3.3 Boundary Integral Equations

1) General Case

Integral representation implies that displacement, $u(y, t)$, $y \in D$ can be obtained when the boundary data $\bar{u}(x, t)$ and $\bar{t}(x, t)$ are supplied. However, we know only either of u or t on the boundary, which is prescribed as boundary conditions. The rest must be sought for beforehand to evaluate the internal displacements and stresses. The unprescribed boundary-values can be obtained by solving the boundary integral equation, which is obtained by taking the limit y to the boundary.

It is noted that for $t \in T^\times \equiv (-\infty, \infty)$, Wheeler and Sternberg [19],

$$U_{ik}(x, t; y, 0) = O\left(\frac{1}{|x - y|}\right), \quad T_{ik}(x, t; y, 0) = O\left(\frac{1}{|x - y|^2}\right)$$

as $x \to y$ and

$$\lim_{\delta \to 0} \int_{s(y; \delta) \subset D} T_{ik}(x, t; y, 0) * f(x, t)\, ds(x) = C_{ik}(y) f(y, t), \qquad (1.49)$$

where $s(y; \delta)$ stands for the part of the surface of a small sphere contained in D, with radius δ and center at y, and $C_{ik} = \delta_{ik}$ for $y \in D$ and $(1/2)\delta_{ik}$ when y is on the smooth surface boundary.

Using these relations, we can reduce the following boundary integral equation from Eq. (1.35)

$$C_{ik}^\varepsilon(y)\, u_i(y, t) = \int_{\partial D} (U_{ik} * t_i - T_{ik} * u_i\, ds(x)$$

$$+ \varrho \int_D U_{ik} * b_i\, dv(x) + \varrho \int_D (v_{0i} U_{ik} + u_{0i} \dot{U}_{ik})\, dv(x). \qquad (1.50)$$

The coefficient C^ε may be obtained in a similar manner as discussed by Hartmann [54].

2) Two-Dimensional Case

In two-dimensional problems, BIEs can be formulated in two ways. One is to use the BIE of three-dimensional form and another path is to formulate BIE with the two-dimensional kernels, Manolis [28].

a) Use of the Three-Dimensional BIE. The three-dimensional kernel (fundamental) solutions are first cast into appropriate form for the plane strain case. Consider an infinitely long cylindrical body or cavity B, in a three-dimensional domain, whose longitudinal axis coincides with the x_3-axis. At time t a wave emenated from a source point y envelopes a spherical region of radius $r = c \cdot t$, where c is the wave speed. If we take the source point to lie on the (x_1, x_2)-plane and assume that the tractions remain constant during the time interval $(K \, \Delta t, (K - 1) \, \Delta t)$, $K = 1, 2, \ldots, N$, the first integral of Eq. (1.50) becomes, Niwa et al. [45]

$$\sum_{K=1}^{N} \int_C U_{\alpha\beta}(\bar{x}, K \, \Delta t; y, 0) \, t_\alpha(\bar{x}, (N - K + 1) \, \Delta t) \, dc(\bar{x}), \qquad (1.51)$$

where

$$U_{\alpha\beta} = \frac{2}{4 \pi \varrho} \left\{ \int_{c_T}^{c_L} \frac{dc}{c^3} \int_{z_{K-1,c}}^{z_{K,c}} \frac{1}{r} (3 r_{,\alpha} r_{,\beta} - \delta_{\alpha\beta}) \, dz \right.$$

$$+ \frac{1}{c_L^2} \int_{z_{K-1,c_L}}^{z_{K,c_L}} \frac{r_{,\alpha} r_{,\beta}}{r} \, dz + \frac{1}{c_T^2} \int_{z_{K-1,c_T}}^{z_{K,c_T}} \frac{1}{r} (\delta_{\alpha\beta} - r_{,\alpha} r_{,\beta}) \, dz \bigg\}$$

and C and $\bar{x}$ are the projections of $B_{K,c} = \{x; x \in B$ and $c(K - 1) \, \Delta T < r \leq c \, K \, \Delta t\}$ and x onto the (x_1, x_2)-plane, and $z_{K,c} = \sqrt{(c \, K \, \Delta t)^2 - (x_\alpha - y_\alpha)(x_\alpha - y_\alpha)}$, $z = x_3 - y_3$ and $r = |x - y|$.

Similarly, the second integral of Eq. (1.50) can be given as

$$\sum_{K=1}^{K} \int_C n_\gamma \{ T_{\alpha\beta\gamma}(\bar{x}, K \, \Delta t; y, 0) \, u_\alpha(\bar{x}, (N - K + 1) \, \Delta t)$$

$$+ Q_{\alpha\beta\gamma}(\bar{x}, K \, \Delta t; y, 0) \, \dot{u}_\alpha(\bar{x}, (N - K + 1) \, \Delta t \} \, dc(\bar{x}), \qquad (1.52)$$

where

$$T_{\alpha\beta\gamma} = \frac{2}{4\pi} \left[-6 c_T^2 \int_{c_T}^{c_L} \frac{dc}{c^3} \int_{z_{K-1,c}}^{z_{K,c}} \frac{1}{r^2} \{ 5 r_{,\alpha} r_{,\beta} r_{,\gamma} - (\delta_{\alpha\gamma} r_{,\beta} + \delta_{\alpha\beta} r_{,\gamma} + \delta_{\beta\gamma} r_{,\alpha}) \} \, dz \right.$$

$$- \int_{z_{K-1,c_L}}^{z_{K,c_L}} \frac{1}{r^2} \left\{ 2 \left(\frac{c_T}{c_L} \right)^2 [6 r_{,\alpha} r_{,\beta} r_{,\gamma} - (\delta_{\alpha\gamma} r_{,\beta} + \delta_{\alpha\beta} r_{,\gamma} + \delta_{\beta\gamma} r_{,\alpha}] \right.$$

$$+ \left(1 - 2 \left(\frac{c_T}{c_L} \right)^2 \right) \delta_{\alpha\gamma} r_{,\beta} \bigg\} \, dz$$

$$+ \int_{z_{K-1,c_T}}^{z_{K,c_T}} \frac{1}{r^2} \{ 2 r_{,\alpha} r_{,\beta} r_{,\gamma} - (2 \delta_{\alpha\gamma} r_{,\beta} + \delta_{\alpha\beta} r_{,\gamma} + \delta_{\beta\gamma} r_{,\alpha}) \} \, dz \bigg],$$

$$Q_{\alpha\beta\gamma} = \frac{2}{4\pi} \left[-\frac{1}{c_L} \int_{z_{K-1,c_L}}^{z_{K,c_L}} \frac{1}{r} \left\{ 2 \left(\frac{c_T}{c_L} \right)^2 r_{,\alpha} r_{,\beta} r_{,\gamma} + \left(1 - 2 \left(\frac{c_T}{c_L} \right)^2 \right) \delta_{\alpha\gamma} r_{,\beta} \right\} \, dz \right.$$

$$+ \frac{1}{c_T} \int_{z_{K-1,c_T}}^{z_{K,c_T}} \frac{1}{r} \{ 2 r_{,\alpha} r_{,\beta} r_{,\gamma} - (\delta_{\alpha\beta} r_{,\gamma} + \delta_{\beta\gamma} r_{,\alpha}) \} \, dz \bigg].$$

Furthermore, the velocity $\dot{u}_\alpha$ is replaced by a backward finite difference in time;

$$\dot{u}_\alpha(\bar{x}, (N-K+1)\,\Delta t) \cong \frac{1}{\Delta t}\{u_\alpha(\bar{x}, (N-K+1)\,\Delta t) - u_\alpha(\bar{x}, (N-K)\,\Delta t\}. \quad (1.53)$$

Finally, we have BIEs for smooth surface as follows (disregarding body force), Niwa et al. [45]

$$\tfrac{1}{2}\,u_\beta(\bar{x}, N\,\Delta t) = u_{0\beta}(\bar{x}, N\,\Delta t; y, 0)$$

$$+ \sum_{K=1}^{N} \int_C \{U_{\alpha\beta}(\bar{x}, K\,\Delta t; y, 0)\,t_\alpha(\bar{x}, (N-K+1)\,\Delta t)$$

$$- n_\gamma [T_{\alpha\beta\gamma}(\bar{x}; K\,\Delta t; y, 0)\,u_\alpha(\bar{x}, (N-K+1)\,\Delta t)$$

$$+ Q_{\alpha\beta\gamma}(\bar{x}, K\,\Delta t; y, 0)\,\dot{u}_\alpha(\bar{x}, (N-K+1)\,\Delta t)]\}\,dc, \quad (1.54)$$

where $u_{0\beta}$ denotes an incident wave field.

Similarly for the anti-plane motion, we have

$$\tfrac{1}{2}\,u_3(\bar{x}, N\,\Delta t) = u_{03}(\bar{x}, N\,\Delta t; y, 0)$$

$$+ \sum_{K=1}^{N} \int_C \{U_{33}(\bar{x}, K\,\Delta t; y, 0)\,t_3(\bar{x}, (N-K+1)\,\Delta t)$$

$$- n_\gamma [T_{33\gamma}(\bar{x}, K\,\Delta t; y, 0)\,u_3(\bar{x}, (N-K+1)\,\Delta t)$$

$$+ Q_{33\gamma}(\bar{x}, K\,\Delta t; y, 0)\,\dot{u}_3(\bar{x}, (N-K+1)\,\Delta t)]\}\,dc \quad (1.55)$$

where

$$U_{33} = \frac{2}{4\pi\varrho\,c_T^2} \int_{z_{K-1,c_T}}^{z_{K,c_T}} \frac{1}{r}\,dz,$$

$$T_{33\gamma} = \frac{2}{4\pi} \int_{z_{K-1,c_T}}^{z_{K,c_T}} \left(\frac{1}{r}\right)_{,\gamma}\,dz,$$

$$Q_{33\gamma} = -\frac{2}{4\pi c_T} \int_{z_{K-1,c_T}}^{z_{K,c_T}} \frac{r_{,\gamma}}{r}\,dz$$

and u_{03} stands for an incident wave field.

b) Use of the Two-Dimensional Kernel. Using the kernels defined by Eq. (1.44) and taking the limit $D \ni x \to x \in C$, we have the two-dimensional BIE from Eqs. (1.47) and (1.48). The BIEs are expressed just by replacing $\varepsilon(y)$ by $C_{\alpha,\beta}(y)$ (free term).

$$C_{\alpha,\beta}^c(y)\,u_\alpha(y, t) = \int_C [V_{\alpha\beta} * t_\alpha - W_{\alpha\beta} * u_\alpha]\,dc(x)$$

$$+ \varrho \int_D V_{\alpha\beta} * b_\alpha\,ds(x)$$

$$+ \varrho \int_D [v_{0\alpha} V_{\alpha\beta} + u_{0\alpha} \dot{V}_{\alpha\beta}]\,ds(x) \quad (1.56)$$

and

$$C^e(y)\, u_3\,(y,\, t) = \int\limits_{C} [V_{33} * t_3 - W_{33} * u_3]\, dc\,(x)$$

$$+ \varrho \int\limits_{D} V_{33} * b_3\, ds\,(x)$$

$$+ \varrho \int\limits_{D} [\dot{v}_{03}\, V_{33} + u_{03}\, \dot{V}_{33}]\, ds\,(x) \qquad (1.57)$$

where $V_{\alpha\beta}$ and $W_{\alpha\beta}$ are given by Eq. (1.44).

It should be noted that in the formulation by use of three-dimensional kernels the integration is carried out over an area segment moving with time, while in the two-dimensional formulation the integration is done over a line segment and over a time. Furthermore, in the former singularities appear only once in the time-marching scheme, whereas in the latter singularities appear at every time step. However, even in the two-dimensional formulation the time integration can be carried out first analytically, and hence singularities appear only once on the first time step, Mansur and Brebbia [46], and Ishihara [55].

1.4 BIE Formulations in the Transformed Domain of Integral Transforms

1.4.1 Fourier Transformed Problems of Elastodynamics

In solving the elastodynamic initial-boundary value problems, we often use the Fourier transform technique. The Fourier transform of a function $f(x, t)$ with respect to t is defined as

$$\bar{f}(x, \omega) = \mathscr{F}\{f(x, t)\} = \frac{1}{2\pi} \int\limits_{-\infty}^{\infty} f(x, t)\, e^{i\omega t}\, dt \qquad (1.58)$$

and its inverse

$$f(x, t) = \int\limits_{-\infty}^{\infty} \bar{f}(x, \omega)\, e^{-i\omega t}\, d\omega$$

where ω is the circular frequency.

The Fourier transform of field equation (1.6) is

$$(\lambda + \mu)\, \nabla\nabla \cdot \bar{u}(x, \omega) + \mu \nabla^2 \bar{u}(x, \omega) + \varrho\, \bar{b}(x, \omega) + \varrho\omega^2\, \bar{u}(x, \omega) = 0,$$

or

$$L\, \bar{u}(x, \omega) \equiv (E + \varrho\omega^2\, 1)\, \bar{u}(x, \omega) = (\lambda\, \nabla\nabla. + \mu \nabla^2\, 1 + \mu \nabla\nabla. + \varrho\omega^2\, 1)\, \bar{u}(x, \omega)$$

$$= \varrho\,[(c_L^2 - 2\, c_T^2)\, \nabla\nabla. + c_T^2\, \nabla^2\, 1 + \omega^2\, 1]\, \bar{u}(x, \omega) = -\, \varrho\, \bar{b}(x, \omega). \qquad (1.59)$$

This equation is amenable to numerical solutions, since hyperbolic partial differential equations are transformed to elliptic ones.

The boundary conditions are also written as

$$\begin{aligned} \bar{u}(x, \omega) &= \bar{\bar{u}}(x, \omega), \quad x \in \partial D_1, \\ \overset{n}{T}\, \bar{u}(x, \omega) &= \bar{\bar{t}}(x, \omega), \quad x \in \partial D_2, \quad \partial D_1 \cup \partial D_2 = \partial D. \end{aligned} \qquad (1.60)$$

Essentially, when integral transforms are used, the solution to a transient elastodynamic problem consists of a series of solutions to a static boundary value problem for a number of discrete parameter ω. This series of solutions must finally be numerically inverted back to the original problem in the form

$$u(x, t) = \int_{-\infty}^{\infty} \tilde{u}(x, \omega)\, e^{-i\omega t}\, d\omega. \tag{1.61}$$

In what follows, we assume that the displacement field is with zero initial conditions and a quiescent past, i.e.,

$$u_0(x) = v_0(x) = 0, \quad u(x, t) = 0 \quad \text{for } t < 0.$$

These assumptions are made for the sake of convenience and no additional difficulty in the solution procedure is encountered even if the assumptions are relaxed.

The Fourier transformed boundary value problems for elastodynamics are classified as follows (for simplicity, we denote D_- (D_+) for the interior (exterior) domain and omit the super tilde ($\sim$), for example $u(x, \omega)$ instead of $\tilde{u}(x, \omega)$);

(i) The first interior (exterior) problem:

$$L\,u(x, \omega) = (E + \varrho\omega^2\, \mathbf{1})\, u(x, \omega) = -\varrho\, b(x, \omega), \quad x \in D_-\ (x \in D_+)$$

$$u(x, \omega) = \bar{u}(x, \omega), \quad x \in \partial D. \tag{1.62}$$

(ii) The second interior (exterior) problem:

$$L\,u(x, \omega) = (E + \varrho\omega^2\, \mathbf{1})\, u(x, \omega) = -\varrho b(x, \omega), \quad x \in D_-\ (x \in D_+)$$

$$\overset{n}{T}u(x, \omega) = \bar{t}(x, \omega), \quad x \in \partial D. \tag{1.63}$$

(iii) The third (mixed) interior (exterior) problem:

$$L\,u(x, \omega) = (E + \varrho\omega^2\, \mathbf{1})\, u(x, \omega) = -\varrho b(x, \omega), \quad x \in D_-\ (x \in D_+)$$

$$u(x, \omega) = \bar{u}(x, \omega), \quad x \in \partial D_1 \tag{1.64}$$

$$\overset{n}{T}u(x, \omega) = \bar{t}(x, \omega), \quad x \in \partial D_2, \quad \partial D_1 \cup \partial D_2 = \partial D.$$

For the exterior problem, the scattered field must satisfy the following radiation conditions to assure the unique solution; Kupradze [10], Doyle [56], and Srivastra and Zischka [57].

$$\frac{\partial u^L}{\partial r} - i k_L\, u^L = o\left(r^{-\frac{n-1}{2}}\right), \quad u^L = o\left(r^{-\frac{n-3}{2}}\right),$$

$$\frac{\partial u^T}{\partial r} - i k_T\, u^T = o\left(r^{-\frac{n-1}{2}}\right), \quad u^T = o\left(r^{-\frac{n-3}{2}}\right) \tag{1.65}$$

as $r = |x| \to \infty$, where $k_L = \omega/c_L$, $k_T = \omega/c_T$ are wave numbers of the longitudinal and transverse waves, n denotes dimension of the space.

Wen $\omega = 0$, Eqs. (1.62) to (1.64) are reduced to the boundary value problems of elastostatics.

In the two-dimensional case, similarly transformed equations and boundary conditions are obtained from Eqs. (1.19) and (1.20) and boundary conditions as follows:

(i) For plane motion:

$$L \, u(x, \omega) = (E + \varrho \omega^2 \, 1) \, u(x, \omega) = - \varrho b(x, \omega), \quad x \in D$$

$$u(x, \omega) = \bar{u}(x, \omega), \quad x \in \partial D_1 \tag{1.66}$$

$$\overset{n}{T} u(x, \omega) = \bar{t}(x, \omega), \quad x \in \partial D_2, \quad \partial D_1 \cup \partial D_2 = \partial D.$$

(ii) For anti-plane motion:

$$\mu (\nabla^2 + k_T^2) \, u_3(x, \omega) = - \varrho b_3, \quad x \in D$$

$$u_3(x, \omega) = \bar{u}_3(x, \omega), \quad x \in \partial D_1 \tag{1.67}$$

$$\overset{n}{T} u_3(x, \omega) = \mu \, n \cdot \nabla u_3(x, \omega) = \bar{t}_3(x, \omega), \quad x \in \partial D_2, \quad \partial D_1 \cup \partial D_2 = \partial D.$$

1.4.2 BIE Formulations of Transformed Problems

1) Potentials and Green's Formula

In what follows, we use the following notations unless otherwise mentioned, Niwa et al. [40, 42].

a) For a limit value

$$\overset{n}{T}_x (S \, \mu)(x^{\pm}) \equiv \lim_{x \to x^{\pm}} \overset{n}{T}_x (S \, \mu)(x), \quad x^{\pm} : D_{\pm} \ni x \to x \in \partial D. \tag{1.68}$$

b) For the integral operators, we use bold upper case letters, $K(x, y; \omega)$ or K, for instance

$$(K \, \mu)(x) = \int_{\partial D} \overset{n}{T}_x \, U(x, y; \omega) \cdot \mu(y) \, ds(y). \tag{1.69}$$

Its adjoint operator is denoted by $K^*(x, y; \omega)$ or K^*

$$(K^* \, \mu)(x) = \int_{\partial D} \overset{n}{T}_y \, \overline{U(x, y; \omega)} \cdot \mu(y) \, ds(y) \tag{1.70}$$

where $\bar{U}$ is the complex conjugate of U.

c) For simple and double layer potentials with densities μ and v distributed over the boundary ∂D, we denote respectively,

$$(S \, \mu)(x) = \int_{\partial D} U(x, y; \omega) \cdot \mu(y) \, ds(y), \quad x \in \mathbb{R}^n \tag{1.71}$$

and

$$(D \, v)(x) = \int_{\partial D} \overset{n}{T}_y \, U(x, y; \omega) \cdot v(y) \, ds(y), \quad x \in \mathbb{R}^n. \tag{1.72}$$

On the boundary,

$$(D \, v)(x) = \int_{\partial D} \overset{n}{T}_y \, U(x, y; \omega) \cdot v(y) \, ds(y) = (\bar{K}^* \, v)(x), \quad x \in \partial D. \tag{1.73}$$

d) Properties of potentials, Kellogg [7], Jaswon and Symm [58].

(i) Simple layer potential $(S\mu)(x)$ is continuous in $\mathbb{R}^n$, and for smooth boundary

$$(S\mu)(x^+) \quad = (S\mu)(x^-) = (S\mu)(x), \quad x \in \partial D \tag{1.74}$$

$$\overset{n}{T}_x(S\mu)(x^+) = -\tfrac{1}{2}\mu(x) + (K\mu)(x), \quad x \in \partial D$$

$$\overset{n}{T}_x(S\mu)(x^-) = \tfrac{1}{2}\mu(x) + (K\mu)(x), \quad x \in \partial D \tag{1.75}$$

$$\overset{n}{T}_x(S\mu)(x^+) - \overset{n}{T}_x(S\mu)(x^-) = -\mu(x), \quad x \in \partial D$$

where $(K\mu)(x)$ is defined by Eq. (1.69).

(ii) Double layer potential

$$(Dv)(x^+) = \tfrac{1}{2}v(x) + (\bar{K}^*v)(x), \quad x \in \partial D$$

$$(Dv)(x^-) = -\tfrac{1}{2}v(x) + (\bar{K}^*v)(x), \quad x \in \partial D \tag{1.76}$$

$$(Dv)(x^+) - (Dv)(x^-) = v(x), \quad x \in \partial D$$

$$\overset{n}{T}_x(Dv)(x^+) = \overset{n}{T}_x(Dv)(x^-) = \overset{n}{T}_x(\bar{K}^*v)(x) = (D_n v)(x), \quad x \in \partial D. \tag{1.77}$$

2) Fundamental Solutions

Fundamental solutions which satisfy the radiation conditions are given as follows;

$$U_{jk}(x, y; \omega) = A(U_1\,\delta_{jk} - U_2\,r_{,j}\,r_{,k}) \tag{$1.78)_1$}$$

(i) $A = \dfrac{1}{4\pi\mu}$,

$$U_1 = \frac{e^{ik_T r}}{r} + \left\{\frac{i}{k_T r} - \frac{1}{(k_T r)^2}\right\}\frac{e^{ik_T r}}{r} - \left(\frac{k_L}{k_T}\right)^2 \left\{\frac{i}{k_L r} - \frac{1}{(k_L r)^2}\right\}\frac{e^{ik_T r}}{r}$$

$$\tag{$1.78)_2$}$$

$$U_2 = \left\{1 + \frac{3i}{k_T r} - \frac{3}{(k_T r)^2}\right\}\frac{e^{ik_T r}}{r} - \left(\frac{k_L}{k_T}\right)^2\left\{1 + \frac{3i}{k_L r} - \frac{3}{(k_L r)^2}\right\}\frac{e^{ik_L r}}{r}$$

$$(\text{for } 3 - D)$$

(ii) $A = \dfrac{i}{4\mu}$,

$$U_1 = H_0^{(1)}(k_T r) - \frac{1}{k_T r}H_1^{(1)}(k_T r) + \left(\frac{k_L}{k_T}\right)^2\frac{1}{k_L r}H_1^{(1)}(k_L r)$$

$$\tag{$1.78)_3$}$$

$$U_2 = -H_2^{(1)}(k_T r) + \left(\frac{k_L}{k_T}\right)^2 H_2^{(1)}(k_L r) \qquad (\text{for } 2 - d),$$

$$T_{jk}(x, y; \omega) = \mu A\left[\left\{\left(\delta_{jk}\frac{\partial r}{\partial n} + n_j\,r_{,k}\right) + \frac{\lambda}{\mu}n_k\,r_{,j}\right\}\frac{dU_1}{dr}\right.$$

$$- \left\{\left(\delta_{jk}\frac{\partial r}{\partial n} + n_j\,r_{,k}\right) + 2\left(n_k\,r_{,j} - 2r_{,j}\,r_{,k}\frac{\partial r}{\partial n}\right) + \alpha\frac{\lambda}{\mu}n_k\,r_{,j}\right\}\frac{U_2}{r}$$

$$\left.- \left\{2r_{,j}\,r_{,k}\frac{\partial r}{\partial n} + \frac{\lambda}{\mu}n_k\,r_{,j}\right\}\frac{dU_2}{dr}\right]. \tag{$1.79)_1$}$$

(i) $\alpha = 2$ and A, U_1, U_2 in Eq. (1.78)$_2$, (i) for $3 - D$ (1.79)$_2$
(ii) $\alpha = 1$ and A, U_1, U_2 in Eq. (1.78)$_3$, (ii) for $2 - D$ (1.79)$_3$

where $H_n^{(1)}(\ )$ is the Hankel function of the first kind and the n-th order, $r_{,j} \equiv \partial r/\partial y_j$, $\partial r/\partial n = (\partial r/\partial y_j)\, n_j$.

Fundamental solutions for anti-plane (Helmholtz) equation are given by

$$U_{33} = \frac{i}{4\mu}\, H_0^{(1)}(k_T r) \tag{1.80}$$

$$T_{33} = -\frac{i k_T}{4}\, H_1^{(1)}(k_T r)\, \frac{\partial r}{\partial n}. \tag{1.81}$$

For reference fundamental solutions of elastostatics are also given below:

$$\overset{\circ}{U}_{jk}(x, y) = \overset{\circ}{A}(\overset{\circ}{U}_1\, \delta_{jk} - \overset{\circ}{U}_2\, r_{,j} r_{,k}), \tag{1.82}_1$$

(i) $\overset{\circ}{A} = \dfrac{1}{8\pi(\lambda + 2\mu)\mu}$, $\overset{\circ}{U}_1 = (\lambda + 3\mu)\dfrac{1}{r}$, $\overset{\circ}{U}_2 = -(\lambda + \mu)\dfrac{1}{r}$ (for $3 - D$) (1.82)$_2$

(ii) $\overset{\circ}{A} = \dfrac{1}{4\pi(\lambda + 2\mu)\mu}$, $\overset{\circ}{U}_1 = (\lambda + 3\mu)\ln\dfrac{1}{r}$, $\overset{\circ}{U}_2 = \lambda + \mu$.   (for $2 - D$) (1.82)$_3$

$\overset{\circ}{T}_{jk}(x, y)$ is given as in Eq. (1.79) with $\overset{\circ}{A}$, $\overset{\circ}{U}_1$ and $\overset{\circ}{U}_2$ of Eqs. (1.82).

Fundamental solutions of the anti-plane (Laplace) equation are given by

$$\overset{\circ}{U}_{33}(x, y) = \frac{1}{2\pi\mu}\, \ln\frac{1}{r}, \tag{1.83}$$

$$\overset{\circ}{T}_{33}(x, y) = \frac{1}{2\pi r}\, \frac{\partial r}{\partial n}. \tag{1.84}$$

These solutions can be reduced from the fundamental solutions of elasto-dynamics by taking the limit of the circular frequency ω to zero. In the two-dimensional case, however, it should be noted that the limit of AU_1 and U_{33} become $\overset{\circ}{A}(\overset{\circ}{U}_1 - (\lambda + 2\mu)\ln k_T - \mu \ln k_L) + \text{const.}$ and $\overset{\circ}{U}_{33} - (\ln k_T)/(2\pi\mu) + \text{const.}$, respectively.

We can also easily recognize that the order of the singularity of the fundamental solutions of elastodynamics is the same as those of elastostatics as $r \to 0$, while keeping ω for a certain value.

3) The Indirect BIE Formulations

We formulate BIE by the use of layer potentials, in which the density is unknown, and physical quantities such as displacement and traction can be obtained indirectly by the aid of the predetermined density. Therefore, we sometimes call this method as "indirect method".

In what follows, as an example, we formulate BIE for the first interior problem.

Assume that the displacement can be expressible by a simple layer potential as

$$u(x) = (S\mu)(x), \quad x \in D_-.$$

Table 1.1. Boundary integral equations and solution forms. $\bar{u}$: Prescribed displacement on ∂D_1, $\bar{t}$: Prescribed traction on ∂D_2, O: Operator, I.P.: Identity pair, A.P.: Adjoint pair, U: Unknown

	Formulation	Integral equation	O	I.P.	A.P.	Solution form	U
First interior problem (F.I.)	$L(D)$	$(\frac{1}{2}I - \bar{K}^*)v = -\bar{u}$	$\bar{K}^*$	I_5	A_4	$u = Dv$	v
	$L(S)$	$S\mu = \bar{u}$	S	I_1		$u = S\mu$	μ
	$\bar{L}(D)$	$(\frac{1}{2}I - K^*)v = -\bar{u}$	K^*		A_1, A_3	$u = \bar{D}v$	v
	$\bar{L}(S)$	$\bar{S}\mu = \bar{u}$	$\bar{S}$			$u = \bar{S}\mu$	μ
	$G(t)$	$(\frac{1}{2}I - K)t = -D_n\bar{u}$	K	I_3	A_1	$u = St - D\bar{t}$	t
	$G(u)$	$St = (\frac{1}{2}I + \bar{K}^*)\bar{u}$	S	I_1			
	$\bar{G}(t)$	$(\frac{1}{2}I - \bar{K})t = -\bar{D}_n\bar{u}$	$\bar{K}$		A_2, A_4	$u = \bar{S}t - \bar{D}\bar{t}$	t
	$\bar{G}(u)$	$St = (\frac{1}{2}I + K^*)\bar{u}$	$\bar{S}$				
Second exterior problem (S.E.)	$L(S)$	$(\frac{1}{2}I - K)\mu = -\bar{t}$	K	I_3	A_3	$u = S\mu$	μ
	$L(D)$	$D_n v = \bar{t}$	D_n	I_2		$u = Dv$	v
	$G(u)$	$(\frac{1}{2}I - \bar{K}^*)u = -S\bar{t}$	$\bar{K}^*$	I_5	A_2	$u = Du - S\bar{t}$	u
	$G(t)$	$D_n u = (\frac{1}{2}I + K)\bar{t}$	D_n	I_2			
Second interior problem (S.I.)	$L(S)$	$(\frac{1}{2}I + K)\mu = \bar{t}$	$-K$	I_4	A_5	$u = S\mu$	μ
	$L(D)$	$D_n v = \bar{t}$	D_n	I_2		$u = Dv$	v
	$\bar{L}(S)$	$(\frac{1}{2}I + \bar{K})\mu = \bar{t}$	$-\bar{K}$		A_7, A_8	$u = \bar{S}\mu$	μ
	$\bar{L}(D)$	$\bar{D}_n v = \bar{t}$	$\bar{D}_n$			$u = \bar{D}v$	v
	$G(u)$	$(\frac{1}{2}I + \bar{K}^*)u = S\bar{t}$	$-K^*$	I_6	A_7	$u = S\bar{t} - Du$	μ
	$G(t)$	$D_n u = -(\frac{1}{2}I - K)\bar{t}$	D_n	I_2			
	$\bar{G}(u)$	$(\frac{1}{2}I + K^*)u = \bar{S}\bar{t}$	$-K^*$		A_5, A_6	$u = \bar{S}\bar{t} - \bar{D}u$	u
	$\bar{G}(t)$	$\bar{D}_n u = -(\frac{1}{2}I - \bar{K})\bar{t}$	$\bar{D}_n$				
First exterior problem (F.E.)	$L(D)$	$(\frac{1}{2}I + \bar{K}^*)v = \bar{u}$	$-\bar{K}^*$	I_6	A_8	$u = Dv$	v
	$L(S)$	$S\mu = \bar{u}$	S	I_1		$u = S\mu$	μ
	$G(t)$	$(\frac{1}{2}I + K)t = D_n\bar{u}$	$-K$	I_4	A_6	$u = D\bar{u} - St$	t
	$G(u)$	$St = -(\frac{1}{2}I - \bar{K}^*)\bar{u}$	S	I_1			

Table 1.1 (continued)

Formulation		Integral equation	I.P.	U
Third (mixed) interior problem (T.I.)	Solution form	$u = (S\,t)_{\partial D_1} - (D\,u)_{\partial D_2} + (S\,t)_{\partial D_2} - (D\,\bar{u})_{\partial D_1}$ in D_-		
	$G(u)$	$(S\,t)_{\partial D_1} - (\bar{K}^*\,u)_{\partial D_2} = \{(\tfrac{1}{2}I + \bar{K}^*)\,\bar{u}\}_{\partial D_1} - (S\,\bar{t})_{\partial D_2}$ on ∂D_1 $(S\,t)_{\partial D_1} - \{(\tfrac{1}{2}I + K^*)\,u\}_{\partial D_2} = (\bar{K}^*\,\bar{u})_{\partial D_1} - (S\,\bar{t})_{\partial D_2}$ on ∂D_2		
	$G(u,t)$	$(S\,t)_{\partial D_1} - (\bar{K}^*\,u)_{\partial D_2} = \{(\tfrac{1}{2}I + \bar{K}^*)\,\bar{u}\}_{\partial D_1} - (S\,\bar{t})_{\partial D_2}$ on ∂D_1 $(K\,t)_{\partial D_1} - (D_n\,u)_{\partial D_2} = (D_n\,\bar{u})_{\partial D_1} + \{(\tfrac{1}{2}I - K)\,\bar{t}\}_{\partial D_2}$ on ∂D_2	I	u t
	$G(t,u)$	$\{(\tfrac{1}{2}I - K)\,t\}_{\partial D_1} + (D_n\,u)_{\partial D_2} = -(D_n\,\bar{u})_{\partial D_1} + (K\,\bar{t})_{\partial D_2}$ on ∂D_1 $(S\,t)_{\partial D_1} - \{(\tfrac{1}{2}I + \bar{K}^*)\,u\}_{\partial D_2} = (\bar{K}^*\,\bar{u})_{\partial D_1} - (S\,\bar{t})_{\partial D_2}$ on ∂D_2		
	$G(t)$	$\{(\tfrac{1}{2}I - K)\,t\}_{\partial D_1} + (D_n\,u)_{\partial D_2} = -(D_n\,\bar{u})_{\partial D_1} + (K\,\bar{t})_{\partial D_2}$ on ∂D_1 $(K\,t)_{\partial D_1} - (D_n\,u)_{\partial D_2} = (D_n\,\bar{u})_{\partial D_1} + \{(\tfrac{1}{2}I - K)\,\bar{t}\}_{\partial D_2}$ on ∂D_2		
Third (mixed) exterior problem (T.E.)	Solution form	$u = (D\,u)_{\partial D_2} - (S\,t)_{\partial D_1} + (D\,\bar{u})_{\partial D_1} - (S\,\bar{t})_{\partial D_2}$ in D_+		
	$G(u)$	$(S\,t)_{\partial D_1} - (\bar{K}^*\,u)_{\partial D_2} = -\{(\tfrac{1}{2}I - \bar{K}^*)\,\bar{u}\}_{\partial D_1} - (S\,\bar{t})_{\partial D_2}$ on ∂D_1 $(S\,t)_{\partial D_1} + \{(\tfrac{1}{2}I - K^*)\,u\}_{\partial D_2} = (\bar{K}^*\,\bar{u})_{\partial D_1} - (S\,\bar{t})_{\partial D_2}$ on ∂D_2		
	$G(u,t)$	$(S\,t)_{\partial D_1} - (\bar{K}^*\,u)_{\partial D_2} = -\{(\tfrac{1}{2}I - \bar{K}^*)\,\bar{u}\}_{\partial D_1} - (S\,\bar{t})_{\partial D_2}$ on ∂D_1 $(K\,t)_{\partial D_1} - (D_n\,u)_{\partial D_2} = (D_n\,\bar{u})_{\partial D_1} - \{(\tfrac{1}{2}I + K)\,\bar{t}\}_{\partial D_2}$ on ∂D_2	I	u t
	$G(t,u)$	$\{(\tfrac{1}{2}I + K)\,t\}_{\partial D_1} - (D_n\,u)_{\partial D_2} = (D_n\,\bar{u})_{\partial D_1} - (K\,\bar{t})_{\partial D_2}$ on ∂D_1 $(S\,t)_{\partial D_1} + \{(\tfrac{1}{2}I - \bar{K}^*)\,u\}_{\partial D_2} = (\bar{K}^*\,\bar{u})_{\partial D_1} - (K\,\bar{t})_{\partial D_2}$ on ∂D_2		
	$G(t)$	$\{(\tfrac{1}{2}I + K)\,t\}_{\partial D_1} - (D_n\,u)_{\partial D_2} = (D_n\,\bar{u})_{\partial D_1} - (K\,\bar{t})_{\partial D_2}$ on ∂D_1 $(K\,t)_{\partial D_1} - (D_n\,u)_{\partial D_2} = (D_n\,\bar{u})_{\partial D_1} - \{(\tfrac{1}{2}I + K)\,\bar{t}\}_{\partial D_2}$ on ∂D_2		

As $D_- \ni x \to x^- \in \partial D$, we have for the prescribed displacement

$$(S\,\mu)(x^-) = \bar{u}(x), \quad x \in \partial D_1 . \tag{1.85}$$

Similarly, assuming

$$u(x) = (\bar{S}\,\mu)(x), \quad x \in D_-$$

we have

$$(\bar{S}\,\mu)(x^-) = \bar{u}(x), \quad x \in \partial D . \tag{1.86}$$

If we assume

$$u(x) = (D\,v)(x), \quad x \in D_-$$

and letting $D_- \ni x \to x^- \in \partial D$, and taking account of properties of the double layer potential, we have

$$(-\tfrac{1}{2}I + \bar{K}^*)\,v(x) = \bar{u}(x), \quad x \in \partial D . \tag{1.87}$$

Similarly, if we start with

$$u(x) = (\bar{D} v)(x), \quad x \in D_-$$

we have

$$(-\tfrac{1}{2} I + K^*) v(x) = \bar{u}(x), \quad x \in \partial D. \tag{1.88}$$

For the first exterior and the second interior or exterior problems, we can similarly formulate BIEs by the use of layer potentials. Such formulations are listed in Table 1.1, Niwa et al. [42] and Kitahara [59].

4) The Direct BIE Formulations

If we take the Fourier transform of the reciprocal relation (1.28), we have

$$\int_{\partial D} t \cdot u' \, ds + \int_D \varrho b \cdot u' \, dv = \int_{\partial D} t' \cdot u \, ds + \int_D \varrho b' \cdot u \, dv, \tag{1.89}$$

where zero initial conditions are assumed for simplicity.

Choosing one state $\mathcal{S} = [u, \overset{n}{T} u, \omega]$ as the actual state and another $\mathcal{S}' = [u', \overset{n}{T} u', \omega]$ for the state corresponding to the fundamental solution, that is $u' \equiv U, t' \equiv U\overset{n}{T} \equiv T$, we have an integral representation (body force is disregarded for simplicity)

$$u(x) = \int_{\partial D} \{U \cdot t - T \cdot u\} \, ds(y) = (S t)(x) - (D u)(x). \tag{1.90}$$

For the interior problem, letting $D_- \ni x \to x^- \in \partial D$ and taking account of the properties of the double layer potentials, we have

$$(\tfrac{1}{2} I + \bar{K}^*) u(x) = (S t)(x), \quad x \in \partial D. \tag{1.91}$$

For the first problem, the BIE becomes

$$(S t)(x) = (\tfrac{1}{2} I + \bar{K}^*) \bar{u}(x), \quad x \in \partial D. \tag{1.92}$$

For the second problem, we have

$$(\tfrac{1}{2} I + \bar{K}^*) u(x) = (S \bar{t})(x), \quad x \in \partial D. \tag{1.93}$$

On the other hand, operating $\overset{n}{T}$ to Eq. (1.90), i.e.

$$\overset{n}{T} u(x) = \overset{n}{T}(S t)(x) - \overset{n}{T}(D u)(x)$$

and taking the limit $D_- \ni x \to x^- \in \partial D$ and using the properties of the simple layer potential, we have

$$(\tfrac{1}{2} I - K) t(x) = - (D_n u)(x). \tag{1.94}$$

For the first problem, we have

$$(\tfrac{1}{2} I - K) t(x) = - (D_n \bar{u})(x), \quad x \in \partial D. \tag{1.95}$$

Similarly, for the second problem, we have

$$(D_n u)(x) = - (\tfrac{1}{2} I - K)\bar{t}(x), \quad x \in \partial D. \tag{1.96}$$

If we start with the conjugate formula,

$$u(x) = \bar{S} t(x) - \bar{D} u(x)$$

we obtain

$$(\bar{S} t)(x) = (\tfrac{1}{2} I + K^*) \, \bar{u}(x), \qquad x \in \partial D, \tag{1.97}$$

$$(\tfrac{1}{2} I - \bar{K}) \, t(x) = - (\bar{D}_n) \, \bar{u}(x), \quad x \in \partial D. \tag{1.98}$$

For the third problem, four types of BIE formulations are possible by using

(i) displacement formula

$$\tfrac{1}{2} u(x) = (S \, t)(x) - (\bar{K}^* u)(x), \quad x \in \partial D_1 \cup \partial D_2,$$

(ii) displacement formula on ∂D_1 and traction formula on ∂D_2

$$\tfrac{1}{2} t(x) = (K \, t)(x) - (D_n \, u)(x), \quad x \in \partial D_2,$$

(iii) traction formula on ∂D_1 and displacement formula on ∂D_2 and
(iv) traction formula both on ∂D_1 and ∂D_2.

For example, we formulate BIE for the interior problem by the aid of (ii). We have

$$(S \, t)(x)_{\partial D_1} - (\bar{K}^* u)(x)_{\partial D_2} = \tfrac{1}{2} \bar{u} + (\bar{K}^* \bar{u})(x)_{\partial D_1} - (S \, \bar{t})(x)_{\partial D_2}, \quad x \in \partial D_1, \tag{1.99}$$

$$(K \, t)(x)_{\partial D_1} - (D_n \, u)(x)_{\partial D_2} = \tfrac{1}{2} \bar{t} - (K \, \bar{t})_{\partial D_1} - (D_n \, \bar{u})(x)_{\partial D_2}, \quad x \in \partial D_2, \tag{1.100}$$

where $(\ )_{\partial D_1}$ and $(\ )_{\partial D_2}$ stand for the integral on ∂D_1 and ∂D_2, respectively. Other formulations can be done similarly.

All the BIE formulated by the above mentioned method are also listed in Table 1.1.

1.4.3 Laplace Transformed Domain BIE

1) Laplace Transformed Elastodynamic Problems

The Laplace transform of a function $f(x, t)$ is defined as

$$\hat{f}(x, s) \equiv \mathscr{L}\{f(x, t)\} = \int_0^\infty f(x, t) \, e^{-st} \, dt \tag{1.101}$$

where s is the Laplace transform parameter.

A Laplace transform of the field equation (1.1) results in the following elliptic differential equation

$$E \, \hat{u} + \varrho \hat{b} - \varrho s^2 \, \hat{u} + \varrho s \, u_0 + \varrho v_0 = 0 \tag{1.102}$$

were u_0, v_0 denote initial values of displacement and velocity.

The Laplace transforms of the boundary conditions and the constitutive equations are simply given by

$$\hat{u} = \hat{\bar{u}}(x, s), \quad x \in \partial D_1, \tag{1.103}$$

$$\hat{t} = \hat{T} \cdot \hat{u}(x, s) = \hat{\sigma} \cdot n = \hat{\bar{t}}(x, s), \quad x \in \partial D_2, \quad \partial D_1 \cup \partial D_2 = \partial D,$$

and

$$\hat{\sigma} = \lambda \, \nabla \cdot \hat{u} \, 1 + \mu \, (\nabla \hat{u} + \hat{u} \nabla)$$

$$= \varrho \, [(c_L^2 - 2 c_T^2) \, \nabla \cdot \hat{u} \, 1 + c_T^2 (\nabla \hat{u} + \hat{u} \nabla)]. \tag{1.104}$$

In the followings, we assume a quiescent past with zero initial conditions and disregard the body force, for the sake of convenience. No additional difficulty arise if these assumptions are relaxed. With these assumptions it is clear that the Laplace transform of the field equation is obtained by just setting parameter s equal to $-i\omega$ in the Fourier transformed Eq. (1.59). The radiation conditions in this case are simply obtained from Eq. (1.65) just by replacing $-i\omega$ by s.

The solution procedure of the elastodynamic problems is the same as that stated in the Fourier transform approach. In order to have the solution of the original problems, solutions obtained from a series of values of parameter s are synthesized after numerical inversion by the inversion integral

$$u(x, t) = \frac{1}{2\pi i} \int_{\beta-i\infty}^{\beta+i\infty} \hat{u}(x, s) \, e^{st} \, ds, \qquad (1.105)$$

where $\beta > 0$ is greater than the real part of all the singularities of $\hat{u}(x, s)$. As for the numerical inversion, some comments will be given later.

2) Fundamental Solutions

The fundamental solutions of Eq. (1.102) can be obtained as follows (for simplicity, the caret ($\hat{\ }$) is omitted hereafter); Cruse and Rizzo [25], Manolis [28], Doyle [60], and Sladek and Sladek [61].

$$U_{jk}(x, s) = A\,(U_1 \, \delta_{jk} - U_2 \, r_{,j} \, r_{,k}), \qquad (1.106)$$

(i) $A = \dfrac{1}{4\pi\mu}$, $\quad U_1 = \dfrac{e^{-sr/c_T}}{r} + \left(\dfrac{c_T}{sr} + \dfrac{c_T^2}{s^2 r^2}\right) \dfrac{e^{-sr/c_T}}{r} - \left(\dfrac{c_L}{sr} + \dfrac{c_L^2}{s^2 r^2}\right) \dfrac{e^{-sr/c_L}}{r}$

$$U_2 = \left(1 + \frac{3c_T}{sr} + \frac{c_T^2}{s^2 r^2}\right) \frac{e^{-sr/c_T}}{r} - \left(\frac{c_T}{c_L}\right)^2 \left(1 + \frac{3c_L}{sr} + \frac{c_L^2}{s^2 r^2}\right) \frac{e^{-sr/c_L}}{r}$$

$$(\text{for } 3 - D),$$

(ii) $A = \dfrac{1}{2\pi\mu}$, $\quad U_1 = K_0\left(\dfrac{sr}{c_T}\right) + \dfrac{c_T}{sr}\left[K_1\left(\dfrac{sr}{c_T}\right) - \dfrac{c_T}{c_L} K_1\left(\dfrac{sr}{c_L}\right)\right]$,

$$U_2 = K_2\left(\frac{sr}{c_T}\right) - \left(\frac{c_T}{c_L}\right)^2 K_2\left(\frac{sr}{c_L}\right) \qquad (\text{for } 2 - D),$$

where $K_n(\ )$ is the modified Bessel function of the second kind and n-th order.

The second solution $T_{jk}(x, s)$ can be expressed by Eq. (1.79) with appropriate A, U_1, U_2 and $\alpha = 2$ or 1 for $3 - D$ or $2 - D$.

For the anti-plane equation we have as follows;

$$U_{33} = \frac{1}{2\pi\mu} K_0\left(\frac{sr}{c_T}\right), \qquad (1.107)$$

$$T_{33} = -\frac{s}{2\pi\mu c_T} K_1\left(\frac{sr}{c_T}\right) \frac{\partial r}{\partial n}. \qquad (1.108)$$

These solutions are easily derived from Eqs. (1.17)−(1.84) by the aid of the formula $2K_\nu(z) = i\pi\, e^{i\nu\pi/2}\, H_\nu^{(1)}(iz)$ and replacing ω by is.

3) BIE Formulations

The BIEs in the Laplace transformed domain can be obtained in a similar manner as mentioned in the derivations of BIEs in the Fourier transformed domain. In this case, transformation parameter ω in the afore-mentioned formula must be replaced by is in order to have appropriate BIEs. For details refer to Cruse and Rizzo [25], Cruse [26], Manolis [28], Manolis and Beskos [29], and Sladek and Sladek [61].

1.5 Integral Equation Formulations for Inhomogeneous Domain

1.5.1 Basic Equations

If the field is inhomogeneous in a sense that Lamé's constants and mass density vary as a function of position, we have to modify the afore-mentioned treatment. In this section, we consider the problem in conjunction with the Fourier transform, Niwa et al. [62]. For the time-space domain formulation, see Tanaka and Tanaka [63].

The Navier-Cauchy equation for inhomogeneous field can be expressed in the Fourier transformed domain as

$$E(\lambda, \mu, \varrho;\, \partial, \omega)\, \tilde{u}(x, \omega) \equiv \nabla(\tilde{\lambda}, \nabla, \tilde{u}) + \nabla \cdot \{\tilde{\mu}\,(\nabla\tilde{u} + \tilde{u}\nabla\} + \varrho\omega^2\,\tilde{u} + \varrho\tilde{b} = 0. \qquad (1.109)$$

Stress and traction are given by

$$\tilde{\sigma}(x, \omega) = \tilde{\lambda}\nabla \cdot \tilde{u}\, 1 + \tilde{\mu}\,(\nabla\tilde{u} + \tilde{u}\,\nabla), \qquad (1.110)$$

$$\tilde{t}(x, \omega) = \overset{n}{T} \cdot \tilde{u}(x, \omega) = \tilde{\lambda}\, n(\nabla \cdot \tilde{u}) + \tilde{\mu}\, n \cdot (\nabla\tilde{u} + \tilde{u}\nabla). \qquad (1.111)$$

Hereafter we omit the super tilde (˜) for simplicity.

If we assume that Lamé's constants λ, μ and mass density can be expressible as a sum of constant value and variation with position, i.e.

$$\lambda(x) = \lambda_0 + \bar{\lambda}(x), \quad \mu(x) = \mu_0 + \bar{\mu}(x), \quad \varrho(x) = \varrho_0 + \bar{\varrho}(x) \qquad (1.112)$$

then we have following two types of field equations from Eq. (1.109),

$$M\bar{u} + F = 0, \qquad (1.113)$$

$$M \equiv (\lambda_0 + \mu_0)\, \nabla\nabla \cdot u + \mu_0\, \nabla^2 u + \varrho_0\, \omega^2\, u,$$

$$F \equiv \nabla(\bar{\lambda}\nabla \cdot u) + \nabla \cdot \{\bar{\mu}\,(\nabla u + u\nabla)\} + \bar{\varrho}\,\omega^2\, u + \varrho b \qquad (1.114)$$

or

$$Nu + H = 0, \qquad (1.115)$$

$$N \equiv (\lambda_0 + \mu_0)\, \nabla\nabla \cdot u + \mu_0\, \nabla^2 u,$$

$$H \equiv \nabla(\bar{\lambda}\nabla \cdot u) + \nabla \cdot \{\bar{\mu}\,(\nabla u + u\nabla)\} + \varrho\omega^2\, u + \varrho b. \qquad (1.116)$$

It is noted here that M is the elastodynamic operator, whereas N is the elastostatic operator.

The first type does not seem amenable from the computational point of view, since the parameter ω may be involved in the fundamental solution and complicated procedure may be need in the domain integral. Hence we restrict ourselves to the second type.

1.5.2 Integral Equation Formulations

Here we will show the direct integral equations. Substituting the actual state $\mathscr{S} = [u, t]$ and the state $\mathscr{S}'' = [u', t']$ corresponding to the fundamental solution into the reciprocal theorem Eq. (1.89), we easily have the following integral representation for the domain $D - \tau(x, \delta)$.

$$\int_{\partial D + s(x;\,\delta)} [\mathring{U}(x, y)\{\overset{n}{T}(\lambda_0, \mu_0; \partial_y)\, u\,(y)\} - \{\mathring{U}(x, y)\,\overset{n}{T}(\lambda_0, \mu_0; \partial_y)\}\, u\,(y)]\, ds\,(y)$$

$$+ \int_{D-\tau(x;\,\delta)} \mathring{U}(x, y)\, H(y)\, dv\,(y) = 0, \quad x \in D \qquad (1.117)$$

where $\mathring{U}$ is the fundamental solution for elastostatics given by Eq. (1.82) and $\tau(x; \delta)$ is a sphere with radius δ and center at x and $s(x; \delta)$ the surface of it.

By the aid of Gauss divergence theorem, the domain integral can be transformed as follows,

$$\int_{D-\tau(x;\,\delta)} \mathring{U}(x, y)\, H(y)\, dv\,(y)$$

$$= \int_{\partial D + s(x;\,\delta)} [\mathring{U}(x, y)\{\overset{n}{T}(\bar{\lambda}, \bar{\mu}; \partial_y)\, u\,(y)\} - \{\mathring{U}(x, y)\,\overset{n}{T}(\bar{\lambda}, \bar{\mu}; \partial_y)\}\, u(y)]\, ds\,(y)$$

$$+ \int_{D-\tau(x;\,\delta)} [\mathring{U}(x, y)\, R(\bar{\lambda}, \bar{\mu}; \partial_y)]\, u\,(y)\, dv\,(y)$$

$$+ \int_{D-\tau(x;\,\delta)} \mathring{U}(x, y)\{\varrho(y)\, \omega^2\, u(y) + \varrho(y)\, b(y)\}\, dv\,(y) \qquad (1.118)$$

where

$$\{\mathring{U}(x, y)\, R\,(\bar{\lambda}, \bar{\mu}; \partial_y)\}_{ij} = [\bar{\lambda}(y)\, \mathring{U}_{ik}(x, y)_{,k}]_{,j} + [\bar{\mu}(y)\{\mathring{U}_{ij}(x, y)_{,k} + \mathring{U}_{ik}(x, y)_{,j}\}]_{,k}$$

with notation $(\ )_{,k} = \partial(\ )/\partial y_k$.

Therefore, substituting Eq. (1.118) into Eq. (1.117) and using the relation

$$\overset{n}{T}(\lambda, \mu; \partial) = \overset{n}{T}(\lambda_0, \mu_0; \partial) + \overset{n}{T}(\bar{\lambda}, \bar{\mu}; \partial),$$

we have

$$\int_{\partial D + s(x;\,\delta)} [\mathring{U}(x, y)\{\overset{n}{T}(\lambda, \mu; \partial_y)\, u\,(y)\} - \{\mathring{U}(x, y)\,\overset{n}{T}(\lambda, \mu; \partial_y)\}\, u\,(y)]\, ds\,(y)$$

$$\int_{D-\tau(x;\,\delta)} \{\mathring{U}(x, y)\, R(\bar{\lambda}, \bar{\mu}; \partial_y)\}\, u\,(y)\, dv\,(y)$$

$$+ \int_{D-\tau(x;\,\delta)} \mathring{U}(x, y)\{\varrho(y)\, \omega^2\, u(y) + \varrho(y)\, b(y)\}\, dv\,(y) = 0, \quad x \in D. \qquad (1.119)$$

When letting $\delta \to 0$, we have for $x \in D_c$

$$\lim_{\delta \to 0} \int_{\tau(x;\delta)} \{ \overset{\circ}{U}(x,y) \, R(\bar{\lambda},\bar{\mu};\partial_y) \} \, u(y) \, dv(y) = 0,$$

$$\lim_{\delta \to 0} \int_{\tau(x;\delta)} \overset{\circ}{U}(x,y) \{ \varrho(y) \, \omega^2 \, u(y) + \varrho(y) \, b(y) \} \, dv(y) = 0,$$

$$\lim_{\delta \to 0} \int_{s(x;\delta)} \overset{\circ}{U}(x,y) \{ \overset{n}{T}(\lambda,\mu;\partial_y) \, u(y) \} \, ds(y) = 0, \tag{1.120}$$

$$\lim_{\delta \to 0} \int_{s(x;\delta)} \{ \overset{\circ}{U}(x,y) \, \overset{n}{T}(\lambda,\mu;\partial_y) \} \, u(y) \, ds(y) = C(x) \, u(x), \quad x \in \partial D.$$

and

$$C(x) = \begin{cases} \dfrac{\mu_0 \, \lambda(x) + 2\lambda_0 \, \mu(x) + 6\mu_0 \, \mu(x)}{3\mu_0 \, (\lambda_0 + 2\mu_0)} & (3-D) \\[4mm] \dfrac{\mu_0 \, \lambda(x) + \lambda_0 \, \mu(x) + 4\mu_0 \, \mu(x)}{2\mu_0 \, (\lambda_0 + 2\mu_0)} & (2-D). \end{cases} \tag{1.121}$$

It is noted that the traction operator $\overset{n}{T}$ involves Lamé's constants $\lambda(y)$ and $\mu(y)$ as a function of y, whereas the fundamental solution $\overset{\circ}{U}(x,y)$ is for the homogeneous solid with constant λ_0 and μ_0.

Furthermore, for $x \in \partial D$, we have the free term

$$\lim_{\delta \to 0} \int_{s(x;\delta)} \overset{\circ}{U}(x,y) \, \overset{n}{T}(\lambda,\mu;\partial_y) \, ds(y) = C^e, \quad x \in \partial D \tag{1.122}$$

where $s(x;\delta)$ is the part of surface of the sphere with radius δ contained in D. For the smooth surface, the free term is given by

$$C^e(x) = \begin{cases} \dfrac{\mu_0 \, \lambda(x) + 2\lambda_0 \, \mu(x) + 6\mu_0 \, \mu(x)}{6\mu_0 \, (\lambda + 2\mu_0)} & (3-D) \\[4mm] \dfrac{\mu_0 \, \lambda(x) + \lambda_0 \, \mu(x) + 4\mu_0 \, \mu(x)}{4\mu_0 \, (\lambda + 2\mu_0)} & (2-D). \end{cases} \tag{1.123}$$

Using the above relations, we can derive the integral equation for an in-homogeneous solid as follows;

$$\varepsilon(x) \, u(x) = \int_{\partial D} \{ \overset{\circ}{U}(x,y) \, t(y) - \overset{\circ}{T}(x,y) \, u(y) \} \, ds(y)$$

$$+ \int_D \{ \overset{\circ}{U}(x,y) \, R(\bar{\lambda},\bar{\mu};\partial_y) \} \, u(y) \, dv(y)$$

$$+ \omega^2 \int_D \varrho(y) \, \overset{\circ}{U}(x,y) \, u(y) \, dv(y) + \int_D \varrho(y) \, \overset{\circ}{U}(x,y) \, b(y) \, dv(y), \tag{1.124}$$

where $\overset{\circ}{T}(x,y) \equiv \overset{\circ}{U}(x,y) \, \overset{n}{T}(\lambda,\mu;\partial_y)$ and

$$\varepsilon(x) = \begin{cases} C(x), & x \in D \\[3mm] \dfrac{C(x)}{2}, & x \in \partial D \\[3mm] 0, & x \in D_C. \end{cases}$$

It is noted that the circular frequency ω is not included in the integral. Thus the formulation of this type may be suitable for dynamic analysis.

If the elastic moduli change abruptly in crossing the surface, say S, we must further modify the formulation. Assume that the domain consists of two sub-domains D_1 and D_2 and S is their common boundary, the rest of the boundary are denoted by ∂D_1 and ∂D_2, respectively. Here, we use subscript 1 and 2 to distinguish the relevant quantities.

Since the elastic moduli and mass density vary continuously in each subdomain, we have for $D_1 \cup \partial D_1 \cup S$

$$\varepsilon_1(x)\,u_1(x) = \int_{\partial D_1 + S} \{\mathring{U}_1(x,y)\,t_1(y) - \mathring{T}_1(x,y)\,u_1(y)\}\,ds(y)$$

$$+ \int_{D_1} \{\mathring{U}_1(x,y)\,R_1(\bar{\lambda}_1, \bar{\mu}_1; \partial_y)\}\,u_1(y)\,dv(y) \qquad (1.125)$$

$$+ \omega^2 \int_{D_1} \varrho_1(y)\,\mathring{U}_1(x,y)\,u_1(y)\,dv(y) + \int_{D_1} \varrho_1(y)\,\mathring{U}_1(x,y)\,b_2(y)\,dv(y)$$

and similar expression with subscript 1 replaced by 2 for $D_2 \cup \partial D_2 \cup S$. These integral equations are coupled by the continuity conditions on the interface S,

$$u_1(x) = u_2(x), \quad t_1(x) + t_2(x) = 0, \quad x \in S. \qquad (1.126)$$

We can solve the coupled integral equations easily by the familiar technique.

The above formulation can be modified further. Since we can presume that the homogeneous part of the each subdomain has the same properties, i.e. $\lambda_{01} = \lambda_{02}$, $\mu_{01} = \mu_{02}$, $\varrho_{01} = \varrho_{02}$, and $\mathring{U}_1(x,y) = \mathring{U}_2(x,y)$ we have the following expression, by simply adding the integral equations for domains 1 and 2 and using the continuity conditions,

$$\{\varepsilon_1(x) + \varepsilon_2(x)\}\,u(x) = \int_{\partial D} \{\mathring{U}(x,y)\,t(y) - \mathring{T}(x,y)\,u(y)\}\,ds(y)$$

$$+ \int_D \{\mathring{U}(x,y)\,R(\bar{\lambda}, \bar{\mu}; \partial_y)\}\,u(y)\,dv(y)$$

$$+ \omega^2 \int_D \varrho(y)\,\mathring{U}(x,y)\,dv(y) + \int_D \varrho(y)\,\mathring{U}(x,y)\,b(y)\,dv(y)$$

$$- \int_S \{\mathring{U}(x,y)\,\overset{n_1}{T}(\Delta\lambda, \Delta\mu; \partial_y)\}\,u(y)\,ds(y) \qquad (1.127)$$

where

$$\varepsilon_1(x) + \varepsilon_2(x) = \begin{cases} C(x), & x \in D \\[2mm] \dfrac{C(x)}{2}, & x \in \partial D \\[3mm] \dfrac{C_1(x) + C_2(x)}{2}, & x \in S \\[3mm] 0, & x \in D_c, \end{cases}$$

$$\Delta\lambda(x) = \lambda_1(x) - \lambda_2(x), \quad \Delta\mu(x) = \mu_1(x) - \mu_2(x)$$

and

$$\{\mathring{U}(x,y)\,\mathring{T}^{n_1}(\varDelta\lambda, \varDelta\mu; \partial_y)\}_{ij}$$

$$= \varDelta\lambda(y)\,\mathring{U}_{ik}(x,y)_{,k}\,(u_1)_j(y) + \varDelta\mu(y)\,(n_1)_k(y)\,\{\mathring{U}_{ij}(x,y)_{,k} + \mathring{U}_{ik}(x,y)_{,j}\}.$$

It is easily shown that this expression is the extended form of Eq. (1.124) and, of course, that for homogeneous case.

1.6 Eigenfrequency Problems

Eigenfrequencies can be obtained as real parameter ω (circular frequency) which assures the existence of a non-singular solution of the homogeneous boundary integral equations derived from Table 1.1 by just putting the right-hand side terms equal to zero. It must be remembered that the parameter ω is implicitly involved in the operator-valued functions. Therefore, we define:
(i) eigenfrequencies of operator $\pm K$ as real values of ω for which the BIE
   $(\frac{1}{2}I \mp K)\,\mu = 0$ has non-trivial solution, and
(ii) eigenfrequencies of the operator S or D_n such the real values of ω that the BIE
   $S\,\mu = 0$ or $D_n\,\mu = 0$ has non-trivial solution. The real values of eigenfrequencies
   are due to the self-adjointness of the Navier-Cauchy operator L.

From the Table 1.1 of homogeneous BIEs, we realize for the interior problems that, Niwa et al. [42],
(i) eigenfrequencies for operators S, $\bar{S}$, K, $\bar{K}$, K^*, and $\bar{K}^*$ coincide each other, and
   that
(ii) eigenfrequencies for operators D_n, $\bar{D}_n$, $-K$, $-\bar{K}$, $-K^*$ and $-\bar{K}^*$ also coincide.

The fact (i), that is, eigenfrequencies of S and K coincide, is verified by taking account that no eigenfrequency and thus no eigenfunction exists for the exterior problems.

In Table 1.1, interrelationships among operators are shown by enclosing by solid lines, broken lines and chain lines, respectively. BIEs enclosed by solid lines are used for determining eigenfrequencies of the first interior problems and those enclosed by broken lines are used for the second interior problems.

It is noted here the followings, Kleinman and Roach [48], Niwa et al. [42] and Shaw [47].
(i) As for the BIEs of the second kind, BIEs for the first (second) exterior problems
   determine the eigenfrequencies of the second (first) interior problems, i.e.
   symbolically

$$\text{F.E.} \leftrightarrow \text{S.I.}, \quad \text{S.E.} \leftrightarrow \text{F.I.}$$

(ii) As for the BIEs of the first kind, BIEs for the first (second) exterior problems
   determine the eigenfrequencies of the first (second) interior problems, i.e.

$$\text{F.E.} \leftrightarrow \text{F.I.}, \quad \text{S.E.} \leftrightarrow \text{S.I.}$$

These facts raise a question whether the solution of the exterior problem obtained from BIE is unique or not. It is easy to show that the uniqueness of the exterior problems breaks down when the circular frequency parameter ω coincides with eigenfrequency of the corresponding interior problems. This inherent problem will be discussed in the next section.

We can also show that the eigenfrequencies of the third interior problems coincide.

1.7 Some Remarks on Inherent Problems of BIEM in Elastodynamics

1.7.1 Fictitious Eigenfrequencies in the Time-Harmonic Elastodynamics

1) Fictitious Eigenfrequencies

It is well known that the ordinary BIE formulation for exterior problem in acoustics fails to yield unique solution at certain wave numbers corresponding to the eigenvalues of the interior region of closed surface, Lamb [64], Kleinman and Roach [48]. The similar fact is also known in elastodynamics, Kupradze [10], and Kobayashi and Nishimura [34]. These special wave numbers are called "characteristic" wavenumber and the corresponding eigenfrequencies are called "fictitious" eigenfrequencies. In numerical procedure, at fictitious eigenfrequencies the coefficient matrix for the algebraic equations obtained from BIE by discretization becomes ill-conditioned and large errors yield.

In view of solving transient exterior problems, for instance soil-structure interaction problems due to seismic waves, we have to pay special attention to the treatment of fictitious eigenfrequencies.

2) Methods to Circumvent Fictitious Eigenfrequencies

Several methods to circumvent the fictitious eigenfrequencies have been proposed. They are classified to the following four types. For simplicity, we consider here the second exterior problem.

a) Formulation by the Aid of the Interior Representation

Noting that the equivalent integral representation to Navier-Cauchy equation together with boundary conditions is a pair of integral expressions

$$u = Du - S\bar{t}, \quad x \in D_+ \tag{1.128}$$

$$0 = Du - S\bar{t}, \quad x \in D_-, \tag{1.129}$$

Copley [65] suggested to use the interior representation (1.129). An unique solution u in D_+ is obtained provided that Eq. (1.129) satisfied at every point in D_-. The concept is also used by Kupradze. Schenk [66] developed this concept and solved the exterior boundary integral equation

$$(\tfrac{1}{2}I - \bar{K}^*)\,u = -S\bar{t} \quad \text{on } \partial D \tag{1.130}$$

together with Eq. (1.129), as a supplementary condition.

Since the number of unknowns is less than that of equations, the system is solved by the least-squares method. However, there is no clear guide how to choose the number and location of points on which supplementary condition works most efficiently.

b) Formulation by Use of Two Integral Representations

Kleinman and Roach [48] proposed to use two boundary integral equations

$$(\tfrac{1}{2} I - \bar{K}^*)\, u = - S\, \bar{t}, \quad \text{S.E.G.} (u) \tag{1.131}$$

$$D_n\, u = (\tfrac{1}{2} I + K)\, \bar{t}, \quad \text{S.E.G.} (t) \tag{1.132}$$

for the second exterior problem. This system of equations can be solvable and gives a unique solution even if parameters ω coincide with eigenfrequencies of the exterior problem, since Eqs. (1.131) and (1.132) have a different set of eigen-frequencies in general. Moreover, the solution of Eq. (1.131) satisfies Eq. (1.132).

Burton and Miller [67] suggested to use the combined integral equations

$$\bar{K}^*\, u - S\, t + \alpha\, (D_n\, u - K\, t) = \tfrac{1}{2} I(u + \alpha\, t) \quad \text{on } \partial D. \tag{1.133}$$

For the suitable choice of coupling constant, $\mathrm{Im}\,(\alpha) \neq 0$ for a real ω, this formulation leads to a unique solution even if ω is an eigenvalue of the first and second interior problems. This method was applied to acoustic problems by Meyer et al. [68] and Terai [69]. A major drawback of this method is that it contains strongly singular integral $D_n\, u$.

On the other hand, Greenspan and Werner [70], Kussmaul [71], and Filippi [72] used a mixed potential method, i.e.

$$u = (S + \alpha D)\, \mu \quad \text{in } D_+ \tag{1.134}$$

$$u = (D + \alpha S)\, \mu \quad \text{in } D_- \tag{1.135}$$

where $\mathrm{Im}\,(\alpha) \neq 0$ for a real ω. With a suitable choice of α leads to the existence of a unique layer density μ.

c) Formulation by Use of a Modified Fundamental Solution

Ursell [73] suggested to use a modified fundamental solution which also satisfies a certain dissipative condition on a circle lying inside the interior domain. The use of his modified fundamental solution overcomes the appearance of eigenfunction, and thus uniqueness is recovered. However, since his fundamental solution involves the computation of an infinite series with complex coefficients, the method is not very effective from the view point of numerical treatment.

Jones [74] modified the method to take into account only a finite number of real and non-zero coefficients instead of an infinite series although the applicability is limited. Jones proposed another method in the same paper to solve the boundary integral equation (1.130) subject to the conditions Eq. (1.129), just as Schenk's method. However, the conditions are satisfied termwise in Jones' method, so his method removes the arbitrariness in the selection of the interior points and provides a definite rule for the number of side conditions.

Jones' method was extended to apply for elastodynamic problems by Kobayashi and Nishimura [75]. In relation of this type of formulation see also Kleinman and Roach [76], and Martin [77].

d) Numerical Technique

In computational point of view, interpolation technique may be efficiently applied to overcome this problem, when fictitious eigenfrequencies are distributed rather sparsely as often encountered in the frequency ranges of practical importance. Since the ordinary BIE formulations work well according to our experience even for such values of ω which deviate from the unfavourable eigenfrequencies ω^* by only a small amount, say $2 \sim 3\%$ of ω^*, some interpolation schemes can be used to evaluate the results for the original frequency, even if it is the corresponding eigenfrequency.

As mentioned in the previous section, eigenfrequencies are obtained as such values of parameter ω for which a system of the homogeneous boundary integral equations has non-trivial solution. The condition for it is well known that the determinant of the coefficient matrix of the discretized equations of the system of BIEs is equal to zero. In the numerical calculation, it is hard to obtain the true root of the determinant, however if the frequency $\dot{\omega}$ is very close to one of the eigenfrequencies the value of the determinant becomes almost zero. This phenomenon can be easily detected during the triangular (L-U) decomposition process by the Crout method, Kobayashi and Nishimura [34].

1.7.2 Half-Plane Problems

1) Reflection of Waves at a Plane Surface

For simplicity we here consider a plane harmonic wave with an amplitude A and propagating in the direction n with constant velocity c

$$u(x, t) = A\,d\,e^{ik(x \cdot n - ct)}, \qquad (1.136)$$

where d is a unit vector defining the direction of motion.

Substituting Eq. (1.136) into the Navier-Cauchy equation, we have

$$(c^2 - c_T^2)\,d - (c_L^2 - c_T^2)\,(n \cdot d)\,n = 0 . \qquad (1.137)$$

Hence the non-trivial solution is possible only when

$$\det\left[(c^2 - c_T^2)\,\delta_{ij} - (c_L^2 - c_T^2)\,n_i\,n_j\right] = 0, \qquad (1.138)$$

so that the plane waves can propagate with two velocities c_L and c_T. If $c = c_L$, then $d = (n \cdot d)\,n$, which implies that the displacement vector is in the direction of propagation, i.e. longitudinal wave. If $c = c_T$, then $n \cdot d = 0$, which implies that the displacement vector is normal to the direction of propagation, i.e. shear wave. Therefore, $\nabla \times u = 0$ for longitudinal wave and $\nabla \cdot u = 0$ for shear wave.

The shear displacement can be any direction in a plane normal to the direction of propagation, but we usually select the (x_1, x_2)-plane to contain the propagation direction n and decompose the displacement into two components, that is one in the (x_1, x_2)-plane and the other normal to it. The former is called "SV-wave" (vertically polarized shear wave) and the latter "SH-wave" (horizontally polarized shear wave).

Next we consider the reflection of waves from the traction free boundary.

a) SH-Waves

An incident SH-wave propagating in the half-space $x_2 < 0$ is represented by

$$u_3^0 = A_0 \, e^{i\eta_0}, \quad \eta_0 = k_0 (x_\alpha \, n_\alpha^0 - c_T \, t). \tag{1.139}$$

A reflected SH-wave is expressed by

$$u_3^1 = A_1 \, e^{i\eta_1}, \quad \eta_1 = k_1 (x_\alpha \, n_\alpha^1 - c_T \, t). \tag{1.140}$$

Since the non-trivial stress components are

$$\sigma_{23}^0 = \mu \, u_{3,2}^0 = i k_0 \mu \, A_0 \, n_2^0 \, e^{i\eta_0}$$
$$\sigma_{23}^1 = \mu \, u_{3,2}^1 = i k_1 \mu \, A_1 \, n_2^1 \, e^{i\eta_1} \tag{1.141}$$

the condition that the plane $x_2 = 0$ is free of tractions, i.e. $\sigma_{23}^0 + \sigma_{23}^1 = 0$ can be satisfied only if

$$k_1 = k_0, \quad n_1^1 = n_1^0, \quad n_2^1 = -n_2^0, \quad A_1 = A_0 . \tag{1.142}$$

b) Reflection of P-Waves

An incident P-wave in $x_2 < 0$ is expressed by

$$\boldsymbol{u}^0 = A_0 \, \boldsymbol{d}^0 \, e^{i\eta_0}, \quad \eta_0 = k_0 (x_\alpha \, n_\alpha^0 - c_0 \, t) \tag{1.143}$$

and reflected waves in $x_2 < 0$ by

$$\boldsymbol{u}^m = A_m \, \boldsymbol{d}^m \, e^{i\eta_m}, \quad \eta_m = k_m (x_\alpha \, n_\alpha^m - c_m \, t) \tag{1.144}$$

where $m = 1$ and 2 imply the P- and SV-waves, thus

$$\boldsymbol{d}^0 = \boldsymbol{n}^0, \quad \boldsymbol{d}^1 = \boldsymbol{n}^1, \quad \boldsymbol{d}^2 = \boldsymbol{i}_3 \times \boldsymbol{n}^2 \quad \text{and} \quad \boldsymbol{d}^2 \cdot \boldsymbol{n}^2 = 0. \tag{1.145}$$

Stress components are given by in general

$$\sigma_{22}^m = i k_m \, [(\lambda + 2\mu) \, d_2^m \, n_2^m + \lambda \, d_1^m \, n_1^m] \, A_m \, e^{i\eta_m},$$
$$\sigma_{21}^m = i k_m \mu \, [d_2^m \, n_1^m + d_1^m \, n_2^m] \, A_m \, e^{i\eta_m} . \tag{1.146}$$

Thus the displacements and stresses on the plane $x_2 = 0$ become

$$\left. \begin{array}{l} u_\alpha^m = A_m \, d_\alpha^m \, e^{i\eta_m^1}, \\[4pt] \sigma_{22}^m = i k_m \, [(\lambda + 2\mu) \, d_2^m \, n_2^m + \lambda \, d_1^m \, n_1^m] \, A_m \, e^{i\eta_m^1}, \\[4pt] \sigma_{21}^m = i k_m \mu \, [d_2^m \, n_1^m + d_1^m \, n_2^m] \, A_m \, e^{i\eta_m^1}, \\[4pt] \eta_m^1 = k_m (x_1 \, n_1^m - c_m \, t). \end{array} \right\} \tag{1.147}$$

On the traction free surface, the sum of the three tractions due to the incident P-wave and reflected P- and SV-waves must vanish, i.e.

$$\sum_{m=0}^{2} \sigma_{22}^m = 0, \quad \sum_{m=0}^{2} \sigma_{21}^m = 0. \tag{1.148}$$

From these conditions, we immediately have the followings;

$$n_0^1 = n_1^1 = n_2^1; \quad k_0\, n_1^0 = k_1\, n_1^1 = k_2\, n_1^2 = k \quad \text{(apparent wave number)}$$

$$k_0\, c_L = k_1\, c_L = k_2\, c_T = \omega \quad \text{(circular frequency)} \tag{1.149}$$

$$(\because c_0 = c_1 = c_L,\ c_2 = c_T)$$

and

$$k_1 = k_0, \quad \frac{k_2}{k_0} = \frac{c_L}{c_T} = \varkappa = \sqrt{\frac{\lambda + 2\mu}{\mu}} = \sqrt{\frac{2(1-v)}{1-2v}},$$

$$n_1^0 = n_1^1, \quad n_1^2 = n_1^0 \frac{k_0}{k_2} = \varkappa^{-1} n_0^1 \quad (v\text{: Poisson's ratio}). \tag{1.150}$$

Thus, using Eq. (1.145), we have also

$$n_2^0 = \sqrt{1 - (n_1^0)^2}, \quad n_2^1 = -\sqrt{1 - (n_1^0)^2}, \quad d_1^2 = -n_2^2, \quad d_2^2 = n_1^2. \tag{1.151}$$

The amplitude of the reflected waves are obtained from Eq. (1.148)

$$\begin{bmatrix} \lambda + 2\mu\,(n_2^1)^2 & 2\mu\,\varkappa\, n_1^2\, n_2^2 \\ 2n_1^1\, n_2^1 & \varkappa\{(n_1^2)^2 - (n_2^2)^2\} \end{bmatrix} \begin{bmatrix} A_1/A_0 \\ A_2/A_0 \end{bmatrix} = \begin{bmatrix} -(\lambda + 2\mu\,(n_2^0)^2) \\ -2n_1^0\, n_2^0 \end{bmatrix}. \tag{1.152}$$

When we substitute

$$n_1^0 = \sin\theta_0, \quad n_2^0 = \cos\theta_0, \quad n_1^1 = \sin\theta_0, \quad n_2^1 = -\cos\theta_0,$$

$$n_1^2 = \sin\theta_2 = \varkappa^{-1}\sin\theta_0, \quad n_2^2 = -\cos\theta_2$$

as in usual expression, we have

$$\frac{A_1}{A_0} = \frac{\sin 2\theta_0 \sin 2\theta_2 - \varkappa^2 \cos^2 2\theta_2}{\sin 2\theta_0 \sin 2\theta_2 + \varkappa^2 \cos^2 2\theta_2}$$

$$\frac{A_2}{A_0} = \frac{2\varkappa \sin 2\theta_0 \cos 2\theta_2}{\sin 2\theta_0 \sin 2\theta_2 + \varkappa^2 \cos^2 2\theta_2}. \tag{1.153}$$

It is noted that when $\sin 2\theta_0 \sin 2\theta_2 = \varkappa^2 \cos^2 2\theta_2$, the incident P-wave is reflected as an SV-wave only, this phenomenon is called "mode conversion", in which case $A_2/A_0 = \varkappa \cot 2\theta_2$.

c) Reflection on SV-Waves

An incident wave is represented by

$$u^0 = A_0\, d^0\, e^{i\eta_0}, \quad \eta_0 = k_0(x_\alpha\, n_\alpha^0 - c_0\, t), \tag{1.154}$$

where $d^0 = i_3 \times n^0$.

Reflected waves are represented as mentioned in (b). In a similar manner as (b), we have

$$k_0\, n_1^0 = k_1\, n_1^1 = k_2\, n_1^2 = k, \quad k_0\, c_T = k_1\, c_L = k_2\, c_T = \omega,$$

$$k_2 = k_0, \quad k_1/k_0 = c_T/c_L = \varkappa^{-1},$$

$$n_1^0 = n_1^2, \quad n_1^1 = \varkappa\, n_1^0, \quad n_2^0 = \sqrt{1 - (n_1^0)^2}, \quad n_2^1 = -\sqrt{1 - (\varkappa\, n_1^0)^2}, \tag{1.155}$$

$$d_1^2 = -n_2^2, \quad d_2^2 = n_1^2.$$

The amplitude ratio of reflected waves are obtained similarly.

$$\frac{A_1}{A_0} = - \frac{\varkappa \sin 4\theta_0}{\sin 2\theta_0 \sin 2\theta_1 + \varkappa^2 \cos^2 2\theta_0},$$

$$\frac{A_2}{A_0} = \frac{\sin 2\theta_0 \sin 2\theta_1 - \varkappa^2 \cos^2 2\theta_0}{\sin 2\theta_0 \sin 2\theta_1 + \varkappa^2 \cos^2 2\theta_0}. \tag{1.156}$$

The reflected P-wave vanishes at $\theta_0 = 0, \frac{\pi}{4}, \frac{\pi}{2}$.
In order to let n_1^1 be less than unity, $n_1^0 = \sin \theta_0$ must be such that

$$\sin \theta_0 \le \varkappa^{-1} \quad \text{or} \quad \theta_0 \le \theta_{cr} = \sin^{-1}(1/\varkappa), \tag{1.157}$$

where θ_{cr} is called the critical angle. If $\theta_0 > \theta_{cr}$, the component n_2^1 becomes

$$n_2^1 = - \sqrt{1 - \varkappa^2 \sin^2 \theta_0} = - i\varkappa\beta, \quad \beta = \sqrt{\sin^2 \theta_0 - \varkappa^{-2}} \tag{1.158}$$

and the reflected P-wave must be evaluated by replacing $\cos \theta_1$ by $- i\varkappa\beta$.

$$u^1 = A^1 d^1 e^{k_0 \beta x_2} e^{i k_0 \sin \theta_0 (x_1 - c_L t/\varkappa \sin \theta_0) - i\alpha},$$

$$A^1 = \frac{A_0 \sin 4\theta_0}{\{\varkappa^2 \cos^4 2\theta_0 + 4(\varkappa^2 \sin^2 \theta_0 - 1) \sin^2 2\theta_0 \sin 2\theta_0\}^{1/2}},$$

$$\tan \alpha = \frac{2(\varkappa^2 \sin^2 \theta_0 - 1)^{1/2} \sin 2\theta_0 \sin \theta_0}{\varkappa \cos^2 2\theta_0}. \tag{1.159}$$

2) Green's Function

From the practical point of view, problems including a half-plane are of importance. It is desirable for this type of problem to satisfy the boundary conditions over the half-plane.

The Green's function which satisfies traction free conditions over the half-plane is given by Kinoshita [78]. Refering to the Cartesian system with its origin at the free surface and x_1- and x_2-axes directing to the right along the horizontal free surface and into the half-plane, respectively, Green's function due to point force acting at $c = (0, c)$ in the direction of x_k-axis, i.e. $\delta_{ik} \delta(x - c)$ is given as follows (for a plane strain state):

$$G_{ik}(x, c) = U_{ik}(x, c) + U_{ik}(x, - c) + \frac{1}{2\pi} \int_{-\infty}^{\infty} A_{ik}(\xi) e^{i\xi x_1} d\xi, \tag{1.160}$$

$$A_{11} = - \frac{\xi^2}{\mu F(\xi) k_T^2 R_T} \{2 R_L R_T e^{-x_2 R_T} - (2\xi^2 - k_T^2) e^{-x_2 R_L}\}$$
$$\cdot \{2 R_L R_T e^{-c R_T} - (2\xi^2 - k_T^2) e^{-c R_L}\}$$

$$A_{21} = - \frac{i\xi}{\mu F(\xi) k_T^2} \{2\xi^2 e^{-x_2 R_T} - 2(\xi^2 - k_T^2) e^{-x_2 R_L}\}$$
$$\cdot \{2 R_L R_T e^{-c R_T} - (2\xi^2 - k_T^2) e^{-c R_L}\}$$

$$A_{12} = - \frac{i\xi}{\mu F(\xi) k_T^2} \{2\xi^2 e^{-x_2 R_L} - (2\xi^2 - k_T^2) e^{-x_2 R_L}\}$$
$$\cdot \{2 R_L R_T e^{-c R_L} - (2\xi^2 - k_T^2) e^{-c R_T}\}$$

$$A_{22} = -\frac{\xi^2}{\mu F(\xi) \, k_T^2 \, R_T} \{2R_L \, R_T \, e^{-x_2 R_L} - 2(\xi^2 - k_T^2) \, e^{-x_2 R_T}\}$$
$$\cdot \{2R_L \, R_T \, e^{-c R_L} - (2\xi^2 - k_T^2) \, e^{-c R_T}\}$$

$$F(\xi) = (2\xi^2 - k_T^2)^2 - 4\xi^2 R_L \, R_T \quad \text{(Rayleigh function)}$$

$$R_L = \sqrt{\xi^2 - k_L^2}, \quad R_T = \sqrt{\xi^2 - k_T^2}$$

where U_{ik} is given by Eq. (1.78).

This Green's function must be evaluated numerically. Using this Green's function, there is no need to evaluate surface integrals on the free boundary of the half-plane. However, of course, we must care for the fictitious eigenfrequencies if this Green's function is used for the half-plane containing some boundaries in it.

Since this rather orthodox way is tedious and CPU time consuming, for practical use we hope to truncate the infinite boundary and to take a only a finite part of it into account. In comparing with the results obtained by the use of the Green's function and those with truncated boundary, we confirmed that the truncation about twice of the incident wavelength gives practically the same results as those obtained by the aid of Green's function, as will be seen later.

With an assumption that the asymptotic behavior of the scattered wave on the free boundary mainly consists of Rayleigh wave, a semi-infinite element can be also devised for the Rayleigh wave, Kobayashi and Nishimura [79].

For the anti-plane problems, Green's function for the half-plane is simply given by

$$G(x, y) = \frac{i}{4\mu} \{H_0^{(1)}(k_T r) + H_0^{(1)}(k_T r')\}, \quad r' = |x' - y|, \qquad (1.161)$$

where x' is the mirror image of x with respect to the traction free boundary. By the use of this Green's function for a fundamental solution, no integral evaluation is required on the free surface. The only drawback of this formulation is the possibility of the fictitious eigenfrequencies when D is an indented half-plane or half-plane having cavities.

1.7.3 Treatment of Singularities

1) Use of Singular Solution and Singular Element

The existence of singularity may pollute the BIE solution. The standard way to improve the solution is, of course, to take the properties of the singularities into account. Eigenfunction expansion may be used at the singularity incorporated with BIEM, Jaswon and Symm [58], Kelly et al. [80].

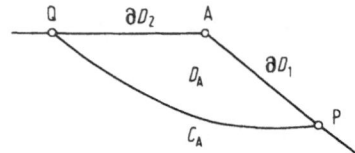

Fig. 1.1.   Singular point A and auxiliary contour C_A

Watson [81] has introduced an auxiliary contour in order to evaluate singular integrals with Hermitian boundary elements in the analysis of fracture mechanics. The similar technique can be also used in acoustics and elastodynamics.

The point is as follows. We formulate BIE in an usual manner, which formally contains the boundary integrals over singular elements AP and AQ, which contains a singular point A as shown in Fig. 1.1. Let assume that displacement and traction over the singular elements consist of sum of the regular part and the singular part, denoted by subscript s. Since the boundary integral equation on $\partial D_1 \cup \partial D_2 \cup C_A$ is given as

$$F(x) \, u_s(x) = \int_{\partial D_1 \cup \partial D_2 \cup C_A} \{U(x, y) \, t_s(y) - T(x, y) \, u_s(y)\} \, ds(y) \qquad (1.162)$$

where

$$F(x) = \begin{cases} 0 & x \bar{\in} D_A \cup \partial D_1 \cup \partial D_2 \cup C_A \\ C_P^e & \text{at } P \\ C_Q^e & \text{at } Q, \end{cases}$$

C^e is the free term evaluated similarly as Eq. (1.49), or (1.122), and u_s is a singular solution which satisfies given boundary conditions and t_s traction derived from u_s, the integrals over the singular elements AP and AQ can be evaluated by the use of integral over an auxiliary contour

$$\int_{\partial D_1 \cup \partial D_2} \{U(x, y) \, t_s(y) - T(x, y) \, u(y)\} \, ds(y)$$
$$= F(x) \, u_s(x) - \int_{C_A} \{U(x, y) \, t_s(y) - u_s(y) \, T(x, y)\} \, ds(y). \qquad (1.163)$$

The idea is also extended to study the singular behaviours of simple layer potential at the external corners, Kobayashi et al. [82]. The singular behaviour of the indirect BIE at external corner are extensively discussed by Hartman [83].

Singular boundary elements are also devised to evaluate boundary integrals, Kelly et al. [80], Lera et al. [84], Atkinson [85]. In this case, the "mixed" element, that is, interpolation functions are different for displacements and tractions, may be well incorporated, Paris et al. [86] and Athanasiadis [87].

2) Use of the Conditions at Element Intersection

In numerical implimentation the boundary is divided into elements. At the intersection of each segment a total of 8 (in 2-D) or 18 (in 3-D) quantities are associated with 2 (or 3) components of traction and displacement at each side of the node. In a well-posed problem, 2 (or 3) components (traction or displacement, or mixed) are prescribed on each of the adjacent elements. 2 (or 3) of the approximated BIEs are relevant to each node. Therefore we have 6 (or 12) relations for 8 (or 18) quantities. 2 (or 6) additionl relations must be supplemented to determine the all relevant quantities.

For simplicity in two-dimensional case, we have the following conditions.

a) Inter-element continuity (or prescribed jump) of displacement provides 2 relations.

$$u_\alpha^1(x_0) = u_\alpha^2(x_0) \quad (\alpha = 1, 2), \qquad (1.164)$$

where x_0 is the position of the intersection node of element 1 and element 2.

b) For the case in which the displacement conditions can not be applicable, Chaudonneret [88] has derived the following two additional independent equations for plane strain state.

(i) The symmetry of stress tensor at the intersection node of element leads to

$$t_\alpha^1 n_\alpha^2 = t_\alpha^2 n_\alpha^1 \quad (\alpha = 1, 2), \tag{1.165}$$

where t_α^m ($m = 1, 2$) and n are traction components and unit outward normal vector of the elements 1 and 2.

(ii) The invariance of trace of strain tensor

$$\varepsilon^1 \alpha \alpha = \varepsilon^2 \alpha \alpha \quad \text{or} \quad \varepsilon_n^1 - \varepsilon_n^2 = \varepsilon_s^2 - \varepsilon_s^1$$

provides the expression

$$t^1 \cdot n^1 - t^2 \cdot n^2 = 2(\lambda + \mu) \left\{ \left(\frac{\partial n_s}{\partial s}\right)^2 - \left(\frac{\partial u_s}{\partial s}\right)^1 \right\}_{x=x_0} \tag{1.166}$$

where n and s denote normal and tangential directions, respectively, in the local coordinates, and ε_n and ε_s stand for normal and tangential strain components on the each element. The displacement gradient usually approximated by appropriate manner, say from the element shape function. This technique is called "double point" method. The method is extended for corner intersection of multiple region by Wardle and Crotty [89] and quadratic element by Rudolphi [90].

In general in two-dimension, six possible types of boundary condition combinations arise on the adjacent elements; (a) tractions-tractions, (b) tractions-mixed, (c) tractions-displacements, (d) mixed-mixed, (e) mixed-displacements, and (f) displacements-displacements. In cases (a), (b), (c) and (d), since two displacement components can be eliminated by the aid of continuity condition of displacement at the intersection, two unknown displacements are determined from BIEs. In cases (e) and (f), two traction components are eliminated by Eqs. (1.165) and (1.166) in order to obtain the solution from BIEs.

The method proposed by Chaudonneret can be extended to the three dimensional cases. The concept of "triple point" (x_1, x_2, x_3) is substituted to the "double point" one. Six additional relations must be supplemented in the three dimensional case. The continuity condition of displacements are $u(x_1) = u(x_2)$, $u(x_2) = u(x_3)$, $u(x_3) = u(x_1)$. If these conditions may not available according to boundary conditions, the symmetry of stress tensor allows to write 3 relations which can be expressed in terms of tractions, and 3 more conditions can be provided by the compatibility between the strain and displacement and by considering every surface associated with triple point.

Similar treatment for potential problems is found in Alarcon et al. [91].

1.7.4 BIE-FE Hybrid Method

It is natural to attempt to combine BIEM with FEM in order to solve more complicated problems. Such hybrid method enjoys merits of both techniques, since FEM is powerful for interior non-homogeneous domain and BIEM can take care of external domain extended to infinity. Therefore, the BIE-FE hybrid method is very useful for the analysis of dynamic soil-structure interaction problems.

General coupling procedure of BIE with other numerical methods is discussed in several papers, Shaw [92], Zienkiewicz et al. [93], Kelly et al. [94], Brebbia [95] and Margulies [96].

The coupling of BIE and FE is simply explained for the reduced two-dimensional problem as follows. If we use FEM for $D_i \cup \partial D_F \cup S$ and BIE for $D_e \cup \partial D_B \cup S$, where S is the common boundary, and ∂D_F and ∂D_B are traction-free.

As for BIE, using the point collocation method we have BIEs satisfied at the collocation points x_i $(i = 1, 2, \ldots, n)$

$$(C + D)\, u(x_i) - (S\, t)(x_i) = 0 \qquad (1.167)$$

where

$$S\, t(x) = \int_C U(x, y)\, t(y)\, dc(y)$$

$$D\, u(x) = \int_C T(x, y)\, u(y)\, dc(y)$$

$$C\, u(x) = \int_C T(x, y)\, u(y)\, \delta(x - y)\, dc(y).$$

We here introduce the shape function matrix

$$N(x) = \begin{bmatrix} N_1 & 0 & \!-\!-\!- & N_n & 0 \\ 0 & N_1 & \!-\!-\!- & 0 & N_n \end{bmatrix} \qquad (1.168)$$

with local support

$$N_i(x_j) = \delta_{ij}$$

displacement and traction are approximated by

$$u(x) = N(x)\, \hat{u}, \quad t(x) = N(x)\, \hat{t} \qquad (1.169)$$

where $\hat{u}, \hat{t}$ denote nodal values.

By the aid of these approximations, Eq. (1.167) is converted into

$$H\, \hat{u} - G\, \hat{t} = 0 \qquad (1.170)$$

where

$$H_{ij} = (C + D)\, N_j(\hat{x}_i) = \left[C + \int_C T(\hat{x}_i, y) \right] N_j(y)\, dc(y)$$

$$G_{ij} = (S)\, N_j(\hat{x}_i) = \int_C U(\hat{x}_i, y)\, N_j(y)\, dc(y).$$

Finally, for $D_e \cup \partial D_B \cup S$, BIE can be written as

$$[H_B \; H_s] \begin{bmatrix} \hat{u}_B \\ \hat{u}_s \end{bmatrix} = [G_B \; G_s] \begin{bmatrix} \hat{t}_B \\ \hat{t}_{sB} \end{bmatrix}. \qquad (1.171)$$

As for FEM, using the same shape function as in BIE, we have for $D_i \cup \partial D_F \cup S$,

$$\begin{bmatrix} K_{FF} & K_{Fs} \\ K_{sF} & K_{ss} \end{bmatrix} \begin{bmatrix} \hat{u}_F \\ \hat{u}_s \end{bmatrix} = \begin{bmatrix} 0 \\ \hat{f}_s \end{bmatrix} = \begin{bmatrix} 0 \\ M\hat{t}_{sF} \end{bmatrix} \qquad (1.172)$$

where K is stiffness matrix and $\hat{f}$ is nodal force defined by

$$K = \int_D [B^T D B - \varrho\, \omega^2\, N^T N]\, ds, \quad \hat{f}_s = \left(\int_C N^T N\, dc \right) \hat{t} = M\hat{t}$$

and B and D are strain matrix and elasticity matrix, i.e.

$$\varepsilon = B u, \quad \sigma = D \varepsilon.$$

By the aid of the coupling conditions of BIE and FE, that is, continuity conditions of displacements and tractions on

$$u_{sF} = u_{sB}, \quad t_{sF} + t_{sB} = 0 \quad \text{on } S \tag{1.173}$$

we have the following system of equations:

$$\begin{bmatrix} K_{FF} & K_{Fs} & 0 & 0 \\ K_{sF} & K_{ss} & M_s & 0 \\ 0 & H_s & -G_s & H_B \end{bmatrix} \begin{bmatrix} \hat{u}_F \\ \hat{u}_s \\ \hat{t}_{sB} \\ \hat{u}_B \end{bmatrix} = \begin{bmatrix} 0 \\ 0 \\ G_B \, \hat{t}_B \end{bmatrix}. \tag{1.174}$$

1.7.5 Miscellaneous in Numerical Treatment

Boundary integral equations in general have to be solved numerically by applying some discretizing technique, say, FEM and method of weighted residuals, Zienkiewicz [97]. In numerical procedures, evaluation of the integral over each element is of basic importance.

Element size must be determined in general in order to express the shortest wave length appropriately, that is, at least four nodal points should be included in the shortest wavelength.

The boundary integrals, whenever non-singular, usually evaluated by Gaussian quadrature formula. Since the evaluation of integrals occupies a significant amount of computational time, it is advised to optimize, if possible, by specifying a certain maximum error bounds on the numerical integrations, Lachat and Watson [98].

When the field points coincide with the node of the same element of integration, ordinary Gaussian quadrature formula cannot be used. In the two-dimensional analysis the integral involving U can be split into a non-singular part and singular part, which has a logarithmic singularity. This particular singular integral can be evaluated by the use of Gaussian quadrature formula with logarithmic weight, Stroud and Secrest [99]. In using this formula, singular point must be placed at the origin.

The singular integral involving T together with free term C is evaluated by applying arbitrary rigid-body displacements, since for existence of such rigid body displacement each coefficient of the on-diagonal blocks must be equal to the sum of all the corresponding off-diagonal coefficients with a change in sign, Cruse [100], Rizzo and Shippy [101], and Lachat and Watson [98].

For the three-dimensional case, integrals involving T can be evaluated by using rigid body displacements, and integrals involving U can be evaluated by the use of Gaussian quadrature formula locating singular point on a common apex of triangular subelements, Rizzo and Shippy [101], and similar technique, Katsikadelis and Armenakas [102].

In the evaluation of integrals in the integral transformed domain, in order to avoid cancellation we had better expand the exponential function (in 3-D), the

Hankel function and the modified Bessel function (in 2-D) into series and delete unfavorable terms, particularly for small argument, Kobayashi and Nishimura [34], Dominguez and Alarcon [51].

The body force integral sometimes can be converted into the surface integral by the use of Gauss' divergence theorem, Stippes and Rizzo [103], Cruse et al. [104] and Danson [105].

The numerical inversion of the Fourier transform is recently accomplished by the use of the Fast Fourier Transform (FFT) algorithm, Cooley and Tukey [106], Brigham [107]. The numerical Laplace inversion transform is also studied in some details, Davies and Martin [108], Narayanan and Beskos [109]. Among many proposed methods, Durbin's method [110] is successfully applied for dynamic response analyses and stress concentration problems by Narayanan and Beskos [109] and Manolis and Beskos [27], and Manolis [28]. Durbin's method is based on the trapezoidal rule applied to the combination of both finite sine and cosine Fourier transforms, whose computations are performed by the aid of the FFT algorithm.

1.8 Application Examples

1.8.1 Transient Response Analysis by the Time-Space Domain BIEM

We have not much information about the applications of BIEM in time-space domain. It has just began to be applied to transient elastodynamic problems. Since in this method time-marching integration is required, the numerical treatment is

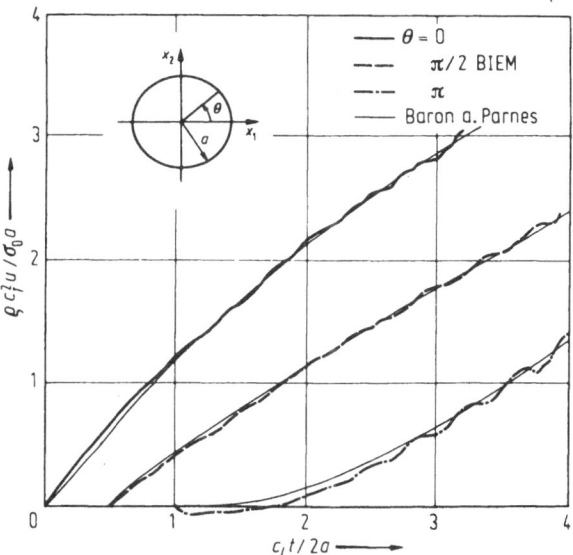

Fig. 1.2. Boundary displacements around a circular opening due to an incident step P-wave compared with Baron and Parnes [111]

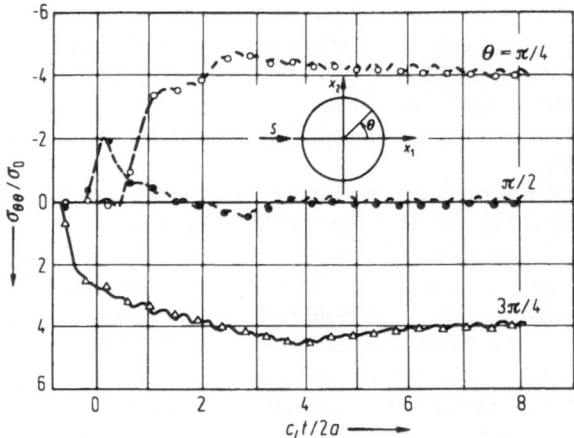

Fig. 1.3. Transient hoop stresses around a circular opening subject to an incident step SV-wave. Curves: Time-space domain BIEM by Fujiki [112], symbols: BIEM with FFT

rather complicated and usually more CPU time consuming than BIEM in the integral transformed domain. However, if we need responses at very early stages excited by impulsive loadings, the time-space domain BIEM may be well suited.

Figure 1.2 shows transient displacement histories at several points on the surface of a circular cylindrical opening contained in an elastic domain, when subjected to a plane step shock

$$u_{01} = \frac{\sigma_0}{3\varrho c_T^2} \{c_L t + (x_1 - a)\} H\{c_L t + (x_1 - a)\}$$

travelling in the direction parallel to x_1-axis, Niwa et al. [45].

The problem was solved by the use of the three-dimensional kernels with plane strain state and Poisson's ratio $\nu = 0.25$. In computation, 24 constant elements were used together with backward finite difference technique for time integration step with time increment Δt such that $c_L \Delta t/a = 0.10$.

The results show fairly good agreement with analytical ones obtained by Baron and Parnes [111].

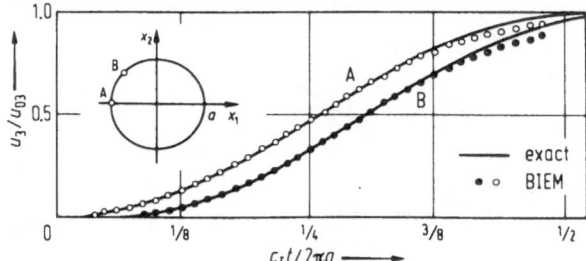

Fig. 1.4. Displacement on the boundary of a circular disc at very early stage due to a shifted sinusoidal incident SH shock wave

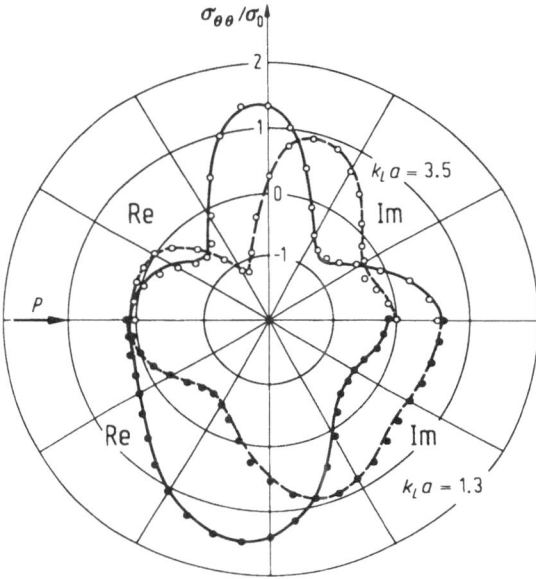

Fig. 1.5. Steady-state hoop stress distribution around a circular opening due to an incident P-wave, Poisson's ratio $v = 0.26$. Curves: Pao [113], symbols: BIEM

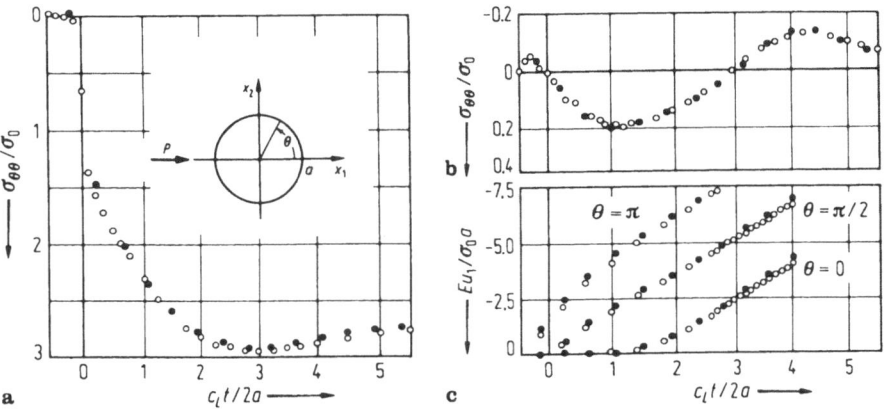

Fig. 1.6 a, b. Transient stresses and displacements around a circular opening due to an incident step P-wave. Open circles: Garnet and Crouzet-Pascal [114], solid circles: BIEM

The similar results due to SV-wave by Fujiki [112], are also compared with those obtained by Fourier transformed domain BIEM, in Fig. 1.3. Both results show a reasonable agreement.

Figure 1.4 shows the early stage transient anti-plane displacements of a unit circular domain subject to an incident wave

$$u_{03} = \{1 - \cos{(c_T(t - t_0) - x_1)}\}\, H\{c_T(t - t_0) - x_1\}.$$

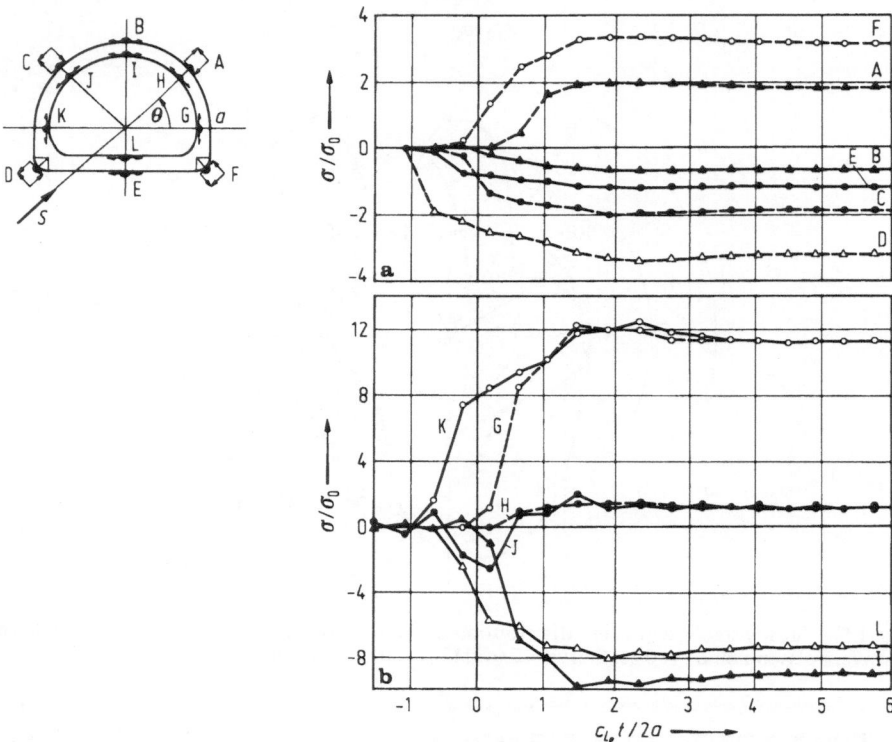

Fig. 1.7. Transient stresses around a horse-shoe-shaped tunnel due to SV-wave. **a** Stresses in the sourrounding medium. **b** Stresses at the inner surface of the lining. Poisson's ratio: $v_e = 1/4$, $v_i = 1/6$, density: $\varrho_i/\varrho_e = 1.0$, Young's modulus: $E_i/E_e = 5$ (i: lining, e: soil)

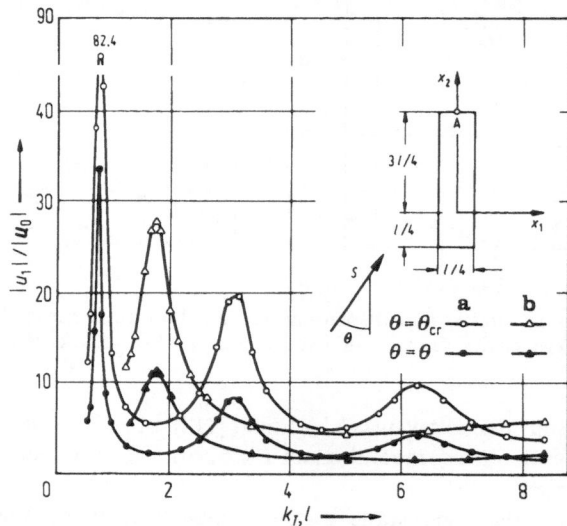

Fig. 1.8. Lateral displacement amplitudes at the top A of a column subject to time-harmonic incident SV-wave; **a** Young's modulus: $E_1/E_2 = 1.0$, density: $\varrho_1/\varrho_2 = 1.0$, **b** $E_1/E_2 = 10$, $\varrho_1/\varrho_2 = 1.0$ (1: column, 2: soil, Poison's ratio $v_1 = v_2 = 1/4$)

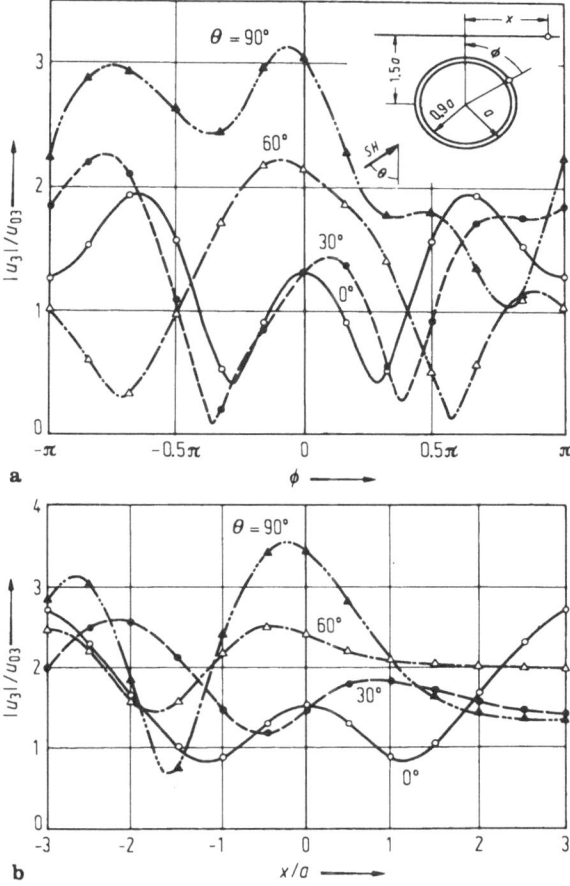

Fig. 1.9. Displacements **a** on the outer surface of lining and **b** on the ground surface due to the incident time-harmonic SH-wave, wave number $k_T a = 1.57$, wave speed ratio $(c_1/c_2)^2 = \varrho_2 \mu_1/(\varrho_1 \mu_2) = 3.0$ (1: lining, 2: soil); symbols: BIEM, curves: analytical solutions

The problem was solved by use of the two-dimensional BIEM, Ishihara [55]. The time integrations were modified first to facilitate computations. In computations, 24 constant elements were used with $\mu = 0.5$, $\varrho = 1.0$, $t_0 = 1.5$ and time increment Δt such that $c_T \Delta t = \pi/40$. The results show good agreement with exact ones at the very early stage, though they gradually deviate from the true values with an increase of time steps. We need further study to improve the numerical procedures. For other examples, see [44 – 46].

1.8.2 Applications of Integral Transformed Domain BIEM

Transient elastodynamic responses as well as time harmonic behaviours so far have been mostly analysed in conjunction with integral transforms. BIEMs in the integral transformed domain are amenable to numerical treatment, since they are

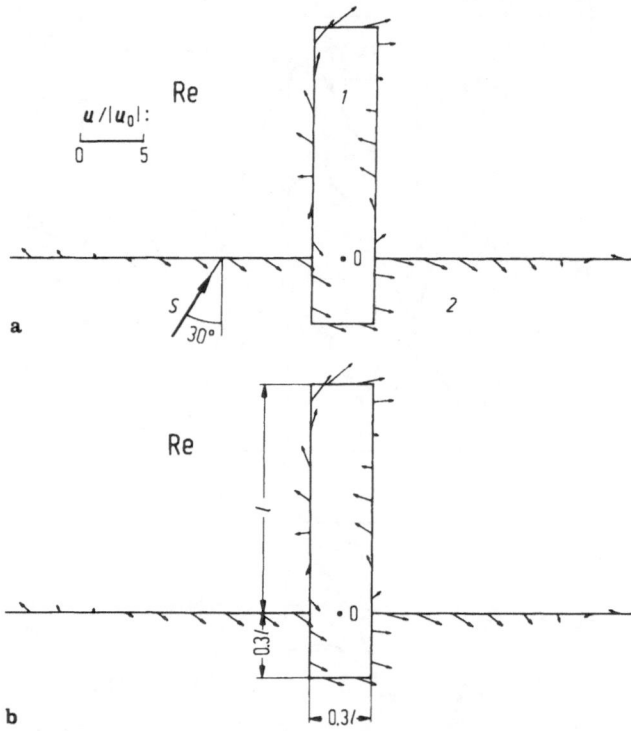

Fig. 1.10. Displacements due to time-harmonic SV-wave by the aid of **a** the half-plane Green's function, and **b** fundamental solution. Incident wave length: $2l$, peak at 0, Young's modulus: $E_1/E_2 = 1.0$, Poisson's ratio: $v_1 = v_2 = 1/4$, density: $\varrho_1/\varrho_2 = 1.0$

of elliptic type boundary value problems, that is, similar as elastostatics. Some fundamental problems have been investigated toward the applications of this method, much of the work still have to be needed, specifically in applications to structure-ground interaction problems and ground motion analysis due to earthquakes. Here we show some basic examples.

Figure 1.5 shows hoop stresses on the surface of an opening (plane strain state) due to sinusoidal incident P-wave. In this case, 24 quadratic boundary elements were employed. BIEM results show an excellent agreement with the analytical solution by Pao [113].

Figure 1.6 shows the displacement and stress time histories at several points on the surface of the opening subject to unit step impulse, where FFT algorithm is used to transform the problem and reconstitution of the transformed solutions. In computations, 24 quadratic elements were also used. The results accord well with analytical ones obtained by Garnet and Crouzet-Pascal [114].

Figure 1.7 shows stress-time history at several points on the lining of a tunnel subject to the incident step SV-wave of oblique incidence. The lining was divided into 24 quadratic elements on inner and outer surfaces, respectively.

Figure 1.8 illustrates the horizontal displacement response curves at the top of the column to the incident SV-wave with different frequencies. In the figure, a

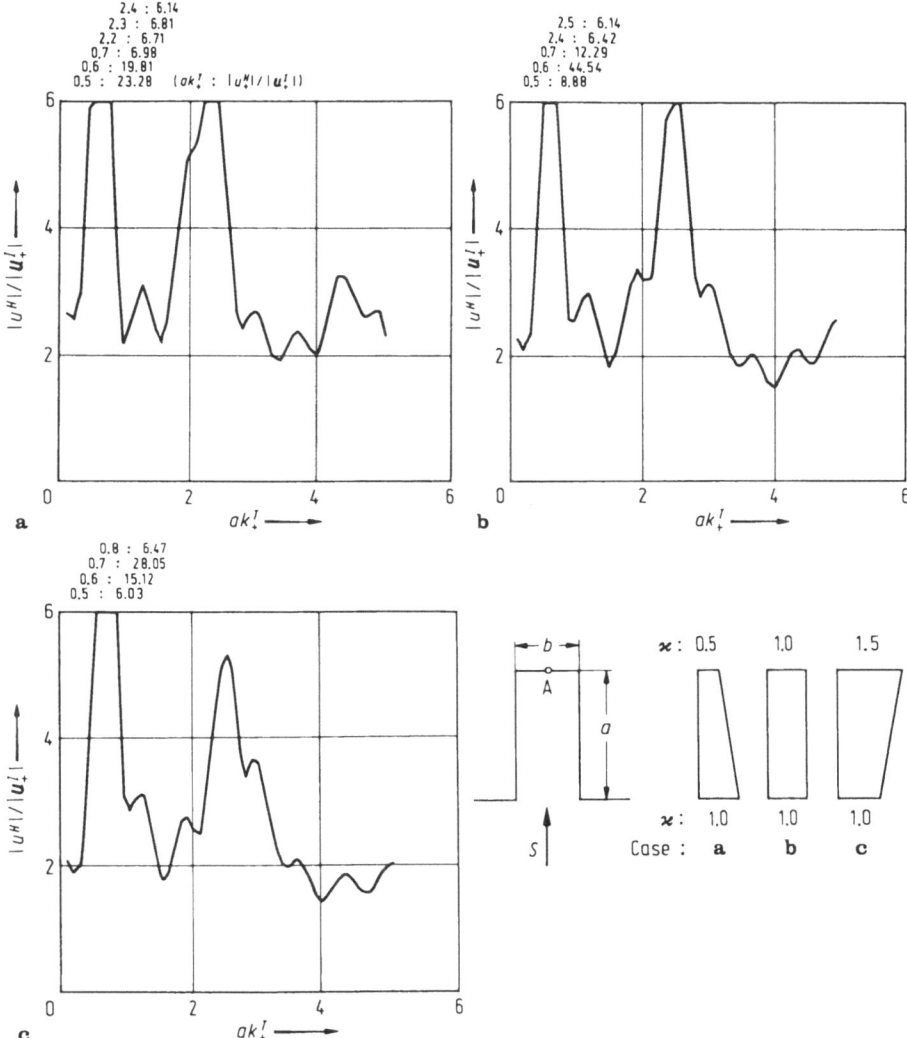

Fig. 1.11 a–c. Horizontal displacement response curves at the top A of an extruded inhomogeneous body subject to time-harmonic SV-wave of vertical incidence. Poisson's ratio: $v_+ = v_- = 1/4$, density: $\varrho_-/\varrho_+ = 1.0$, shear modulus: $\varkappa = \mu_-/\mu_+$ (+: half-plane, −: extruded body)

comparison is also made for the responses due to the different incident angles, that is, vertical and critical incident angles. Their response curves are similar, but the amplitude for the latter is much larger than the former.

Figure 1.9 shows the anti-plane displacement amplitudes on the outer surface of the linning and on the ground surface due to time-harmonic SH-waves. The analysis was done with 120 constant boundary elements and utilizing the half-plane Green's function. The results are in good accordance with analytical ones.

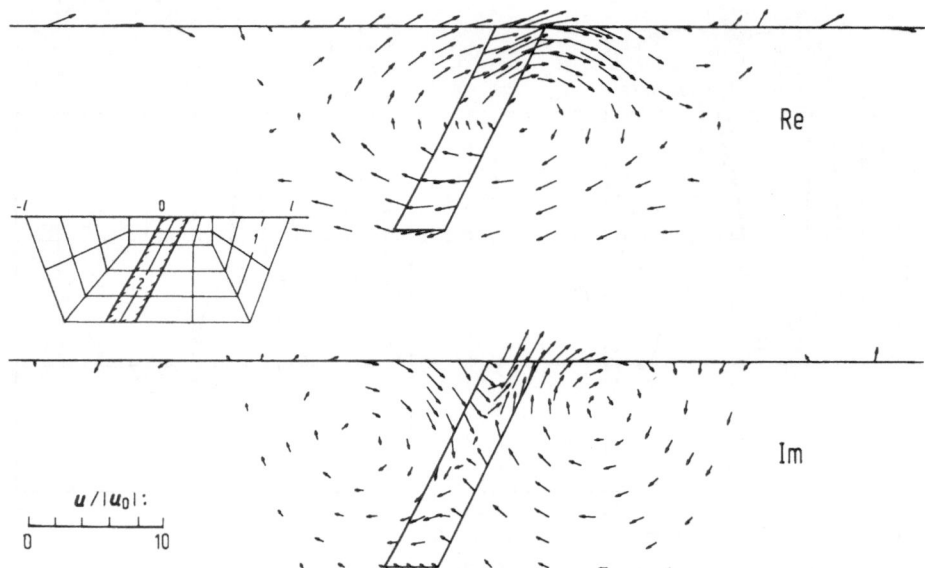

Fig. 1.12. Displacements of non-homogeneous ground subject to the time-harmonic SV-wave of vertical incidence by BIE-FE hybrid method. Incident wave length: $\pi l/2$, Young's modulus: $E_1/E_2 = 10.0$, Poisson's ratio: $\nu_1/\nu_2 = 1.0$, density: $\varrho_1/\varrho_2 = 1.0$

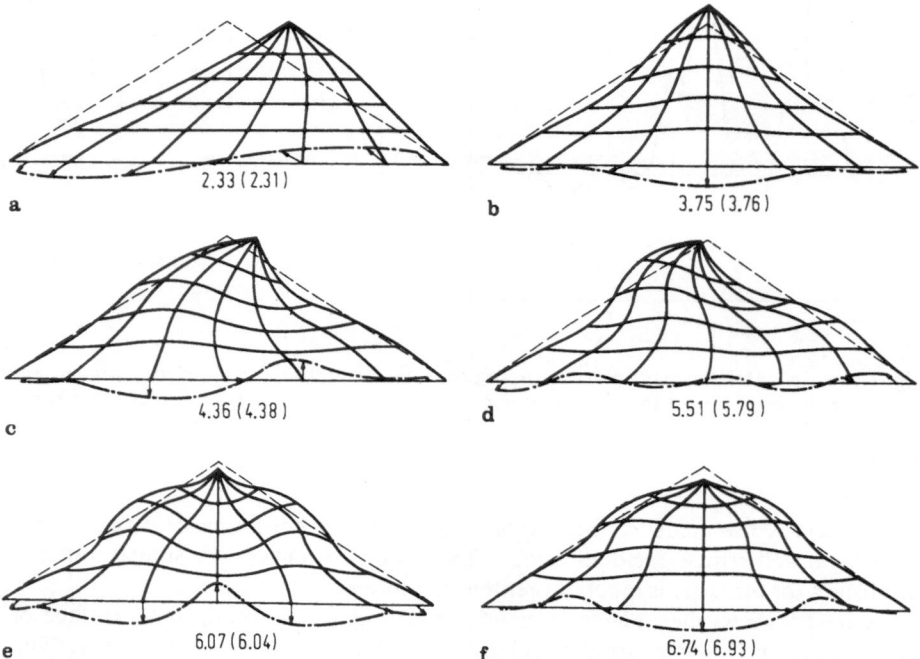

Fig. 1.13a–f. Eigenmodes and eigendensities (reactions from the ground). $k_T = \omega/c_T$ by BIEM and (Clough and Chopra [116]) ω: eigenfrequency

Figure 1.10a illustrates the displacements subject to time-harmonic SV-wave with an incident angle 30° by the use of the half-plane Green's function by Kinoshita [78], while Fig. 1.10b shows the result for the same problem obtained by the use of fundamental solution, in which the truncation of the half-plane boundary is made for twice of the incident wave length. Both results show excellent agreement. Hence, it may be concluded that the truncation of the boundary by twice of the incident wave length is good enough for practical purposes.

Figure 1.11 shows the horizontal displacement response curves at the top of an extruded inhomogeneous body subject to SV-wave of vertical incidence, Niwa et at. [62].

In Fig. 1.12 we show displacements of the non-homogeneous ground due to the incident SV-wave. In the analysis BIE-FE hybrid method was used by employing quadratic elements, Kishima [115].

In the final example, eigenmodes and eigendensities (reactions from the base) corresponding to eigenfrequencies are shown in Fig. 1.13 in comparison with those obtained by Clough and Chopra [116] by FEM.

Other examples are found in [24−42, 49−51, 59, 62, 65−72, 75, 78, 79, 82, 93−95] and Wong and Jennings [117], Sanchez-Sesma and Rosenblueth [118], Alarcon et al. [119], Dravinski [120, 121, 122], Toki and Sato [123], Nardini and Brebbia [124], Kobori and Shinozaki [125], Matsuoka et al. [126], Sato et al. [127], Kobayashi [128], and Mita and Takanashi [129].

1.9 Concluding Remarks

In this chapter, we have seen some fundamentals of BIEM and the related subjects. Much progress have been made recently in this field. We are now in the stage of practical applications. However, some important problems still remain.

The first problem is to develop an efficient algorithm to determine eigenfrequencies. The second is to develop powerful techniques to circumvent the fictitious eigenfrequencies. The third is to formulate efficient BIEM which manages singularity and develop hopefully a simple procedure for it.

Some steps to challenge these problems have been introduced in this chapter. Further researches will find out more effective and versatile ways in the applications of BIEM to initial-boundary value problems.

References

1 Green, G., see *Hydrodynamics*. H. Lamb, § 43, § 58, Dover, 1954
2 Betti, E., see *A Treatise on the Mathematical Theory of Elasticity*. A.E. Love, § 121, Dover, 1944
3 Somigliana, C., see A.E. Love, loc. cit. § 169
4 Kirchhoff, G., see A.E. Love, loc. cit. § 210.
5 Baker, B.B. and Copson, E.T., *The Mathematical Theory of Huygens's Principle*. 2nd ed. Oxford Univ. Press, 1950
6 Fredholm, I., see *Three-dimensional Problems of Mathematical Theory of Elasticity and Thermoelasticity*. V.D. Kupradze (ed.), North-Holland, 1979

7 Kellogg, O.D., *Foundations of Potential Theory.* Springer-Verlag, 1929

8 Muskhelishivili, N.I., *Singular Integral Equations* (transl. by Radok, J.R.M.). Noordhoff, 1953

9 Mikhlin, S.G., *Multidimensional Singular Integrals and Integral Equations.* Pergamon, 1965

10 Kupradze, V.D., *Potential Methods in the Theory of Elasticity.* Israel Progr. Sci. Transl., Jerusalem, 1965

11 Kupradze, V.D. (ed.), *Three-dimensional Problems of the Mathematical Theory of Elasticity and Thermoelasticity.* North-Holland, 1979

12 Love, A.E., *A Treatise of the Mathematical Theory of Elasticity.* Dover, 1944

13 Morse, P.M. and Feshbach, H., *Method of Theoretical Physics.* McGraw-Hill, 1953

14 Eringen, A.C. and Suhubi, E.S., *Elastodynamics.* Academic Press, 1975

15 Achenbach, J.D., *Wave Propagation in Elastic Solids.* North Holland, 1973

16 Graff, K.F., *Wave Motion in Elastic Solids.* Oxford Univ. Press, 1975

17 Pao, Y.H. and Mow, C.C., *Diffraction of Elastic Waves and Dynamic Stress Concentrations.* Crane-Russak, 1972

18 Ewing, W.M., Jardetzky, W.S., and Press, F., *Elastic Waves in Layered Media.* McGraw-Hill, 1957

19 Wheeler, C.T. and Sternberg, E., Some theorems in classical elastodynamics. Arch. Rat. Mech. Anal. **31**, 51–90, 1968

20 Friedman, M.B. and Shaw, R.P., Diffraction of pulses by cylindrical obstacles of arbitrary cross-section. J. Appl. Mech. **29**, 40–46, 1962

21 Chen, L.H. and Schweikert, J., Sound radiation from an arbitrary body. J. Acoust. Soc. Am. **35**, 1626–1632, 1963

22 Banaugh, R.P. and Goldsmith, W., Diffraction of steady acoustic waves by surfaces of arbitrary shape. J. Acoust. Soc. Am. **35**, 1590–1601, 1963

23 Banaugh, R.P. and Goldsmith, W., Diffraction of steady elastic waves by surfaces of arbitrary shape. J. Appl. Mech. **29**, 589–597, 1963

24 Rizzo, F., An integral equation approach to boundary value problems of classical elastostatics. Quart. Appl. Math. **25**, 83–95, 1967

25 Cruse, J.A., and Rizzo, F., A direct formulation and numerical solution of the general transient elastodynamic problems, I. J. Math. Anal. Appl. **22**, 244–259, 1968

26 Cruse, J.A., A direct formulation and numerical solution of the general transient elastodynamic problems, II. J. Math. Anal. Appl. **22**, 341–355, 1968

27 Manolis, G.D. and Beskos, D.E., Dynamic stress concentration studies by boundary integrals and Laplace transform. Int. J. num. Meth. Engng. **17**, 573–599, 1981

28 Manolis, G.D., A comparative study on three boundary element method approaches to problems in elastodynamics. Int. J. num. Meth. Engng. **19**, 73–91, 1983

29 Manolis, G.D. and Beskos, D.E., Dynamic response of lined tunnels by an isoparametric boundary element method. Comp. Meth. Appl. Mech. Engng. **36**, 291–307, 1983

30 Niwa, Y., Kobayashi, S., and Yokota, Y., Application of integral equation methods to the determination of static and steady-state dynamic stresses around cavities of arbitrary shape. Proc. Japan Soc. Civil Eng. **195**, 27–35 (in Japanese), 1971

31 Niwa, Y., Kobayashi, S., and Azuma, N., An analysis of transient stresses produced around cavities of an arbitrary shape during the passage of travelling wave. Memo. Fac. Eng., Kyoto Univ. **37**, 28–46, 1975

32 Niwa, Y., Kobayashi, S., and Fukui, T., Application of integral equation method to some geomechanical problems. Proc. 2nd Int. Conf. Num. Meth. Geomech., 120–131, ASCE, 1976

33 Kobayashi, S. and Nishimura, N., Dynamic analysis of underground structures by the integral equation method. Proc. 4th Int. Conf. Num. Meth. Geomech. **1**, 401–409, Balkema, 1982

34 Kobayashi, S. and Nishimura, N., Transient Stress Analysis of Tunnels and Cavities of Arbitrary Shape due to Travelling Waves. Chapt. 7 of *Developments in Boundary Element Methods* – **2**. P.K. Banerjee and R.P. Shaw (eds.), Applied Science, 1982

35 Tai, G.R.C. and Shaw, R.P., Helmholtz equation eigenvalues and eigenmodes for arbitrary domains. J. Acoust. Soc. Am. **56**, 796–804, 1974

36 De Mey, G., Calculation of eigenvalues of Helmholtz equation by an integral equation. Int. J. num. Meth. Engng. **10**, 59−66, 1976

37 Augirre-Ramirez, G. and Wong, J. P., Eigenvalues of the Helmholtz equation by boundary integral equation-finite element method. Proc. Int. Symp. Innovative Num. Anal. in Appl. Eng. Sci., CETIM, Versailles, 107−110, 1977

38 Hutchinson, J.R., Determination of membrane vibrational characteristics by the boundary integral equation methods. *Recent Advances in Boundary Element Methods*. C.A. Brebbia (ed.), 301−316, Pentech Pr., 1978

39 Vivoli, J. and Filippi, P., Eigenfrequencies of thin plates and layer potentials. J. Acoust. Soc. Am. **55**, 562−567, 1974

40 Niwa, Y., Kobayashi, S., and Kitahara, M., Applications of integral equation method to eigenvalue problems of elasticity. Proc. Japan Soc. Civil Eng. **285**, 17−28 (in Japanese), 1979

41 Niwa, Y., Kobayashi, S., and Kitahara, M., Eigenfrequency analysis of a plate by the integral equation method. Theoret. Appl. Mech. **29**, 287−307, Univ. of Tokyo Press, 1981

42 Niwa, Y., Kobayashi, S., and Kitahara, M., Determination of Eigenvalues by Boundary Element Methods. Chapt. 6 of *Developments in Boundary Element Methods − 2*. P.K. Banerjee and R.P. Shaw (eds.), Applied Science, 1982

43 Cole, D.M., Kosloff, D.D., and Minster, J.B., A numerical boundary integral equation method for elastodynamics, I. Bul. Seismo. Soc. Am. **68**, 1331−1357, 1978

44 Mansur, W.J. and Brebbia, C.A., Numerical implimentation of the boundary element method for two dimensional transient scalar wave propagation problems. Appl. Math. Model. **6**, 299−306, 1982

45 Niwa, Y., Fukui, T., Kato, S., and Fujiki, K., An application of the integral equation method to two-dimensional elastodynamics. *Theoret. and Appl. Mech.* **28**, 281−290, Univ. of Tokyo Pr., 1980

46 Mansur, W.J. and Brebbia, C.A., Transient elastodynamics using a time-stepping technique. *Boundary Elements*. C.A. Brebbia, T. Futagami, and M. Tanaka (eds.), 677−698, Springer, 1983

47 Shaw, R.P., Boundary Integral Equation Methods Applied to Wave Problems, Chapt. 6 of *Developments in Boundary Element Methods − 1*. P.K. Banerjee and R. Butterfield (eds.), Applied Science Pub., 1979

48 Kleinman, R.E., and Roach, G.F., Boundary integral equations for the three dimensional Helmholtz equation. SIAM Review **16**, 214−236, 1974

49 Banerjee, P.K. and Butterfield, R., *Boundary Element Methods in Engineering Science*. McGraw-Hill, 1981

50 Brebbia, C.A. and Walker, S., *Boundary Element Techniques in Engineering*. Newness-Butterworths, 1980

51 Dominguez, J. and Alarcon, E., Elastodynamics. Chapt. 7 of *Progress in Boundary Element Methods − 1*. C.A. Brebbia (ed.), Pentech Pr., 1981

52 Sternberg, E. and Gurtin, M.E., On the completeness of certain stress function in the linear theory of elasticity. Proc. 4th U.S. Nat. Congr. Appl. Mech. **2**, 793−797, 1962, ASME

53 Sommerfeld, A., *Partial Differential Equation in Physics*. Academic Pr., 1949

54 Hartman, F., Computing the C-matrix in non-smooth boundary points. *New Development in Boundary element Methods*. C.A. Brebbia (ed.), 367−379, CML Pub., 1980

55 Ishihara, Y., An approach to BIE solution for Wave Equation. Graduation Work, Fac. of Eng. (in Japanese), 1983

56 Doyle, J.M., Radiation conditions in elasticity. ZAMP **16**, 527−531, 1965

57 Srivastava, S. and Zischka, K.A., Radiation conditions and uniqueness theorem for n-dimensional wave equation in an infinite domain. J. Math. Anal. Appl. **45**, 764−776, 1974

58 Jawson, M.A. and Symm, G., *Integral Equation Methods in Potential Theory and Elastostatics*. Academic Pr., 1977

59 Kitahara, M., *Boundary Integral Equation Methods to Eigenvalue Problems of Elastodynamics and Thin Plates*. Elsevier, 1985

60 Doyle, J.M., Integration of the Laplace transformed equations of classical elastokinetics. J. Math. Anal. Appl. **13**, 118–131, 1966
61 Sladek, V. and Sladek, J., Boundary integral equation method in thermoelasticity, part I: General analysis. Appl. Math. Model. **7**, 241–253, 1983
62 Niwa, Y., Kitahara, M., and Hirose, S., Elastodynamic problems for inhomogeneous bodies. *Boundary Elements.* C.A. Brebbia, T. Futagami, and M. Tanaka (eds.), 751–763, Springer, 1983
63 Tanaka, M. and Tanaka, K., On boundary-volume element discretization of inhomogeneous elastodynamic problems. Appl. Math. Model. **5**, 194–198, 1981
64 Lamb, H., *Hydrodynamics*, sect. 290, Dover, 1954
65 Copley, L.G., Integral equation method for radiation from vibrating bodies. J. Acoust. Soc. Am. **41**, 807–816, 1967
66 Schenk, H.A., Improved integral formulation for acoustic radiation problems. J. Acoust. Soc. Am. **44**, 45–58, 1968
67 Burton, A.J. and Miller, G.F., The application of integral equation methods to the numerical solution of some exterior boundary-value problems. Proc. Roy. Soc. London, Ser. **A, 323**, 201–210, 1971
68 Meyer, W.L., Bell, W.A., and Zinn, B.T., Boundary integral solutions of three dimensional acoustic radiation problems. J. sound and vib. **59**, 245–262, 1978
69 Terai, T., On calculation of sound fields around three dimensional objects by integral equation methods. J. sound and vib. **69**, 71–100, 1980
70 Greenspan, D. and Werner, P., A numerical method for the exterior Dirichlet problem for reduced wave equation. Arch. Rat. Mech. Anal. **23**, 288–316, 1966
71 Kussmaul, R., Ein numerisches Verfahren zur Lösung des Neumannschen Außenraumproblems für die Helmholtzsche Schwingungsgleichung. Computing **4**, 246–262, 1979
72 Filippi, P., Layer potentials and acoustic diffraction. J. sound. and vib. **54**, 473–500, 1977
73 Ursell, F., On the exterior problems of acoustics. Proc. Cambridge Phil. Soc. **74**, 117–125, 1973
74 Jones, D.S., Integral equations for the exterior acoustic problem. Q. Jl. Mech. Appl. Math. **27**, 129–142, 1974
75 Kobayashi, S. and Nishimura, N., On the indeterminancy of BIE solutions for the exterior problems of time-harmonic elastodynamics and incompressible elastostatics. *Boundary Element Methods in Engineering.* C.A. Brebbia (ed.), 282–296, Springer, 1982
76 Kleinman, R.E. and Roach, G.F., On modified Green functions in exterior problems for the Helmholtz equation. Proc. Roy. Soc. London **A 383**, 313–332, 1982
77 Martin, R.A., On the null-field equations for the exterior problems of acoustics. Q. Jl. Mech. Appl. Math. **33**, 385–396, 1980
78 Kinoshita, M., Structure-Ground Dynamic Interaction Analysis by BIEM using Half-Plane Green's Function. M. Sci. Thesis, Facul. of Eng., Kyoto Univ., 1983
79 Kobayashi, S. and Nishimura, N., Analysis of dynamic soil-structure interactions by boundary integral equation method. Num. Meth. in Eng., Proc. 3rd Int. Symp. Num. Meth. Eng. Paris **1**, 353–362, Pluralis, 1983
80 Kelly, D.W., Mustoe, G.W., and Zienkiewicz, O.C., On a hierarchical order for trial functions in numerical procedures based on satisfaction of the governing equations. *Recent Advances in Boundary Element Methods.* C.A. Brebbia (ed.), 359–373, Pentech Pr., 1978
81 Watson, J.O., Hermitian cubic boundary elements for plane problems of fracture mechanics. Res. Mech. **4**, 23–42, 1982
82 Kobayashi, S., Nishimura, N., and Kawakami, T., Simple layer potential method for domains having external corners. Appl. Math. Model. **8**, 61–65, 1984
83 Hartman, F., The complementary problem of finite elastic bodies. *New Developments in Boundary Element Methods.* C.A. Brebbia (ed.), 229–246, 1980
84 Lera, S.G., Paris, E., and Alarcon, E., Treatment of singularities in 2-D domains using BIEM. Appl. Math. Model. **6**, 111–118, 1982
85 Atkinson, C., Fracture Mechanics Stress Analysis. Chapt. 3. *Progress in Boundary Element Methods — 2.* C.A. Brebbia (ed.), Pentech Pr., 1983

86 Paris, F., Martin, A., and Alarcon, E., Potential Theory. Chapt. 3 of *Progress in Boundary Element Methods* — 1. C.A. Brebbia (ed.), Pentech Pr., 1981

87 Athanasiadis, G., Torsion prismatischer Stäbe nach der Singularitatenmethode. Ing. Arch. **49**, 89 – 96, 1980

88 Chaudonneret, M., On the discontinuity of the stress vector in the boundary integral equation method for elastic analysis. *Recent Advances in Boundary Element Methods*. C.A. Brebbia (ed.), 233 – 249, Pentech Pr., 1978

89 Wardle, L.J. and Crotty, J.M., Two-dimensional boundary integral equation analysis for non-homogeneous mining applications. *Recent Advances in Boundary Element Methods*. C.A. Brebbia (ed.), 233 – 249, Pentech Pr., 1978

90 Rudolphi, T.J., An implimentation of the boundary element method for zoned media with stress discontinuities. Int. J. num. Meth. Engng. **19**, 1 – 15, 1983

91 Alarcon, E., Martin, A., and Paris, F., Boundary elements in potential and elasticity theory. Computers and Structures **10**, 351 – 362, 1979

92 Shaw, R.P., Coupling boundary integral equation method to other numerical techniques. *Recent Advances in Boundary Element Methods*. C.A. Brebbia (ed.), 137 – 147, Pentech Press. 1978

93 Zienkiewicz, O.C., Kelly, D.W., and Bettess, P., Marriage a la mode — The best of both worlds (Finite Element and Boundary Integrals). Chapt. 5 of *Energy Methods in Finite Element Analysis*. Glowinski, R., Rodin, E.Y., and Zienkiewicz, O.C. (eds.), Wiley, 1979

94 Kelly, D.W., Mustoe, G.G.W., and Zienkiewicz, O.C., Coupling Boundary Element Methods with Other Numerical Methods. Chapt. 10 of *Development in Boundary Element Methods* — 1. P.K. Banerjee and R.P. Butterfield (eds.), 1979

95 Brebbia, C.A., *The Boundary Element Methods for Engineers*. Pentech Pr., 1978

96 Margulies, M., Combination of the boundary element and finite element methods. Chapt. 8 of *Progress in Boundary Element Methods* — 1. C.A. Brebbia (ed.), 1981

97 Zienkiewicz, O.C., *The Finite Element Method*. 3rd ed., McGraw-Hill, 1977

98 Lachat, J.C. and Watson, J.D., Effective numerical treatment of boundary integral equations; A formulation for three-dimensional elastostatics. Int. num. Meth. Engng. **10**, 991 – 1005, 1976

99 Stroud, A.H. and Secrest, D., *Gaussian Quadrature Formula*. Prentice-Hall, 1966

100 Cruse, T.A., An improved boundary-integral equation method for three-dimensional elastic stress analysis. Computers and Structures **4**, 741 – 754, 1974

101 Rizzo, F.J. and Shippy, D.J., An advanced boundary integral equation method for three-dimensional thermoelasticity. Int. J. num. Meth. Engng. **11**, 1753 – 1768, 1977

102 Katsikadis, J.T. and Armenakas, A.Z., Numerical evaluation of double integrals with a logarithmic or Cauchy-type singularity. J. Appl. Mech. **50**, 682 – 684, 1983

103 Stippes, M. and Rizzo, F.J., A note on the body force integral of classical elastostatics. ZAMP **28**, 339 – 341, 1977

104 Cruse, T.A., Snow, D.W., and Wilson, R.B., Numerical solution in axisymmetri elasticity. Computers and Structures **7**, 445 – 451, 1977

105 Danson, D., Linear Isotropic Elasticity with Body Forces. Chapt. 4 of *Progress in Boundary Element Methods* — 2. C.A. Brebbia (ed.), 1983

106 Cooley, J.W. and Tukey, J.W., An algorithm for machine calculation of complex Fourier series. Math. Comp. **19**, 297 – 301, 1965

107 Brigham, E.O., *The Fast Fourier Transform*. Prentice-Hall, 1974

108 Davies, B. and Martin, B., Numerical inversion of the Laplace transform: a survey and comparison of methods. J. Comput. Phys. **33**, 1 – 32, 1979

109 Narayanan, G.V. and Beskos, D.E., Numerical operational methods for time-dependent linear problems. Int. J. num. Meth. Engng. **18**, 1829 – 1854, 1982

110 Durbin, F., Numerical inversion of Laplace transforms: an efficient improvement to Dubner and Abate's method. Computer J. **17**, 371 – 376, 1974

111 Baron, M.L. and Parnes, R., Displacements and velocities produced by the diffraction of pressure wave by a cylindrical cavity in an elastic medium. J. Appl. Mech. **29**, 385 – 395, 1962

112 Fujiki, K., Analysis of the transient stresses around underground cavities by the integral equation method. M. Sci. Thesis, Kyoto Univ., 1980

113 Pao, Y.-H., Dynamic stress concentration in an elastic plate. J. Appl. Mech. **29**, 299−305, 1962

114 Garnet, H. and Crouzet-Pascal, J., Transient response of a circular cylinder of arbitrary thickness, in an elastic medium, to a plane dilatational wave. J. Appl. Mech. **33**, 521−531, 1966

115 Kishima, T., Dynamic response analysis of non-homogeneous ground by BIE-FE hybrid method. Graduation Work, Fac. of Eng., Kyoto Univ. (in Japanese), 1984

116 Clough, R.W. and Chopra, A.K., Earthquake stress analysis in earth dams. Proc. ASCE **92, EM 2**, 197−211, 1966

117 Wong, H.L. and Jennings, P.C., Effects of Canyon Topography on strong ground motion. Bul. Seismo. Soc. Am. **65**, 1239−1257, 1975

118 Sanchez-Sesma, F.J. and Rosenblueth, E., Ground motions at canyons of arbitrary shapes under incident SH waves. Earthq. Eng. Struct. Dyn. **7**, 441−450, 1979

119 Alarcon, E., Dominguez, J., and Cano, F., Dynamic stiffness of foundations. *New Developments in Boundary Element Methods.* C.A. Brebbia (ed.), 264−280, 1980

120 Dravinski, M., Scattering of SH waves by subsurface topography. Proc. ASCE, **EM 1**, 1−17, 1982

121 Dravinski, M., Scattering of elastic waves by an alluvial valley. Proc. ASCE, **EM 1**, 19−31, 1982

122 Dravinski, M., Scattering of plane harmonic SH wave by dipping layers of arbitrary shape. Bul. Seismo. Soc. Am. **73**, 1303−1319, 1983

123 Toki, K. and Sato, T., Seismic response analyses of ground with irregular profiles by the boundary element method. *Boundary Elements.* C.A. Brebbia, T. Futagami, and M. Tanaka (eds.), 699−708, Springer, 1983

124 Nardini, D. and Brebbia, C.A., Transient dynamic analysis by the boundary element method. ibid., 719−730

125 Kobori, T. and Shinozaki, Y., Applications of the boundary integral equation method to dynamic soil-structure interaction analysis under topographic site condition. ibid., 731−740

126 Matsuoka, O., Yokoi, T., and Torii, K., The fundamental solution for periodically oscillating line torsional loads on a circular ring interior of a semi-infinite solid and its application. ibid., 741−749

127 Sato, T., Kawase, H., and Yoshida, J., Dynamic response analysis of rigid foundations subjected to seismic waves by boundary element method. ibid., 765−774

128 Kobayashi, S., Some problems of the boundary integral equation method in elasto-dynamics. ibid., 775−784

129 Mita, A. and Takanashi, W., Dynamic soil-structure interaction analysis by hybrid method. ibid., 785−794

Chapter 2

Elastic Potentials in BIE Formulations

by R. P. Shaw

2.1 Introduction

This chapter is an expansion of a paper, Shaw [1], given at a recent BEM meeting. The aim of this work is to examine some of the several boundary integral equation formulations available for problems of linear elasticity. In particular, emphasis will be placed on a comparison of the advantages and disadvantages of the widely used displacement-traction formulation, e.g. Cruse and Rizzo [2], based on reciprocity theorems of elasticity and the displacement potential formulations, e.g. Banaugh and Goldsmith [3], which are actually introduced earlier, at least for elastodynamic problems.

It is clear that elastodynamic problems should, in principle at least, be more difficult than elastostatic problems due to the presence of an additional independent variable, the time t, in transient problems or at least an additional parameter, the frequency p in time harmonic problems. In practice, this has not always been the case. This discrepancy lies in two basic distinctions between elastostatics and elastodynamics. Time harmonic problems satisfy Helmholtz equations when expressed in terms of elastic potentials; the fundamental solutions to these equations are considerably simpler than those to the Navier equation required in the displacement-traction formulation in elastodynamics. This advantage does not carry over directly to elastostatics. Even more advantageous is the concept of retarded time values in transient problems which allow the resulting approximating system of algebraic equations obtained from a BEM solution procedure to the original BIE to be successive rather than simultaneous in character, e.g. Friedman and Shaw [4] for acoustics and Cole et al. [5] for elastodynamics. Thus it is not surprising that elastodynamic problems appear to be the first treated by BIE and BEM approaches. Note that a distinction is drawn here between the BIE formulation which is an exact representation of the original problem and the BEM numerical solution procedure which is only one, albeit the most popular by far, method of approximate solution of the BIE. This point will be addressed again in a later section.

The chapter will be laid out in sections as follows. The first section will consider elastodynamic problems with both displacement-traction and elastic potential formulations. The advantages and disadvantages of both approaches will be addressed for both transient and time harmonic problems. This is followed by a section on elastostatics where a special discussion of the reduction of elastic potential formulations from dynamic to static problems will be given. Next a brief discussion of various solution procedures, including but not limited to BEM, will be given. Finally a summary of concepts and suggestions for further study will be given. While this chapter is not meant to be a complete review of the field, it is hoped that the references given will be sufficiently representative to allow these concepts and suggestions to be seen as clearly based on the existing literature.

2.2 Elastodynamic Formulations

The governing equation of linear isotropic elastodynamics is the Navier equation, as found in any of the standard elastodynamic texts, e.g. [6, 7];

$$\varrho\,\ddot{\bar{u}} = (\lambda + \mu)\,\nabla\,(\nabla \cdot \bar{u}) + \mu\nabla^2\bar{u} = (\lambda + 2\mu)\,\nabla\,(\nabla \cdot \bar{u}) - \mu\nabla \times (\nabla \times \bar{u}), \qquad (2.1)$$

where $\bar{u}$ is the displacement field, λ and μ are the Lamé elastic parameters and ϱ is density. An elastodynamic problem further requires a set of initial conditions, e.g. an initial displacement field and an initial velocity field, and a prescribed boundary condition at all bounding surface points. For infinite domains, there may also be a radiation condition on the displacement field "at infinity". Typical boundary conditions require a prescribed displacement on some portion of the boundary surface, S_1, traction over some other portion, S_2, and/or an elastic constraint over the remaining surface, S_3, where any of these surface segments may be zero but where their sum adds to the complete surface.

Although it is possible to work directly in the time domain, e.g. [8], many common formulations of elastodynamic problems are based on either a Laplace transformation, e.g. [9], or a Fourier transformation, e.g. [10], of Eq. (2.1). These transformations bring the transient formulation to a form either similar to or identical to the time harmonic case respectively. This of course still leaves the inversion problem to be faced.

Now consider some BIE representations of this problem. The most commonly used representation in engineering analysis is one based on reciprocity theorems and expressed directly in terms of surface displacements and tractions. Such an approach has a clear advantage over all others in that it works directly with the physically significant variables which inevitably are prescribed as boundary conditions. It pays a price however in the complexity of the integral kernels required, i.e. the "fundamental solutions". Consider an actual displacement-traction BIE for a field point, $P(x_i)$, on a smooth surface, without body forces for convenience:

$$\tfrac{1}{2}\,u_j(P) = \int_S \{t_i(Q)\,U_{ij}(P, Q) - u_i(Q)\,T_{ij}(P, Q)\}\,dS(Q), \qquad (2.2)$$

where t_i is the surface traction. The point $Q(x_{0i})$ represents the integration variable and U_{ij} and T_{ij} are the "fundamental solutions", "integral kernels" or "Green's tensors". While there are several forms which these could take, the most common is the Stokes solution due to a point load. Appropriate forms for these terms are given in [8, 9, 10] for the direct time dependent, the Laplace transform and the Fourier transform forms respectively.

The analogous elastic potential formulation actually has two distinct forms. The most common is the Lamé displacement potential based on a Helmholtz decomposition of the displacement:

$$\bar{u} = \nabla \phi + \nabla \times \bar{\psi} \qquad (2.3\,a)$$

with an auxiliary condition usually taken as

$$\nabla \cdot \bar{\psi} = 0 \qquad (2.3\,b)$$

which leads to wave equations on the two displacement potentials ϕ and $\bar{\psi}$,

$$\nabla^2 \phi = \ddot{\phi}/c_D^2, \qquad (2.4\,a)$$

$$\nabla^2 \phi + k_D^2 \phi = 0, \qquad (2.4\,b)$$

$$\nabla^2 \bar{\psi} = \ddot{\bar{\psi}}/c_R^2, \qquad (2.5\,a)$$

$$\nabla^2 \bar{\psi} + k_R^2 \bar{\psi} = 0, \qquad (2.5\,b)$$

where the two wave speeds, $c_D = \left(\dfrac{\lambda + 2\mu}{\varrho}\right)^{1/2}$ and $c_R = \left(\dfrac{\mu}{\varrho}\right)^{1/2}$, represent the speed of propagation of dilatational and rotational (or pressure and shear) waves respectively in the time dependent cases, (a) and $k_D = p/c_D$ and $k_R = p/c_R$ in the time harmonic cases, (b). This in fact is the form used in most of the classical elastodynamic solutions as found in [6, 7]. Alternatively, a Stokes formulation based on dilatation and rotation is found in Love [11],

$$\theta = \nabla \cdot \bar{u}, \qquad (2.6)$$

$$\bar{\omega} = \tfrac{1}{2} \nabla \times \bar{u} \qquad (2.7)$$

leads to the same wave equations on these variables. These two systems are clearly interconnected, i.e.

$$\theta = \nabla^2 \phi = \ddot{\phi}/c_D^2, \qquad (2.8)$$

$$\bar{\omega} = -\tfrac{1}{2} \nabla^2 \bar{\psi} = -\ddot{\bar{\psi}}/2c_R^2. \qquad (2.9)$$

The BIE formulations of these potential representations are actually identical to those for the acoustic potential, i.e. those for the wave equation or, in the time harmonic case, the Helmholtz equation which are well known, e.g. a review article by Shaw [12]. Typically, they are

$$\frac{1}{2} \phi(\bar{x}, t) = \frac{1}{4\pi} \int_S \left\{ \frac{1}{R} \frac{\partial \phi(\bar{x}_0, t_{0D})}{\partial n_0} + \left[\frac{1}{c_D R} \frac{\partial \phi(\bar{x}_0, t_{0D})}{\partial t_{0D}} + \frac{\phi(\bar{x}_0, t_{0D})}{R^2} \right] \frac{\partial R}{\partial n_0} \right\} dS(\bar{x}_0) \qquad (2.10)$$

$$\frac{1}{2} \bar{\psi}(\bar{x}, t) = \frac{1}{4\pi} \int_S \left\{ \frac{1}{R} \frac{\partial \bar{\psi}(\bar{x}_0, t_{0R})}{\partial n_0} + \left[\frac{1}{c_R R} \frac{\partial \bar{\psi}(\bar{x}_0, t_{0R})}{\partial t_{0R}} + \frac{\bar{\psi}(\bar{x}_0, t_{0R})}{R^2} \right] \frac{\partial R}{\partial n_0} \right\} dS(\bar{x}_0) \qquad (2.11)$$

for the time dependent case where R is the distance between the field point and the integration point and $t_0 = t - R/c$ is a retarded time due to the finite speed of propagation of wave effects, in this case at two different wave speeds, c_D in Eq. (2.10) and c_R in Eq. (2.11) respectively. The corresponding Helmholtz formulation is

$$\frac{1}{2} \phi(\bar{x}) = \frac{1}{4\pi} \int_S \left\{ \frac{\exp(ik_D R)}{R} \frac{\partial \phi(\bar{x}_0)}{\partial n_0} - \phi(\bar{x}_0) \frac{\partial}{\partial n_0} \left[\frac{\exp(ik_D R)}{R} \right] \right\} dS(\bar{x}_0), \qquad (2.12)$$

$$\frac{1}{2}\,\bar{\psi}\,(\bar{x}) = \frac{1}{4\pi} \int_S \left\{ \frac{\exp\,(ik_R R)}{R} \frac{\partial \bar{\psi}\,(\bar{x}_0)}{\partial n_0} - \bar{\psi}\,(\bar{x}_0) \frac{\partial}{\partial n_0} \left[\frac{\exp\,(ik_R R)}{R} \right] \right\} dS\,(\bar{x}_0). \tag{2.13}$$

This form also applies to the Fourier transformed case as well with a slight modification for the Laplace transform, i.e. (s) replacing $(-ip)$. As the point of this discussion is not so much to solve specific problems as to discuss the characteristics of their formulations, no further detail need be given here, save to refer to the forms required for the typical boundary conditions, which are given in [13] for the Lamé potentials and clearly involve tangential as well as normal derivatives of the potentials. This complicates the problem if only nodal values of the potentials are computed as would be the case in a BEM method.

At this point it is worth noting that the preference for one formulation over another may be related to the background of the researcher as much as to the actual superiority of one approach to another. Researchers coming from a structural analysis background seem to generally prefer the displacement-traction formulation and corresponding BEM solution technique due to its apparent similarity to the well known finite element method while those researchers coming from a fluids and/or acoustics background appear to be drawn to the potential formulations and, in many cases, to non-BEM solution methods. This can be seen very clearly in the question of "significant variable". If the solution is sought for engineering purposes, the actual stress and displacement fields are the significant variables. However, as would be the case in seismic prospecting, etc., if the significant variable were a time, e.g. first arrival time, the prefered formulation might well be one which distinguished clearly between the two types of elastic waves, i.e. an elastic potential formulation.

2.3 Elastostatic Formulations

Although elastostatic problems exist as a class in their own right, it is of interest to examine the manner in which they may be obtained from elastodynamic problems as the time dependence is gradually removed. This may be of significance for low frequency elastodynamics problems. This question does not appear to have been specifically addressed in the displacement-traction formulation but has appeared in the elastic potential formulations as a serious dilemma. In the displacement-traction formulation, the elastostatic Green's tensors are well known and widely used, e.g. [14].

The analogous reduction in the Lamé potential formulation immediately runs into difficulty. Were the transition smooth as the frequency p approaches zero in the time harmonic case for example, Eqs. (2.4b) and (2.5b) would reduce to Laplace equations with displacements expressed as a sum of a scalar and a vector harmonic displacement potential. This is well known to not be the case as Sternberg [15] has shown, i.e. the Lamé displacement potentials maintain a time dependence even though the displacement itself is static.

Recently, Kaul and his coworkers [16, 17] developed this point further by using the Stokes elastic potential formulation. In this form, the difficulty may be seen immediately. The Navier equation for a time harmonic or Fourier transformed problem may be solved for the displacement field as

$$\bar{u}(\bar{x};p) = \frac{-c_D^2 \nabla \theta(\bar{x};p) + 2c_R^2 \nabla \times \bar{\omega}(\bar{x};p)}{p^2}. \tag{2.14}$$

Clearly as p approaches zero, i.e. in the transition from elastodynamics to elastostatics, a catastrophe (in mathematical terms) occurs. To have a bounded displacement field, the limiting values of dilatation and rotation must be related;

$$\theta_0(\bar{x}) = \lim_{p \to 0} \theta(\bar{x};p), \tag{2.15}$$

$$\bar{\omega}_0(\bar{x}) = \lim_{p \to 0} \bar{\omega}(\bar{x};p) \tag{2.16}$$

and

$$c_D^2 \nabla \theta_0(\bar{x}) = 2c_R^2 \nabla \times \bar{\omega}_0(\bar{x}). \tag{2.17}$$

By L'Hopital's rule, the displacement field may be expressed as

$$\bar{u}(\bar{x}) = \lim_{p \to 0} \bar{u}(\bar{x};p) = -c_D^2 \nabla \theta_1(\bar{x}) + 2c_R^2 \nabla \times \bar{\omega}_1(\bar{x}) \tag{2.18}$$

where θ_1 and ω_1 are new functions defined as

$$\theta_1(\bar{x}) = \lim_{p \to 0} \frac{\partial \theta(\bar{x};p)}{\partial(p^2)}, \tag{2.19}$$

$$\bar{\omega}_1(\bar{x}) = \lim_{p \to 0} \frac{\partial \bar{\omega}(\bar{x};p)}{\partial(p^2)}. \tag{2.20}$$

Manipulating these equations gives

$$\nabla^2 \theta_1(\bar{x}) = -\theta_0(\bar{x})/c_D^2, \tag{2.21}$$

$$\nabla^2 \bar{\omega}_1(\bar{x}) = -\bar{\omega}_0(\bar{x})/c_R^2, \tag{2.22}$$

$$\nabla^2 \theta_0(\bar{x}) = 0, \tag{2.23}$$

$$\nabla^2 \bar{\omega}_0(\bar{x}) = 0 \tag{2.24}$$

together with the auxiliary conditions

$$\nabla \cdot \bar{\omega}_0 = 0, \tag{2.25}$$

$$\nabla \cdot \bar{\omega}_1 = 0 \tag{2.26}$$

and Eq. (2.17).

This solution formulation may be further manipulated until the basic dependent variables are expressed as harmonic potentials, analogous to the Neuber-Papkovich potentials, e.g. [18]. Work is presently underway to examine the usefulness of this approach in a BIE formulation, [19], and no further development will be given here save to say that the relationship of these new potentials to the usually prescribed boundary conditions remains as awkward as ever.

2.4 Solution Methods

The most popular solution method for BIE in all fields is that of BEM. Indeed, many authors prefer to go directly to BEM formulations of problems. This approach is essentially the method of weighted residuals applied directly to the governing equations with the trial or basis functions defined as nonzero only over a small subsection of the original boundary. Clearly the method of weighted residuals is far more general than that actually used in most BEM solution procedures. BEM actually follows the line of reasoning used in finite element methods in discretizing the domain which in this case is only the boundary rather than the total volume. In fact it is common to use isoparametric elements, e.g. using the same shape functions to describe the geometry as to approximate the dependent variables, just as in finite elements. The solution is then expressed in terms of nodal values, as in the finite element case, which are then solved for by a system of algebraic equations. This approach is widely known and will not be discussed further except in comparison to some other approaches.

There are a wide range of alternative solution methods to BIE; some of these are problem dependent in the sense that they are appropriate only for certain classes of problems. Others hold in general. One classical approach is based on the Picard iteration method and has been used in the solution of an acoustic problem by Chertock [20]. This approach is appropriate for Fredholm integral equations of the second kind where the unknown appears outside as well as inside the integral. An assumption is made for the zeroth order approximation to this unknown, usually that it may be taken to be zero. This then is used under the integral to yield the first order solution explicitly. The procedure is repeated until the change in solution from one iteration to the next is acceptably small. Questions of convergence do arise, but the biggest drawback appears to be the computational effort required since the quadratures will undoubtedly have to be carried out numerically.

Another approach is applicable to problems which differ only slightly from those whose solutions are known. In these cases, an asymptotic expansion in terms of some small physical parameter may be possible such that the basic problem for each order of the expansion is expressed in terms of the known solution procedure, e.g. Shaw [21] for acoustic radiation from a submerged elastic shell bounded by nonconcentric circular cylinders or [22] for elastic plate vibrations.

The most common alternative however would be one based on a Galerkin approach, or a weighted residual approach using continuous trial functions, i.e. functions which are defined as essentially nonzero over the entire bounding surface. In this case, the unknowns would be the coefficients multiplying the trial functions. Since the form of the dependence of the dependent variables on the coordinates describing the surface would be known, tangential derivatives arising in the appropriate form of the boundary conditions when expressed in terms of elastic potentials could be readily carried out. A form of this approach, essentially a Fourier series, was used by Sharma [23] in the solution of an axisymmetric elastodynamics problem.

Determining the coefficients in a Galerkin approach is essentially a least mean square "best fit"; this is facilitated by but not dependent on the basis functions being orthogonal over the domain in question.

2.5 Comments and Suggestions

The purpose of this chapter has been to review and compare some of the several methods available for the solution of elasticity problems by BIE methods. The most significant point to be made is that there exist several alternative formulations and solution methods and there may not be ONE superior approach to all problems of elastodynamics and elastostatics. While comparisons have been made between displacement-traction formulation, transient elastodynamic BEM solutions based on direct time domain, Laplace transform and Fourier transform methods by Manolis [24], there still seems to be room for further examination of the alternative elastic potential formulation and alternative solution methods. While it is probably true that the best general purpose computer code would be one based on BEM solutions to displacement-traction BIE formulations, e.g. BEASY [25], it is also probably true that some specific classes of problems may be better addressed by the alternatives described above. Thus the final suggestion to be made here is that those involved in basic research consider these alternatives in light of the comments made above.

References

1 Shaw, R.P., Alternative solution methods in elastic BIE problems. Presented at the 6th BEM conference, Q.E.II, July, 1984
2 Cruse, T.A. and Rizzo, F.J., A direct formulation and numerical solution of the general transient elastodynamic problem, I. J. Math. Anal. Appl. **22**, 244–259, 1968
3 Banaugh, R.P. and Goldsmith, W., Diffraction of steady elastic waves by surfaces of arbitrary shape. J. Appl. Mech. **30**, 589–597, 1963
4 Friedman, M.B. and Shaw, R.P., Diffraction of a plane shock wave by an arbitrary rigid obstacle. J. Appl. Mech. **29**, 40–46, 1962
5 Cole, D.M., Kosloff, D.D., and Minster, J.B., A numerical boundary integral equation for elastodynamics, I. Bull. Seis. Soc. Amer. **68**, 1331–1357, 1978
6 Miklowitz, J., *The Theory of Elastic Waves and Waveguides*. North-Holland Press, N.Y., 1978
7 Achenbach, J.D., *Wave Propagation in Elastic Solids*. North-Holland Press, N.Y., 1975
8 Karabalis, D.L. and Beskos, D.E., Dynamic response of 3D rigid surface foundations by the time domain boundary element method. Earth Engr. Struc. Dyn. **12**, 73–94, 1984
9 Manolis, G.D. and Beskos, D.E., Dynamic stress concentration studies by the integral equation method. Proc. 2nd Int. Symp. Innovative Num. Anal., ed. Shaw, R.P., et al., Univ. of Va. Press, 459–463, 1980
10 Kobayashi, S. and Nishimura, N., Transient analysis of tunnels and caverns of arbitrary shape due to travelling waves. *Developments in Boundary Element Methods*. 2, ed. Banerjee, P.K. and Shaw, R.P., Chap. 7, App. Sci. Pub., London, 1982
11 Love, A.E.H., *The Mathematical Theory of Elasticity*. 4th ed., Dover Pub., N.Y., 1944
12 Shaw, R.P., Boundary integral equation methods applied to wave problems. *Developments in Boundary Element Methods*. 1, ed. Banerjee, P.K. and Butterfield, R., Chap. 6, App. Sci. Pub., London, 1980
13 Banaugh, R.P., Application of integral representations of displacement potentials in elastodynamics. Bull. Seis. Soc. Amer. **54**, 1073–1086, 1964
14 Brebbia, C.A., *The Boundary Element Method for Engineers*. Pentech Press, London, 1978
15 Sternberg, E., On the integration of the equations of motion in the classical theory of elasticity. Arch. Rat. Mech. Anal. **6**, 34–50, 1960
16 Kircher, T.A. and Kaul, R.K., On the concept of generalized potentials in the classical theory of elasticity. FEAS Rep. No. DA8325075, SUNY at Buffalo, 1983

17 Ghorieshi, A.J. and Kaul, R.K., On the exact solution of problems in elastostatic in terms of harmonic functions. FEAS Rep. (unnumbered), SUNY at Buffalo, 1984
18 Saada, A.S., *Elasticity; Theory and Applications.* Pergamon Press, N.Y., 1974
19 Shaw, R.P. and Kaul, R.K., The use of harmonic potentials in elastostatic BIE formulations. In preparation, 1985
20 Chertock, G., Convergence of iterative solutions to integral equations for sound radiation. Quar. Appl. Math. **26**, 268–271, 1968
21 Shaw, R.P., Time harmonic acoustic radiation from a submerged elastic shell defined by nonconcentric circular cylinders. J. Acous. Soc. Amer. **64**, 311–317, 1978
22 Shaw, R.P., Elastic plate vibrations by boundary integral equations; part I – Infinite plates. Res. Mech. **4**, 83–88, 1982
23 Sharma, D.L., Scattering of steady elastic waves by surfaces of arbitrary shape. Bull. Seis. Soc. Amer. **57**, 795–812, 1967
24 Manolis, G.D., A comparative study on three boundary element method approaches to problems in elastodynamics. Int. J. Num. Meth. Eng. **19**, 73–91, 1983
25 *BEASY – A Boundary Element Method Code.* Computational Mechanics Centre, Ashurst Lodge, Ashurst, U.K.

Chapter 3

Time Dependent Non-Linear Potential Problems

by P. Skerget and C. A. Brebbia

3.1 Introduction

In a previous publication [4] the authors have presented the formulation for time independent non-linear potential problems based on the Kirchhoff's transform. The transform was used to convert a nonlinear material problem into a linear one using different functions to define the conductivity. The method was applied with mixed and nonlinear radiation boundary conditions. This approach allowed for the solution of the nonlinear potential problems without need to define any internal cells.

In this chapter the use of Kirchoff's transform is extended to deal with time dependent problems. This requires a time integration which can be carried out in two different forms as pointed out by Wrobel and Brebbia [1]. The first approach consists in referring always to the initial conditions at time zero and in most cases does not require any domain integrations; but the number of operations is considerable. The second approach requires the computation of the initial condition − i.e. domain-integrals at the beginning of each new step. Although internal cells are now required, the technique is computationally very efficient and will be the one discussed here. In spite of this the first approach can be of interest when the domain extends to infinity and it could be useful to implement it with the theory described in this chapter. The interested reader is directed to [3]. Other applications of Wrobel and Brebbia technique can be seen in Chap. 2 of this volume.

In what follows the general theory for time dependent non-linear potential problems will be presented and in particular:

(i) The time dependent diffusion equation with non-linear conductivity has been expressed in integral form using Kirchoff's transform. The transformation effectively reduces the space dependent non linearities to the boundary. Cells are then only required for the computation of initial conditions if using the second approach. Notice that these cells are not needed when referring the problem always to the initial time zero (1st approach above see [3]).

(ii) A time and space dependent fundamental solution has been applied. This solution previously used in linear parabolic integral equation systems (see [1]) gives extremely accurate and stable results even for large time steps. The resulting integral equations also incorporate mixed and radiation type of boundary conditions in addition to the essential and natural conditions. Other relationships on the boundary could be easily incorporated.

(iii) Several applications are presented to illustrate how the method can be used to solve practical engineering problems.

The chapter demonstrates how the use of Kirchoff's transform can be extended to time dependent potential problems for which the conductivity is a function of the potential or temperature.

3.2 Governing Equations

Let us consider that our problem is governed by the diffusion equation for homogeneous, isotropic media with constant conductivity

$$\frac{\partial u}{\partial t} = a \nabla^2 u + p/c \varrho \tag{3.1}$$

where c and ϱ are specific heat and density, a is diffusity $(a = k_0/c \varrho)$ and p is a heat source.

The linear boundary conditions corresponding to Eq. (3.1) are

$$\text{Essential} \quad u = \bar{u} \qquad \text{on } \Gamma_1$$

$$\text{Natural} \quad k_0 \frac{\partial u}{\partial n} = -q \qquad \text{on } \Gamma_2 \tag{3.2}$$

$$\text{Mixed} \quad k_0 \frac{\partial u}{\partial n} = -h_0(u - u_f) \quad \text{on } \Gamma_3$$

where h_0 is a constant heat transfer coefficient and u_f is the temperature of the surrounding media. For some special problems, such as cavities, the mixed boundary conditions can be expressed in a different manner, i.e.

$$k_0 \frac{\partial u}{\partial n} = -h_0(u - u_r) \quad \text{on } \Gamma_3 \tag{3.3}$$

where u_r is now the unknown temperature on the opposite boundary.

If non-linear radiation type boundary conditions exist they can be written as,

$$k_0 \frac{\partial u}{\partial n} = -h(u - u_f) - \varepsilon \sigma(u^4 - u_s^4) \quad \text{on } \Gamma_3 \tag{3.4}$$

where σ is the Stefan-Boltzman constant, ε is temperature dependent emissivity between the surface Γ_3 and the environment at the temperature u_s. h is generally a temperature dependent heat transfer coefficient.

The initial conditions needed for the solution of Eq. (3.1) are

$$u = u_i \quad \text{in } \Omega \text{ at } t = t_i. \tag{3.5}$$

Sometimes when higher order time-interpolation functions are introduced, the fluxes on the boundary have to be prescribed, i.e.

$$q = 2_i \quad \text{on } \Gamma \text{ at } t = t_i. \tag{3.6}$$

When the conductivity k depends on the potential the governing equation (3.1) becomes nonlinear, i.e.

$$c \varrho \frac{\partial u}{\partial t} = \frac{\partial}{\partial x} \left(k \frac{\partial u}{\partial x} \right) + \frac{\partial}{\partial y} \left(k \frac{\partial u}{\partial y} \right) + \frac{\partial}{\partial z} \left(k \frac{\partial u}{\partial z} \right) + p. \tag{3.7}$$

For these cases as demonstrated in [4] the Kirchhoff's transform provides an economic and efficient way of solving the equations without the need to define internal cells. The transform is defined as,

$$\Psi = K[u] = \int_{u_0}^{u} k(u) \, du \tag{3.8}$$

where u_0 is arbitrary reference value. Equation (3.7) can be transformed into a linear one in terms of the new variable $\Psi(u)$, that is

$$\frac{\partial \Psi}{\partial t} = a \nabla^2 \Psi + a p. \tag{3.9}$$

Thus the form of Eq. (3.1) is preserved.

The transformed boundary conditions (3.2) corresponding to Eq. (3.9) are:

$$\text{Essential} \quad \Psi = \bar{\Psi} = K[\bar{u}] \quad \text{on } \Gamma_1$$
$$\text{Natural} \quad \frac{\partial \Psi}{\partial n} = -\bar{q} \quad \text{on } \Gamma_2. \tag{3.10}$$

Notice that these are still linear relationship. The transformation of mixed boundary conditions instead gives slightly non-linear equations, i.e.

$$\frac{\partial \Psi}{\partial n} = -h_0 (K^{-1}[\Psi] - u_f) \quad \text{on } \Gamma_3 \tag{3.11}$$

or in terms of Eq. (3.3)

$$\frac{\partial \Psi}{\partial n} = -h_0 (K^{-1}[\Psi] - K^{-1}[\Psi_r]) \quad \text{on } \Gamma_3 \tag{3.12}$$

where K^{-1} is the inverse Kirchoff's transform.

Upon transformation the radiation boundary conditions become even more complex and can be written as,

$$\frac{\partial \Psi}{\partial n} = -h (K^{-1}[\Psi] - u_f) - \varepsilon \sigma (K^{-1}[\Psi]^4 - u_s^4) \quad \text{on } \Gamma_3. \tag{3.13}$$

The initial conditions are now,

$$\Psi = \Psi_i = K[u_i] \quad \text{in } \Omega \text{ at } t = t_i \tag{3.14}$$

and

$$q = q_i \quad \text{on } \Gamma \text{ at } t = t_i. \tag{3.15}$$

3.3 Homogeneous Parabolic Equation

For unsteady heat conduction problems with constant conductivity a boundary integral equation relating boundary values for potentials and its normal derivatives over the boundary Γ can be obtained by weighting the governing equation and boundary conditions by a function $\overset{*}{u}$. As the problem is time dependent the equation has to be weighted also with respect to time [3]. This yields the following weighted residual statement,

$$\int\limits_{t_0}^{t_F}\int\limits_{\Omega}\left[a\,\nabla^2 u\,(s,\,t)-\frac{\partial u\,(s,\,t)}{\partial t}\right]\overset{*}{u}(\xi,\,t_F;\,s,\,t)\,d\Omega\,(s)\,dt$$

$$=\int\limits_{t_0}^{t_F}\int\limits_{\Gamma_2}a\,[q\,(s,\,t)-\bar q\,(s,\,t)]\,\overset{*}{u}(\xi,\,t_F;\,s,\,t)\,d\Gamma\,(s)\,dt$$

$$-\int\limits_{t_0}^{t_F}\int\limits_{\Gamma_1}a\,[u\,(s,\,t)-\bar u\,(s,\,t)]\,\overset{*}{q}(\xi,\,t_F;\,s,\,t)\,d\Gamma\,(s)\,dt \qquad (3.16)$$

where t_0 is the initial and t_F is the final time,

$$\overset{*}{q}(\xi,\,t_F;\,s,\,t)=\partial\overset{*}{u}(\xi,\,t_F;\,s,\,t)/\partial n\,(s)\quad\text{and}\quad t_0\leq t\leq t_F.$$

Integrating by parts twice the Laplacian with respect to x_i gives,

$$\int\limits_{t_0}^{t_F}\int\limits_{\Omega}a\,\nabla^2\overset{*}{u}(\xi,\,t_F;\,s,\,t)\,u\,(s,\,t)\,d\Omega\,(s)\,dt-\int\limits_{t_0}^{t_F}\int\limits_{\Omega}\frac{\partial u\,(s,\,t)}{\partial t}\overset{*}{u}(\xi,\,t_F;\,s,\,t)\,d\Omega\,(s)\,dt$$

$$=\int\limits_{t_0}^{t_F}\int\limits_{\Gamma}a\,u\,(s,\,t)\,\overset{*}{q}(\xi,\,t_F;\,s,\,t)\,d\Gamma\,(s)\,dt-\int\limits_{t_0}^{t_F}\int\limits_{\Gamma}a\,q\,(s,\,t)\,\overset{*}{u}(\xi,\,t_F;\,s,\,t)\,d\Gamma\,(s)\,dt. \qquad (3.17)$$

Integrating now by parts the time derivative produces the following expression,

$$\int\limits_{t_0}^{t_F}\int\limits_{\Omega}\left[a\,\nabla^2\overset{*}{u}(\xi,\,t_F;\,s,\,t)+\frac{\partial\overset{*}{u}(\xi,\,t_F;\,s,\,t)}{\partial t}\right]u\,(s,\,t)\,d\Omega\,(s)\,dt$$

$$-\left[\int\limits_{\Omega}\overset{*}{u}(\xi,\,t_F;\,s,\,t)\,u\,(s,\,t)\,d\Omega\,(s)\right]_{t=t_0}^{t=t_F}$$

$$=\int\limits_{t_0}^{t_F}\int\limits_{\Gamma}a\,u\,(s,\,t)\,\overset{*}{q}(\xi,\,t_F;\,s,\,t)\,d\Gamma\,(s)\,dt$$

$$-\int\limits_{t_0}^{t_F}\int\limits_{\Gamma}a\,q\,(s,\,t)\,\overset{*}{u}(\xi,\,t_F;\,s,\,t)\,d\Gamma\,(s)\,dt. \qquad (3.18)$$

Let $\overset{*}{u}$ be the fundamental solution of the parabolic differential equation defined by [1].

$$a\,\nabla^2\overset{*}{u}(\xi,\,t_F;\,s,\,t)+\frac{\partial\overset{*}{u}(\xi,\,t_F;\,s,\,t)}{\partial t}+\delta(\xi,\,s)\,\delta(t_F,\,t)=0 \qquad (3.19)$$

where δ are Dirac delta functions given by

$$\overset{*}{u}(\xi,\,t_F;\,s,\,t)=\begin{cases}0 &,\;t>t_F\\[2ex]\dfrac{1}{(4\,\pi\,a\,\tau)\,d/2}\exp\left(-\dfrac{r^2\,(\xi,\,s)}{4\,a\,\tau}\right),&\;t<t_F\end{cases} \qquad (3.20)$$

where $\tau = t_F - t$; and d is the spatial dimension of the problem. The fundamental solution represents the effect of the unit point source applied at the point ξ at time t_F on the reference point s at time t in an infinite region and possesses the following properties.

$$a \nabla^2 \overset{*}{u}(\xi, t_F; s, t) + \frac{\partial \overset{*}{u}(\xi, \tau_F; s, t)}{\partial t} = 0 \quad \text{in } \Omega \text{ and } t < t_F \tag{3.21}$$

$$\int_\Omega u(s, t)\, \overset{*}{u}(\xi, t_F; s, t)\, d\Omega(s) = u(\xi, t_F) \quad \text{for } t = t_F. \tag{3.22}$$

For the two dimensional problems which will be discussed in this chapter the fundamental solution and its normal derivative are given by

$$\overset{*}{u}(\xi, t_F; s, t) = \frac{1}{4 \pi a \tau} \exp\left(-\frac{r^2(\xi, s)}{4 a \tau}\right)$$

$$\overset{*}{q}(\xi, t_F; s, t) = \frac{d(\xi, s)}{8 \pi a^2 \tau^2} \exp\left(-\frac{r^2(\xi, s)}{4 a \tau}\right) \tag{3.23}$$

where $d(\xi, s) = (x(\xi) - x(s)) \cdot n_x(s) + (Y(\xi) - Y(s)) \cdot n_Y(s)$ and $n_x(s)$, $n_Y(s)$ are the direction cosinus of the normal at the point s.

The form of $\overset{*}{u}$ and $\overset{*}{q}$ can similarly be deduced for 3-D cases.

In order to investigate the singularity that occurs in the integrals in Eq. (3.18) at time $t = t_F$ at the peak of the Dirac delta function, some may subtract or add to the upper limit of the integrals an arbitrarily small quantity ε. The first integral on the left hand side of (3.18) is zero for $t_0 \leq t < t_F - \varepsilon$ because of Eq. (3.21). Taking the limit as $\varepsilon \to 0$ and accounting for condition (3.22) Eq. (3.18) becomes,

$$u(\xi, t_F) + \int_{t_0}^{t_F} \int_\Gamma a\, u(s, t)\, \overset{*}{q}(\xi, t_F; s, t)\, d\Gamma(s)\, dt \tag{3.24}$$

$$= \int_{t_0}^{t_F} \int_\Gamma a\, q(s, t)\, \overset{*}{u}(\xi, t_F; s, t)\, d\Gamma(s)\, dt + \int_\Omega u_0(s, t = t_0)\, \overset{*}{u}(\xi, t_F; s, t = t_0)\, d\Omega(s).$$

Equation (3.24) is valid for any point inside the domain, but in order to obtain a boundary integral equation, we have to take point ξ to the boundary yielding the following boundary integral equation,

$$c(\xi)\, u(\xi, t_F) + \int_{t_0}^{t_F} \int_\Gamma a\, u(s, t)\, \overset{*}{q}(\xi, t_F; s, t)\, d\Gamma(s)\, dt \tag{3.25}$$

$$= \int_{t_0}^{t_F} \int_\Gamma a\, q(s, t)\, \overset{*}{u}(\xi, t_F; s, t)\, d\Gamma(s)\, dt + \int_\Omega u_0(s, t = t_0)\, \overset{*}{u}(\xi, t_F; s, t = t_0)\, d\Omega(s).$$

This relationship can also be written in incremental form, i.e. from t_{F-1} to t_F, as

$$c(\xi)\, u(\xi, t_F) + \int_{t_{F-1}}^{t_F} \int_\Gamma a\, u(s, t)\, \overset{*}{q}(\xi, t_F; s, t)\, d\Gamma(s)\, dt$$

$$= \int_{t_{F-1}}^{t_F} \int_\Gamma a\, q(s, t)\, \overset{*}{u}(\xi, t_F; s, t)\, d\Gamma(s)\, dt$$

$$+ \int_\Omega u_{F-1}(s, t = t_{F-1})\, \overset{*}{u}(\xi, t_F; s, t = t_{F-1})\, d\Omega(s). \tag{3.26}$$

Two different time-marching schemes can now be employed using Eqs. (3.25) or (3.26). Both adopt a time-stepping technique where u and q are assumed to vary within each time step according to constant linear of higher order interpolation functions.

For the first scheme, Eq. (3.25), the time integration process starts at time $t = t_0$. Any domain integral at $t = t_0$ may be transformed into equivalent boundary integral for harmonic type initial conditions or it vanishes for the case $u_0 = 0$ [6]. In this way no domain integrals are required during the marching process.

The second scheme effectively marches on time by translating the axis in such a way that the initial time zero refers now to t_{F-1}. The solution progresses in a more conventional manner by carrying the influence of the previous steps on the initial condition — i.e. it produces a domain integral. The result is a boundary integral relationship similar to the one obtained for elliptic equations including a domain integral for the initial conditions at the beginning of each time step. This scheme will be the only one applied here. For further information about the other approach the reader is referred to Wrobel and Brebbia [1].

The values of the variables u and q are assumed to vary within each boundary element and each time step according to the space, φ, and time, Ψ, interpolation functions i.e.;

$$u(\eta, t) = \varphi^T \Psi U_m^n$$
$$q(\eta, t) = \varphi^T \Psi Q_m^n. \qquad (3.27)$$

The index n refers to the number of boundary nodes within each element for which the nodal values of u and q are associated and the index m refers to the degree of variation of the function Ψ, i.e. $m = 1$ if Ψ is constant, $m = 1, 2$ if Ψ is linear etc.

Let us assume that the boundary Γ is divided into E elements with N_E boundary nodes and domain Ω discretised into C cells with N_I internal points. The time dimension is also discretised into F time steps and one can write the discretised form of (3.26) as,

$$c(\xi) u(\xi, t_F) + a \sum_{e=1}^{E} \left[\int_{\Gamma_e} \varphi^T \int_{t_{F-1}}^{t_F} \overset{*}{q}(\xi, t_F; s, t) \Psi \, dt \, d\Gamma(s) \right] U_m^n$$

$$= a \sum_{e=1}^{E} \left[\int_{\Gamma_e} \varphi^T \int_{t_{F-1}}^{t_F} \overset{*}{u}(\xi, t_F; s, t) \Psi \, dt \, d\Gamma(s) \right] Q_m^n$$

$$+ \sum_{c=1}^{C} \int_{\Omega_c} \overset{*}{u}(\xi, t_F; s, t_{F-1}) u(s, t_{F-1}) \, d\Omega(s). \qquad (3.28)$$

One can also represent the geometrical variation by expressing the coordinates in terms of the local coordinates η as

$$x(\eta) = \varphi^T x^n \quad \text{and similarly for } Y \text{ and } Z \text{ coordinates.} \qquad (3.29)$$

Since the interpolation functions φ are expressed in terms of local coordinates, it is necessary to transform the elements of the surface differential $d\Gamma$ from the global cartesian system to the local system of coordinates as,

$$d\Gamma = |J| \, d\eta \qquad \text{for two-dimensional problems}$$
$$d\Gamma = |J| \, d\eta_1 \, d\eta_2 \quad \text{for three-dimensional problems} \qquad (3.30)$$

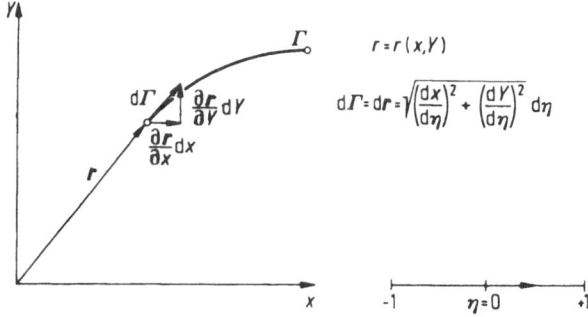

Fig. 3.1.   Geometrical definitions for boundary

For two-dimensional problems the Jacobian can be expressed simply, see Fig. 3.1.

3.4 Constant and Linear Time Interpolation

Constant Time Interpolation

Assuming that the functions u and q remain constant on time, one can write $\Psi = 1$ for this case. Applying Eq. (3.28) to all N_E boundary nodes, a final system of equations can be easily obtained,

$$\mathbf{H}\,\mathbf{U}_F = \mathbf{G}\,\mathbf{Q}_F + \mathbf{B}\,\mathbf{U}_{F-1}. \tag{3.31}$$

The coefficients of the matrices $\hat{\mathbf{H}}$, $\mathbf{G}$ and $\mathbf{B}$ are formed of a series of terms resulting from integration over the elements. These terms can be independently obtained (discontinuous elements) or combination of terms (continuous elements).

$$h_e^n = a \int_{\Gamma_e} \varphi^n \int_{t_{F-1}}^{t_F} \overset{*}{q}\,(\xi, t_F; s, t)\, dt\, d\Gamma\,(s)$$

$$g_e^n = a \int_{\Gamma_e} \varphi^n \int_{t_{F-1}}^{t_F} \overset{*}{u}\,(\xi, t_F; s, t)\, dt\, d\Gamma\,(s) \tag{3.32}$$

$$B_c = \int_{\Omega_c} \overset{*}{u}(\xi, t_F; s, t_{F-1})\, u\,(s, t_{F-1})\, d\Omega\,(s)$$

where the index n refers to the number of boundary nodes within each element. Note that $H_{ij} = \hat{H}_{ij} + c_i\,\delta_{ij}$.

The time integrals in (3.32) can be evaluated analytically as

$$\int_{t_{F-1}}^{t_F} \overset{*}{u}(\xi, t_F; s, t)\, dt = \frac{1}{4\pi a}\,E_1\,(x_{F-1})$$

$$\int_{t_{F-1}}^{t_F} \overset{*}{q}(\xi, t_F; s, t)\, dt = \frac{d}{2\pi a\, r^2}\,\exp\,(-x_{F-1}) \tag{3.33}$$

and the argument x_{F-1} is

$$x_{F-1} = \frac{r^2}{4a(t_F - t_{F-1})} \qquad (3.34)$$

where E_1 is the exponential — integral function defined by

$$E_1(x) = \int_x^\infty \frac{e^{-t}}{t}\, dt. \qquad (3.35)$$

Space Integration

The remaining step in the numerical solution of Eq. (3.28) is the computation of the space integrals. Although the space interpolation function φ can be taken as constant, linear or higher order, only discontinuous linear elements of the type shown in Fig. 3.2 are used in this chapter.

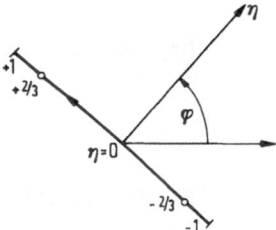

Fig. 3.2.   Linear discontinuous element

Where the space interpolation functions φ^n and vectors $\mathbf{u}^n$ and $\mathbf{q}^n$ are,

$$\varphi^n = \frac{1}{4}\begin{bmatrix} 2 - 3\,\eta \\ 2 + 3\,\eta \end{bmatrix}; \quad \mathbf{u}^n = \begin{bmatrix} u^1 \\ u^2 \end{bmatrix}; \quad \mathbf{q}^n = \begin{bmatrix} q^1 \\ q^2 \end{bmatrix}. \qquad (3.36)$$

The regular integrals h_e^n and g_e^n, for the case when the source point is not on the element, can be evaluated by standard gaussian formulae, i.e.

$$h_e^n = \frac{d_{ie}}{4\pi} \sum_{k=1}^K \frac{1}{r_{ik}^2} \exp(-x_{F-1})_k\, \varphi_k^n\, W_k$$

$$g_e^n = \frac{l_e}{8\pi} \sum_{k=1}^K E_1(x_{F-1})_k\, \varphi_k^n\, W_k \qquad (3.37)$$

where $d_{ie} = (x^i - x^1)(Y^2 - Y^1) + (Y^i - Y^1)(x^1 - x^2)$, $n = 1, 2$ and K represents the number of integration points.

For convenience during numerical computation the exponential integral function is approximated by rational and polynomials expansions [8],

$$E_1(x) = -\ln x + a_0 + a_1 x + a_2 x^2 + a_3 x^3 + a_4 x^4 + a_5 x^5 + \varepsilon(x)$$
$$0 < x \le 1; \quad |\varepsilon(x)| < 2 \cdot 10^{-7} \qquad (3.38)$$

or

$$E_1(x) = \frac{1}{x\, e^x} \cdot \frac{x^4 + b_1 x^3 + b_2 x^2 + b_3 x + b_4}{x^4 + c_1 x^3 + c_2 x^2 + c_3 x + c_4} + \varepsilon(x) \qquad (3.39)$$

$$1 \le x \le \infty; \quad [\varepsilon(x)] < 2 \cdot 10^{-8}.$$

When the source point lies on the element of integration, the integrals g_e^n contain a logarithmic singularity. One has then to compute the following integrals

$$g_e^1 = \frac{l_e}{8\pi} \int_{-1}^{+1} E_1(x_{F-1}) \frac{1}{4} (2-3\eta) \, d\eta$$

$$g_e^2 = \frac{l_e}{8\pi} \int_{-1}^{+1} E_1(x_{F-1}) \frac{1}{4} (2+3\eta) \, d\eta. \qquad (3.40)$$

In this case expanding the exponential integral in series [8],

$$E_1(x) = -c - \ln x + \sum_{n=1}^{\infty} (-1)^{n-1} \frac{x^n}{n \, n!} \qquad (3.41)$$

where $c = 0.5772156649$, the integrals (3.40) can be evaluated analytically for elements other than quadratic or higher order.

For straight-line elements the integrals h_e^n are identically zero due to the orthogonality between r and n. Since the source point ξ lies on the smooth boundary the c coefficient [e.g. (3.28)] is $c(\xi) = \frac{1}{2}$.

The coefficient of matrix **B** are formed by integrating over the cells. The domain is discretised in triangles such that the Cartesian coordinates x of points within each element are expressed by the following equations [5]

$$x(\eta) = x_1 \eta_1 + x_2 \eta_2 + x_3 \eta_3$$
$$Y(\eta) = Y_1 \eta_1 + Y_2 \eta_2 + Y_3 \eta_3 \qquad (3.42)$$

where $\eta_1, \eta_2, \eta_3 = 1 - \eta_1 - \eta_2$ are area coordinates – Fig. 3.3.

The transformation of the element of surface $d\Omega$ from the global Cartesian system of coordinates to the local system of coordinates is given by,

$$d\Omega = |J| \, d\eta_1 \, d\eta_2 \qquad (3.43)$$

where the Jacobian $|J|$ equals twice the area of the triangle.

Two alternative ways can be postulated for computing the domain integrals. First, one can express the function u in terms of nodal values and appropriate interpolation functions, as in finite elements, i.e.

$$u(\eta) = \varphi^T u^n \qquad (3.44)$$

where $\sum \varphi_i = 1$ and the index n is associated to the nodal points. Here we take the linear discontinuous element which is shown in the Fig. 3.4 where the interpola-

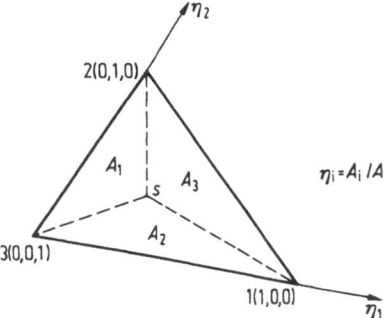

Fig. 3.3.   Definition of area coordinates

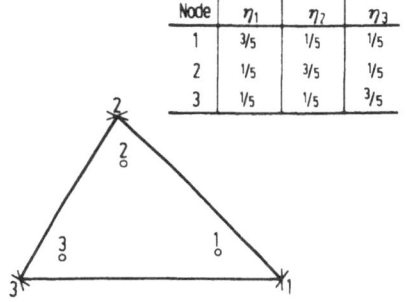

Node	η_1	η_2	η_3
1	3/5	1/5	1/5
2	1/5	3/5	1/5
3	1/5	1/5	3/5

Fig. 3.4.   Linear discontinuous element

tion functions φ^n and vector $\mathbf{u}^n$ are given by,

$$\varphi^n(\eta) = \frac{1}{2} \begin{Bmatrix} 4\eta_1 - \eta_2 - \eta_3 \\ 4\eta_2 - \eta_3 - \eta_1 \\ 4\eta_3 - \eta_1 - \eta_2 \end{Bmatrix} ; \quad \mathbf{u}^n = \begin{bmatrix} u^1 \\ u^2 \\ u^3 \end{bmatrix} \qquad (3.45)$$

and the domain integrals are of the form,

$$B_c^n = \int_{\Omega_c} \varphi^n \overset{*}{u}(\xi, t_F; s, t_{F-1})\, d\Omega(s) = \int_0^1 \left[\int_0^{1-\eta_2} \varphi^n \overset{*}{u}(\xi, t_F; s, t_{F-1})\, |J|\, d\eta_1 \right] d\eta_2. \qquad (3.46)$$

It is generally more convenient to compute the domain integrals, which are regular, by using a suitable numerical scheme. For triangular cells Hammer's integration formulae can be used,

$$B_c^n = A_c \sum_{k=1}^K \frac{1}{4\pi a(t_F - t_{F-1})} \exp\left(-\frac{r_{ik}^2}{4a(t_F - t_{F-1})}\right) W_k\, \varphi_k^n \qquad (3.47)$$

where A_c is the area of the triangle and k represents the number of integration points.

Alternatively, no interpolation functions are introduced and the values for function u are computed directly at each integration points yielding,

$$B_c = \int_{\Omega_c} \overset{*}{u}(\xi, t_F; s, t_{F-1})\, u(s, t_{F-1})\, d\Omega(s) = \int_1^0 \left[\int_0^{1-\eta_2} \overset{*}{u}(\xi, t_F; s, t_{F-1})\, u(s, t_{F-1})\, |J|\, d\eta_1 \right] d\eta_2$$

$$\qquad (3.48)$$

again Hammer's formulae can be used

$$B_c = A_c \sum_{k=1}^K \frac{1}{4\pi a(t_F - t_{F-1})} \exp\left(-\frac{r_{ik}^2}{4a(t_F - t_{F-1})}\right) W_k\, u_{(F-1)k}. \qquad (3.49)$$

The formulae with seven integration points has been applied for the examples shown in this chapter. The domain element is represented in the Fig. 3.5.

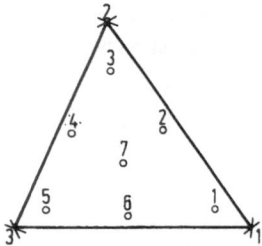

Fig. 3.5. Internal cell

It can be seen from Eqs. (3.37), (3.46) and (3.49), that the elements of matrices $\mathbf{G}$, $\mathbf{H}$ and $\mathbf{B}$ depend on geometrical data, properties of the solid and the magnitude of the time step. Thus, adopting a constant time step throughout the analysis, they can be computed only once and stored.

Linear Time Interpolation

Let us now assume a linear variation on time within each time step for u and q according to the following interpolation functions,

$$\Psi_1 = \frac{t_F - t}{\Delta t_F}; \quad \Psi_2 = \frac{t - t_{F-1}}{\Delta t_F} \quad (3.50)$$

where $\Delta t_F = t_F - t_{F-1}$. Applying Eq. (3.28) to all N_E boundary nodes, the following equations can be written,

$$\mathbf{H}_1 \mathbf{U}_{F-1} + \mathbf{H}_2 \mathbf{U}_F = \mathbf{G}_1 \mathbf{Q}_{F-1} + \mathbf{G}_2 \mathbf{Q}_F + \mathbf{B} \mathbf{U}_{F-1} \quad (3.51)$$

where the elements of the matrices are given by terms such as,

$$h_{e1}^n = \frac{a}{\Delta t_F \Gamma_e} \int_{\Gamma_e} \varphi^n \int_{t_{F-1}}^{t_F} (t_F - t) \overset{*}{q}(\xi, t_F; s, t) \, dt \, d\Gamma(s)$$

$$h_{e2}^n = \frac{a}{\Delta t_F \Gamma_e} \int_{\Gamma_e} \varphi^n \int_{t_{F-1}}^{t_F} (t - t_{F-1}) \overset{*}{q}(\xi, t_F; s, t) \, dt \, d\Gamma(s)$$

$$g_{e1}^n = \frac{a}{\Delta t_F \Gamma_e} \int_{\Gamma_e} \varphi^n \int_{t_{F-1}}^{t_F} (t_F - t) \overset{*}{u}(\xi, t_F; s, t) \, dt \, d\Gamma(s) \quad (3.52)$$

$$g_{e2}^n = \frac{a}{\Delta t_F \Gamma_e} \int_{\Gamma_e} \varphi^n \int_{t_{F-1}}^{t_F} (t - t_{F-1}) \overset{*}{u}(\xi, t_F; s, t) \, dt \, d\Gamma(s)$$

$$B_c = \int_{\Omega_c} \overset{*}{u}(\xi, t_F; s, t_{F-1}) \, u(s, t = t_{F-1}) \, d\Omega(s)$$

where index n refers to the number of boundary nodes within each element and $H_{ij,2} = \hat{H}_{ij,2} + c_i \delta_{ij}$.

The time integrals in (3.52) can be computed analytically as

$$\int_{t_{F-1}}^{t_F} (t_F - t) \overset{*}{u}(\xi, t_F; s, t) \, dt = \frac{\Delta t_F}{4 \pi a} [\exp(- x_{F-1}) - X_{F-1} E_1(x_{F-1})]$$

$$\int_{t_{F-1}}^{t_F} (t - t_{F-1}) \overset{*}{u}(\xi, t_F; s, t) \, dt = \frac{\Delta t_F}{4 \pi a} [E_1(x_{F-1}) + x_{F-1} E_1(x_{F-1}) - \exp(- x_{F-1})]$$

$$\int_{t_{F-1}}^{t_F} (t_F - t) \overset{*}{u}(\xi, t_F; s, t) \, dt = \frac{d}{8 \pi a^2} E_1(x_{F-1})$$

$$\int_{t_{F-1}}^{t_F} (t - t_{F-1}) \overset{*}{q}(\xi, t_F; s, t) \, dt = \frac{d}{8 \pi a^2} [\exp(- x_{F-1})/x_{F-1} - E_1(x_{F-1})]. \quad (3.53)$$

Space Integration

Let us assume a quadratic space variation for the functions u and q with the geometry still represented by a straight line segment — Fig. 3.6.

The space interpolation functions φ and vectors $\mathbf{u}^n$, and $\mathbf{q}^n$ are,

$$\varphi^n(\eta) = \begin{bmatrix} \frac{1}{2}\eta(\eta - 1) \\ (1 - \eta^2) \\ \frac{1}{2}\eta(\eta + 1) \end{bmatrix}, \quad \mathbf{u}^n = \begin{bmatrix} u^1 \\ u^2 \\ u^3 \end{bmatrix}, \quad \mathbf{q}^n = \begin{bmatrix} q^1 \\ q^2 \\ q^3 \end{bmatrix}. \quad (3.54)$$

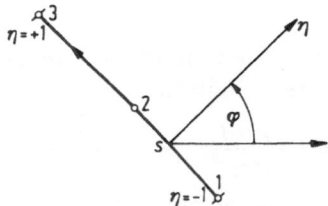

Fig. 3.6. Quadratic element

When the source points do not lie on the element of integration, integrals h_{e1}^n, h_{e2}^n, g_{e1}^n and g_{e2}^n can be evaluated by standard Gauss formulas,

$$h_{e1}^n = \frac{d_{ie}}{16\pi a \Delta t_F} \sum_{k=1}^{K} E_1(x_{F-1})_k \varphi_k^n W_k$$

$$h_{e2}^n = \frac{d_{ie}}{16\pi a \Delta t_F} \sum_{k=1}^{K} [\exp(-x_{F-1})/x_{F-1} - E_1(x_{F-1})]_k \varphi_k^n W_k$$

$$g_{e1}^n = \frac{l_e}{8\pi} \sum_{k=1}^{K} [\exp(-x_{F-1}) - x_{F-1} E_1(x_{F-1})]_k \varphi_k^n W_k$$

$$g_{e2}^n = \frac{l_e}{8\pi} \sum_{k=1}^{K} [E_1(x_{F-1}) + x_{F-1} E_1(x_{F-1}) - \exp(-x_{F-1})]_k \varphi_k^n W_k.$$

(3.55)

When the source point lies on the element under consideration, the integrals g_{e2}^n contain a logarithmic singularity. One has to solve the following integrals,

$$g_{e2}^n = \frac{l_e}{8\pi} \left[\int_{-1}^{+1} E_1(x_{F-1}) \varphi^n d\eta + \int_{-1}^{+1} x_{F-1} E_1(x_{F-1}) \varphi^n d\eta - \int_{-1}^{+1} \exp(-x_{F-1}) \varphi^n d\eta \right]$$

$$n = 1, 2, 3. \quad (3.56)$$

Notice that only the first integral above is singular and that it is the same form as the integrals in Eq. (3.40). Thus it can be evaluated in closed form as previously described. The remaining integrals, which are non-singular, are computed by using standard Gaussian formulas.

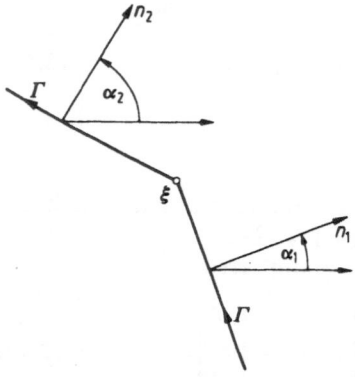

Fig. 3.7. Computation of weight $c(\xi)$

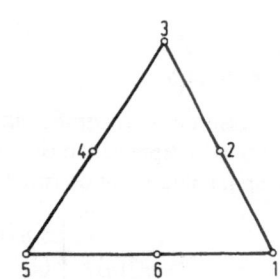

Fig. 3.8. Quadratic triangle internal cell

For straight line elements the integrals h_{e1}^n and h_{e2}^n are identically zero due to the orthogonality between r and n.

The $c(\xi)$ coefficients for the mid-nodes are $\frac{1}{2}$ because the boundary is smooth, but for the extreme-nodes they are computed by,

$$c(\xi) = (\pi + \alpha_1 - \alpha_2)/2\pi. \tag{3.57}$$

Once again the domain is modelled by discretising it into triangles, but a quadratic variation of the function u is now assumed – Fig. 3.8.

The interpolation functions φ^n and vector $\mathbf{u}^n$ are given by,

$$\varphi^n(\eta) = \begin{bmatrix} \eta_1(2\eta_1 - 1) \\ 4\eta_1\eta_2 \\ \eta_2(2\eta_2 - 1) \\ 4\eta_2\eta_3 \\ \eta_3(2\eta_3 - 1) \\ 4\eta_3\eta_1 \end{bmatrix} \quad ; \quad \mathbf{u}^n = \begin{bmatrix} u^1 \\ u^2 \\ u^3 \\ u^4 \\ u^5 \\ u^6 \end{bmatrix} \tag{3.58}$$

and the domain integrals are of the form (3.46) and can be computed by applying formula (3.47).

3.5 Non-Linear Boundary Conditions for the Case of Constant Conductivity

The governing integral equation relating the boundary values for potential and its normal derivatives is described by (3.28).

We will now develop the boundary integral equations for the other boundary condition cases, using the elements constant in time and linear in space described in 3.4. For this kind of boundary element, the integral equation can be written as,

$$c(\xi)\,u(\xi, t_F) + \sum_{e=1}^{E} [h_e^1\ h_e^2] \begin{bmatrix} u^1 \\ u^2 \end{bmatrix}_{t_F} = \sum_{e=1}^{E} [g_e^1\ g_e^2] \begin{bmatrix} q^1 \\ q^2 \end{bmatrix}_{t_F} + B(\xi, t_{F-1}). \tag{3.59}$$

Let us first consider the mixed type boundary conditions described by Eq. (3.2), by substituting $q_{t_r}^n$ by,

$$q_{t_r}^n = -\frac{h_0^n}{k_0}(u^n - u_f^n)_{t_r}; \quad n = 1, 2. \tag{3.60}$$

These conditions apply on the Γ_3 part of the boundary so that the whole boundary is now $\Gamma = \Gamma_1 + \Gamma_2 + \Gamma_3$, yielding the following boundary integral equation,

$$c(\xi)\,u(\xi, t_F) + \sum_{e=1}^{E} [h_e^1\ h_e^2] \begin{bmatrix} u^1 \\ u^2 \end{bmatrix}_{t_F} + \frac{1}{k_0} \sum^{E_3} [g_e^1\ g_e^2] \begin{bmatrix} h_0^1 & u^1 \\ h_0^2 & u^2 \end{bmatrix}_{t_F}$$

$$= \sum^{E_{1+2}} [g_e^1\ g_e^2] \begin{bmatrix} q^1 \\ q^2 \end{bmatrix}_{t_F} + \frac{1}{k_0} \sum^{E_3} [g_e^1\ g_e^2] \begin{bmatrix} h_0^1 & u_f^1 \\ h_0^2 & u_f^2 \end{bmatrix}_{t_F} + B(\xi, t_{F-1}). \tag{3.61}$$

Applying Eqs. (3.61) to all boundary nodes the following system of equations is obtained,

$$(\mathbf{H} + \mathbf{CG})\, \mathbf{U}_F = \mathbf{G}(\mathbf{Q} + \mathbf{CU}_f)_F + \mathbf{B}\,\mathbf{U}_{F-1} \tag{3.62}$$

Equations (3.62) can be reordered in such a way that all the unknowns are on the left hand side and the following linear set of equations is thus obtained,

$$\mathbf{A}\mathbf{X} = \mathbf{F}_0. \tag{3.63}$$

Similarly, one can write integral equation for the boundary conditions described by equations (3.3) introducing

$$q^n_{t_r} = -\frac{h^n_0}{k_0}(u^n - u^n_f)_{t_r}; \quad n = 1, 2 \tag{3.64}$$

on the Γ_3 part of the boundary where u^n_f represent unknown nodal values for temperatures on the opposite boundary, yielding the following formulae,

$$c(\xi)\, u(\xi, t_F) + \sum^{E} [h^1_e\, h^2_e] \begin{bmatrix} u^1 \\ u^2 \end{bmatrix}_{t_r} + \frac{1}{k_0} \sum^{E_3} [g^1_e\, g^2_e] \begin{bmatrix} h^1_0 & u^1 \\ h^2_0 & u^2 \end{bmatrix}_{t_r} - \frac{1}{k_0} \sum^{E_3} [g^1_e\, g^2_e] \begin{bmatrix} h^1_0 & u^1_f \\ h^2_0 & u^2_f \end{bmatrix}_{t_r}$$

$$= \sum^{E_{1+2}} [g^1_e\, g^2_e] \begin{bmatrix} q^1 \\ q^2 \end{bmatrix}_{t_r} + B(\xi, t_{F-1}). \tag{3.65}$$

Applying Eqs. (3.65) to all boundary nodes, one can write,

$$(\mathbf{H} + \mathbf{CG})\, \mathbf{U}_F - \mathbf{CG}\,\mathbf{U}_{fF} = \mathbf{GQ}_F + \mathbf{B}\,\mathbf{U}_{F-1}. \tag{3.66}$$

Again a linear set of equations (3.63) is obtained.

Thus, for the constant conductivity k_0 the system of equation is linear if the boundary conditions are linear, i.e. (3.2) or (3.3).

The only nonlinearities considered here are due to the temperature dependent heat transfer coefficient and the radiation described by Eq. (3.4). One can divide the heat transfer coefficient into constant and temperature dependent part as,

$$h(u) = h_0 + h_N(u) \tag{3.67}$$

substituting the value $q^n_{t_r}$ on the Γ_3 part of the boundary by,

$$q^n_{t_r} = -\frac{h^n_0}{k_0}(u^n - u^n_f)_{t_r} - \frac{h^n_N}{k_0}(u^n - u^n_f)_{t_r} - \frac{\varepsilon^n \sigma}{k_0}(u^4 - u^4_s)^n_{t_r}. \tag{3.68}$$

This yields the following discretised form for the integral equation,

$$c(\xi)\, u(\xi, t_F) + \sum^{E} [h^1_e\, h^2_e] \begin{bmatrix} u^1 \\ u^2 \end{bmatrix}_{t_r} + \frac{1}{k_0} \sum^{E_3} [g^1_e\, g^2_e] \begin{bmatrix} h^1_0 & u^1 \\ h^2_0 & u^2 \end{bmatrix}_{t_r}$$

$$= \sum^{E_{1+2}} [g^1_e\, g^2_e] \begin{bmatrix} q^1 \\ q^2 \end{bmatrix}_{t_r} + \frac{1}{k_0} \sum^{E_3} [g^1_e\, g^2_e] \begin{bmatrix} h^1_0 & u^1_f \\ h^2_0 & u^2_f \end{bmatrix}_{t_r} + B(\xi, t_{F-1})$$

$$+ \frac{1}{k_0} \sum^{E_3} [g^1_e\, g^2_e] \begin{bmatrix} h^1_N (u_f - u)^1 \\ h^2_N (u_f - u)^2 \end{bmatrix}_{t_r} + \frac{\sigma}{k_0} \sum^{E_3} [g^1_e\, g^2_e] \begin{bmatrix} \varepsilon^1 (u^4_s - u^4)^1 \\ \varepsilon^2 (u^4_s - u^4)^2 \end{bmatrix}_{t_r}. \tag{3.69}$$

Equation (3.70) consists of two parts, the first is linear and identical to (3.61), the second part is non linear. The final system of equations can then be expressed as,

$$\mathbf{A X}_N = \mathbf{F}_0 + \mathbf{F}_N \qquad (3.70)$$

the non linear part $\mathbf{F}_N$ caused by thermal radiation is highly non linear and appropriate methods for solving this system of equations need to be employed. Brown's method has been used by the authors and proved to converge very quickly. The functions $\mathbf{Y}$ whose roots have to be found, can be obtained by rewriting Eq. (3.70) as,

$$\mathbf{Y} = \mathbf{0} = \mathbf{X}_N - \mathbf{X}_L - \mathbf{A}^{-1} \mathbf{F}_N \qquad (3.71)$$

where $\mathbf{X}_L = \mathbf{A}^{-1} \mathbf{F}_0$ is linear solution, $\mathbf{A}^{-1}$ is inverse matrix and $\mathbf{X}_N$ is non linear solution.

3.6 Non-Linear Boundary Conditions for the Case of Temperature Dependent Conductivity

Let the heat conductivity be a function of temperature. Now boundary conditions of the mixed-type introduce non linearities which generally are not too severe. One can write a boundary integral equation similar to (3.59) by changing the variable u to Ψ taking $k_0 = 1$, i.e.

$$c(\xi)\, \Psi(\xi, t_F) + \sum_{e=1}^{E} [h_e^1\ h_e^2] \begin{bmatrix} \psi^1 \\ \psi^2 \end{bmatrix}_{t_F} = \sum_{e=1}^{E} [g_e^1\ g_e^2] \begin{bmatrix} q^1 \\ q^2 \end{bmatrix}_{t_F} + B(\xi, t_{F-1}) \qquad (3.72)$$

and Eq. (3.61) can be written as,

$$c(\xi)\, \Psi(\xi, t_F) + \sum^{E} [h_e^1\ h_e^2] \begin{bmatrix} \psi^1 \\ \psi^2 \end{bmatrix}_{t_F} + \sum^{E_3} [g_e^1\ g_e^2] \begin{bmatrix} h_0^1 & K^{-1} [\Psi]^1 \\ h_0^2 & K^{-1} [\Psi]^2 \end{bmatrix}_{t_F}$$

$$= \sum^{E_{1+2}} [g_e^1\ g_e^2] \begin{bmatrix} q^1 \\ q^2 \end{bmatrix}_{t_F} + \sum^{E_3} [g_e^1\ g_e^2] \begin{bmatrix} h_0^1 & u_f^1 \\ h_0^2 & u_f^2 \end{bmatrix}_{t_F} + B(\xi, t_{F-1}). \qquad (3.73)$$

The integral formula is non linear due to the application of Kirchoff's transform. One can now separate Eq. (3.73) into a linear and non linear part introducing the following identity,

$$h_0\, K^{-1} [\Psi] = h_0\, \Psi - (h_0\, \Psi - h_0\, K^{-1} [\Psi]). \qquad (3.74)$$

Equation (3.73) can be rewritten as,

$$c(\xi)\, \Psi(\xi, t_F) + \sum^{E} [h_e^1\ h_e^2] \begin{bmatrix} \psi^1 \\ \psi^2 \end{bmatrix}_{t_F} + \sum^{E_3} [g_e^1\ g_e^2] \begin{bmatrix} h_0^1 & \psi^1 \\ h_0^2 & \psi^2 \end{bmatrix}_{t_F}$$

$$= \sum^{E_{1+2}} [g_e^1\ g_e^2] \begin{bmatrix} q^1 \\ q^2 \end{bmatrix}_{t_F} + \sum^{E_3} [g_e^1\ g_e^2] \begin{bmatrix} h_0^1 & u_f^1 \\ h_0^2 & u_f^2 \end{bmatrix}_{t_F} + B(\xi, t_{F-1})$$

$$+ \sum^{E_3} [g_e^1\ g_e^2] \begin{bmatrix} h_0^1 (\Psi - \bar{K}^1 [\Psi]^1 \\ h_0^2 (\Psi - \bar{K}^1 [\Psi]^2 \end{bmatrix}_{t_F}. \qquad (3.75)$$

The linear part of the above equation is identical to Eq. (3.61), so that one can write the final system as,

$$A\,X_N = F_0 + F_N \tag{3.76}$$

which can be solved by a simple iterative process.

Following the same ideas a boundary integral equation can be developed for the heat transfer boundary conditions (3.12) such that,

$$
c(\xi)\,\Psi(\xi, t_F) + \sum_{}^{E} [h_e^1\, h_e^2] \begin{bmatrix} \Psi^1 \\ \Psi^2 \end{bmatrix}_{t_F} + \sum_{}^{E_3} [g_e^1\, g_e^2] \begin{bmatrix} h_0^2 & \Psi^1 \\ h_0^3 & \Psi^2 \end{bmatrix}_{t_F} - \sum_{}^{E_3} [g_e^1\, g_e^2] \begin{bmatrix} h_0^1 & \Psi_r^1 \\ h_0^2 & \Psi_r^2 \end{bmatrix}_{t_F}
$$

$$
= \sum_{}^{E_{1+2}} [g_e^1\, g_e^2] \begin{bmatrix} q^1 \\ q^2 \end{bmatrix}_{t_F} + B(\xi, t_{F-1}) + \sum_{}^{E_3} [g_e^1\, g_e^2] \begin{bmatrix} h_0^1(\Psi - \bar{K}^1[\Psi])^1 \\ h_0^2(\Psi - \bar{K}^1[\Psi])^2 \end{bmatrix}_{t_F}
$$

$$
- \sum_{}^{E_3} [g_e^1\, g_e^2] \begin{bmatrix} h_0^1(\Psi_r - \bar{K}^1[\Psi_r])^1 \\ h_0^2(\Psi_r - \bar{K}^1[\Psi_r])^2 \end{bmatrix}_{t_F} . \tag{3.77}
$$

The temperature dependent heat transfer coefficient can be easily considered applying Eq. (3.67).

Finally, one can also develop a formulation for boundary conditions of the type (3.13), following the same ideas previously described,

$$
c(\xi)\,\Psi(\xi, t_F) + \sum_{}^{E} [h_e^1\, h_e^2] \begin{bmatrix} \Psi^1 \\ \Psi^2 \end{bmatrix}_{t_F} + \sum_{}^{E_3} [g_e^1\, g_e^2] \begin{bmatrix} h_0^1 & \Psi^1 \\ h_0^2 & \Psi^2 \end{bmatrix}_{t_F} .
$$

$$
= \sum_{}^{E_{1+2}} [g_e^1\, g_e^2] \begin{bmatrix} q^1 \\ q^2 \end{bmatrix}_{t_F} + \sum_{}^{E_3} [g_e^1\, g_e^2] \begin{bmatrix} h_0^1 & u_f^1 \\ h_0^2 & u_f^2 \end{bmatrix}_{t_F} + B(\xi, t_{F-1})
$$

$$
+ \sum_{}^{E_3} [g_e^1\, g_e^2] \begin{bmatrix} h_0^1(\Psi - \bar{K}^1[\Psi])^1 \\ h_0^2(\Psi - \bar{K}^1[\Psi])^2 \end{bmatrix}_{t_F} + \sum_{}^{E_3} [g_e^1\, g_e^2] \begin{bmatrix} h_N^1(u_f - \bar{K}^1[\Psi])^1 \\ h_N^2(u_f - \bar{K}^1[\Psi])^2 \end{bmatrix}_{t_F}
$$

$$
+ \sigma \sum_{}^{E_3} [g_e^1\, g_e^2] \begin{bmatrix} \varepsilon^1\{u_s^4 - (\bar{K}^1[\Psi])^4\}^1 \\ \varepsilon^2\{u_s^4 - (\bar{K}^1[\Psi])^4\}^2 \end{bmatrix}_{t_F} . \tag{3.78}
$$

The linear part in (3.78) is again the same as in Eq. (3.61). The only difference is in the non linear part. The final system of equations can be written as (3.70) and once again solved using Brown's method.

3.7 Applications

The applications presented here concern the solution of time dependent conduction problems with constant and potential dependent conductivity. They were analysed using two types of elements (i) linear discontinuous elements with constant interpolation in time and (ii) quadratic continuous elements with linear interpolation in time.

Example 1. *Rectangular plate with natural and essential boundary conditions and non-linear conductivity.* The first example analyses heat conduction in a rectangular plate — Fig. 3.9. Because of symmetry only half of the plate has to be considered. This is discretised using ten discontinuous linear elements in a space and constant on time. the total number of nodal points is twenty with eight internal cells showing in the figure. The material constants are $c = \varrho = a = 1$ and conductivity varies exponentially

$$k = 1. \quad \exp\left(0.3\,\frac{u - 0.5}{0.5}\right).$$

Boundary conditions are prescribed fluxes on two boundaries and temperatures on one boundary. Initial conditions are $u = 0$ on the whole plate. The problem is rendered into a linear one by using Kirchoff's transform and solved on time using time step $\tau = t_F - t_{F-1} = 1.0$. The numerical results (Fig. 3.10) converge to the

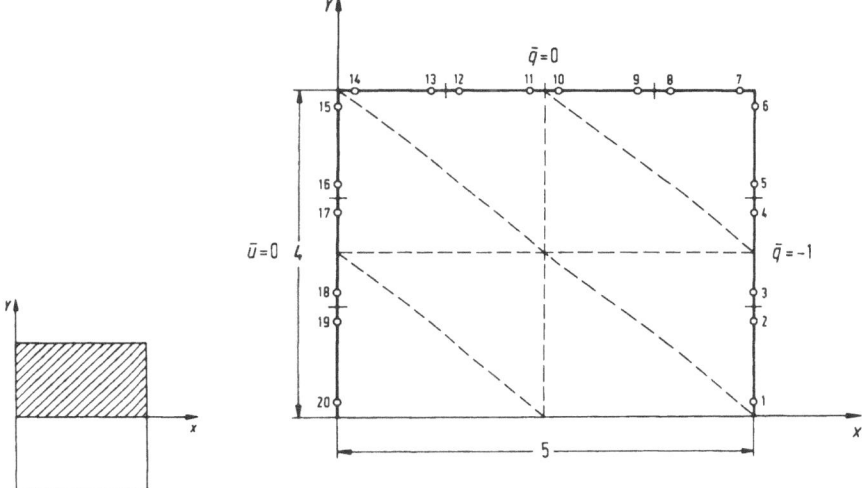

Fig. 3.9. B.E.M. discretisation (10 elements, 20 nodes, 8 cells, 56 points)

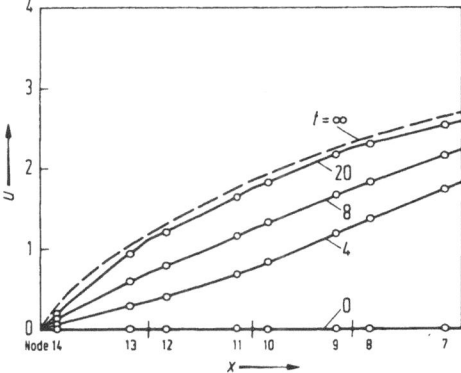

Fig. 3.10. Temperatures at boundary nodes

Table 3.1. Potentials at boundary nodes

No.	$u(t = 4)$	$u(t = 8)$	$u(t = 12)$	$u(t = 16)$	$u(t = 20)$	$u(t = \infty)$
1	1.79	2.17	2.38	2.50	2.57	2.699
2	1.79	2.17	2.38	2.50	2.57	2.699
3	1.79	2.18	2.38	2.50	2.57	2.699
4	1.80	2.18	2.38	2.50	2.57	2.699
5	1.81	2.18	2.38	2.50	2.57	2.699
6	1.81	2.18	2.38	2.50	2.57	2.699
7	1.71	2.11	2.32	2.44	2.51	2.642
8	1.34	1.79	2.03	2.17	2.25	2.395
9	1.15	1.63	1.88	2.02	2.11	2.254
10	0.81	1.30	1.55	1.69	1.78	1.936
11	0.65	1.12	1.37	1.52	1.60	1.748
12	0.37	0.76	0.97	1.09	1.17	1.301
13	0.26	0.56	0.74	0.84	0.91	1.018
14	0.05	0.12	0.17	0.20	0.22	0.312

steady state solution which can be found analytically by the expression

$$u(t = \infty) = 0.5 + \frac{5}{3} \ln\left(\frac{3}{5} x + e^{-0.3}\right).$$

The boundary element results behave well and no obvious jumps exist in spite of using discontinuous elements. The discontinuous elements have also given satisfactory results and they have become popular in engineering practice [9].

Example 2. *Rectangular plate with non-linear boundary conditions and non-linear conductivity.* Because of symmetry only one quarter need to be considered – shown in Fig. 3.11. The example was studied using linear discontinuous elements in

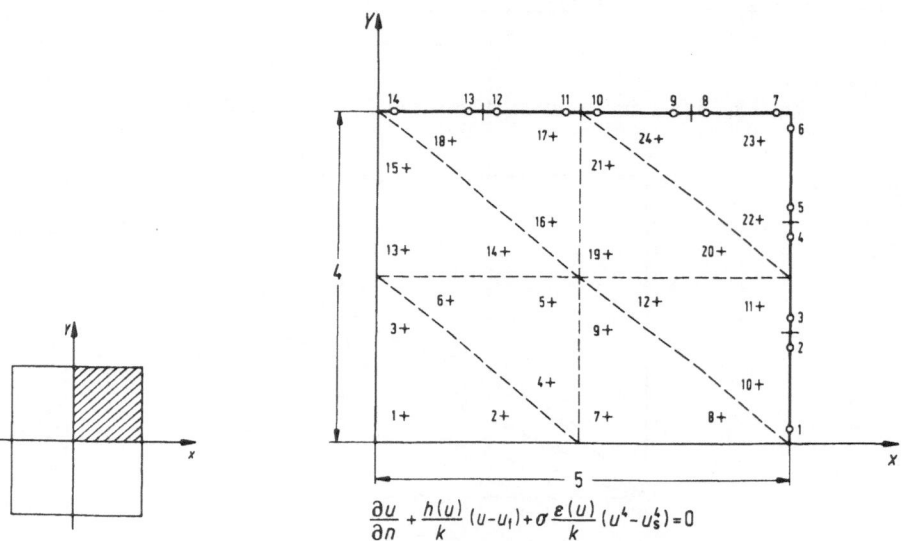

$$\frac{\partial u}{\partial n} + \frac{h(u)}{k}(u - u_1) + \sigma \frac{e(u)}{k}(u^4 - u_s^4) = 0$$

Fig. 3.11. B.E.M. discretisation (7 elements, 14 nodes, 8 cells, 54 points)

space and constant on time. The domain was defined by eight discontinuous internal cells over which the potential varies linearly in function of the nodal values indicated in the figure. Again all quantities were normalised, taking specific heat $c = 1$, density $\varrho = 1$, and diffusivity $a = 1$. All over the external boundary radiation boundary conditions were applied. The conductivity, heat transfer coefficient and emmisivity are defined by

$$k(u) = 1. \quad \exp\left(0.2\, \frac{u - 0.5}{0.5}\right)$$

$$h(u) = 5. \quad + 0.2 \exp\left(0.2\, \frac{u - 0.5}{0.5}\right)$$

$$\varepsilon(u) = 0.5 + 0.1 \exp\left(0.2\, \frac{u - 0.5}{0.5}\right).$$

The initial conditions are $u = 0$ everywhere in domain. The problem is now non linear due to mixed and radiation boundary conditions. Kirchoff's transform was applied in conjunction with Brown's interation method.

The time step was taken $\tau = t_F - t_{F-1} = 0.5$ and in each time step only three interations were needed to converge the results to required accuracy. As the surrounding media is assumed to be $u_f = u_s = 1$ the steady state solution everywhere is equal to 1. Notice that the results for boundary nodes (Table 3.2) converge

Table 3.2. Potentials at boundary nodes and internal points (see Fig. 3. 11)

No.	$u(t = 2)$	$u(t = 4)$	$u(t = 6)$	$u(t = \infty)$
Boundary nodes				
1	0.917	0.953	0.970	1.0
2	0.921	0.957	0.972	1.0
3	0.926	0.960	0.975	1.0
4	0.942	0.970	0.981	1.0
5	0.953	0.977	0.985	1.0
6	0.982	0.991	0.995	1.0
7	0.983	0.992	0.995	1.0
8	0.956	0.978	0.986	1.0
9	0.945	0.972	0.982	1.0
10	0.928	0.961	0.975	1.0
11	0.924	0.957	0.972	1.0
12	0.918	0.951	0.968	1.0
13	0.917	0.949	0.966	1.0
14	0.916	0.947	0.964	1.0
Internal points				
1	0.093	0.373	0.592	1.0
5	0.281	0.555	0.714	1.0
10	0.756	0.862	0.913	1.0
15	0.639	0.772	0.849	1.0
19	0.544	0.744	0.837	1.0
23	0.931	0.966	0.979	1.0

rapidly to the stationary value. The temperature at the internal points instead appear to converge more slowly to the stationary solution.

Example 3. *Rectangular plate with a cavity with natural and non-linear boundary conditions and non-linear conductivity.* Only one quarter of the problem depicted in the Fig. 3.12 needs to be discretised because of symmetry. Twenty linear discontinuous boundary elements in space and constant on time wer employed. A specific heat $c = 840$, density $\varrho = 2200$, diffusivity $a = 0.83874\ E\text{-}06$ were taken and the conductivity was defined by

$$k(u) = 1. \quad \exp\left(0.2\,\frac{u - 500}{500}\right).$$

Values for the heat transfer coefficient $h_0 = 40$ with $u_f = 500$ and emmsivity $\varepsilon_0 = 0.7$ with $u_s = 1000$ were taken. Initial conditions were $u_0 = 320$ everywhere. This problem is highly non linear because of the non linear boundary conditions in addition to temperature dependent conductivity. It was solved using Brown's iterative method. Convergence was achieved in all cases in three cycles. Notice that very high accuracy was demanded (convergence criterion $0.1\ E\text{-}3$, number of significant digits desired 6) in comparison with usual engineering accuracy. Twenty two internal cells were considered with seven integration points each. The time step was taken $\tau = t_F - t_{F-1} = 2000$.

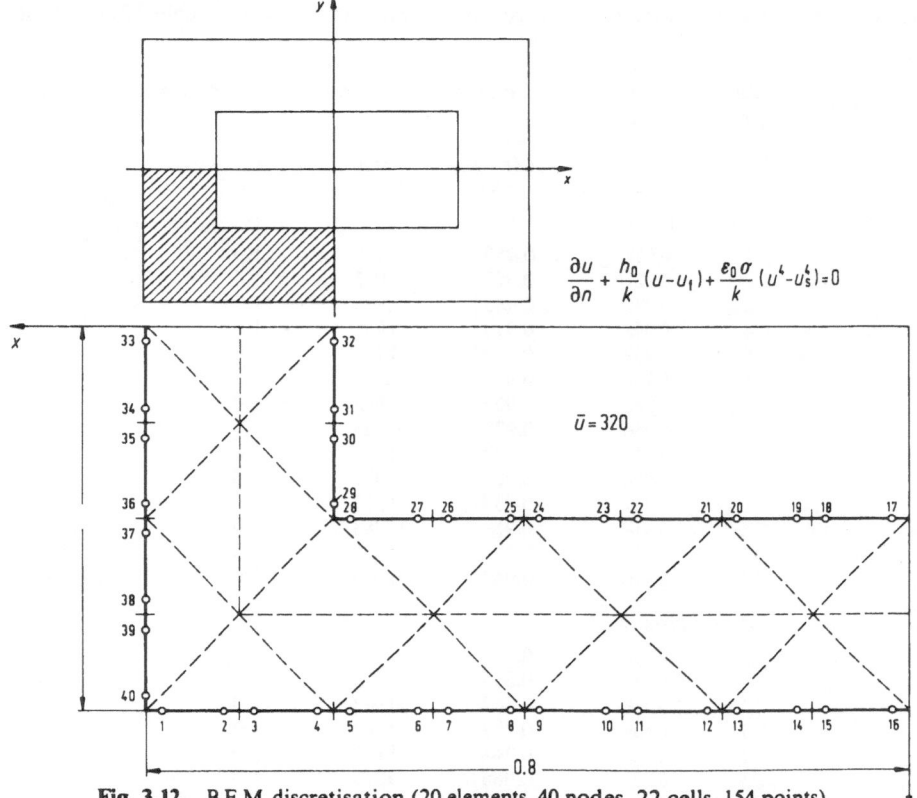

$$\frac{\partial u}{\partial n} + \frac{h_0}{k}\,(u - u_1) + \frac{\varepsilon_0\,\sigma}{k}\,(u^4 - u_s^4) = 0$$

$\bar{u} = 320$

Fig. 3.12. B.E.M. discretisation (20 elements, 40 nodes, 22 cells, 154 points)

Table 3.3. Potentials at boundary nodes (see Fig. 3.12)

No.	x	Y	$u\,(t = 6000)$	$u\,(t = 12000)$	$u\,(t = 18000)$
17	− 0.016	− 0.200	848	860	863
18	− 0.083	− 0.200	848	860	863
19	− 0.116	− 0.200	848	860	863
20	− 0.183	− 0.200	848	860	863
21	− 0.216	− 0.200	848	860	863
22	− 0.283	− 0.200	848	860	863
23	− 0.316	− 0.200	848	860	863
24	− 0.383	− 0.200	848	860	863
25	− 0.416	− 0.200	848	860	863
26	− 0.483	− 0.200	848	859	862
27	− 0.516	− 0.200	847	858	861
28	− 0.583	− 0.200	828	841	846
29	− 0.600	− 0.183	828	841	846
30	− 0.600	− 0.116	847	858	861
31	− 0.600	− 0.083	847	858	862
32	− 0.600	− 0.016	848	859	862

Results for the boundary nodes tended to converge to the steady state solution reported by [2] as shown in Table 3.3.

Example 4. *Concrete section.* Figure 3.13 depicts part of the section of concrete bridge with a rectangular cavity subjected to thermal loading. The section bridge is assumed to be subjected to complex radiation and mixed boundary conditions.

On the top surface the radiation condition can be defined

$$I\,[\mathrm{W/m^2}] = \frac{43.2 \times 10^6}{16 \times 3600} \sin \frac{\pi\,(t - 4)}{16}.$$

For the bottom surface the similar condition applies

$$I\,[\mathrm{W/m^2}] = \frac{4.0 \times 10^6}{16 \times 3600} \sin \frac{\pi\,(t - 5)}{16}$$

where intensity of radiation I is in $[\mathrm{W/m^2}]$ and time t [h].

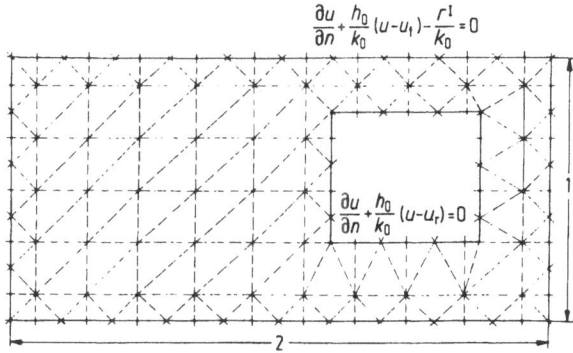

Fig. 3.13. B.E.M. discretisation (82 continuous quadratic elements, 164 nodes, 164 continuous quadratic cells, 246 points)

On the vertical surface on the left a linear variation of intensity is assumed for the radiaton, i.e.

$$
\begin{aligned}
\text{at} \quad t &= 4 \to I = 0 \\
\text{at} \quad t &= 10 \to I = 400 \\
\text{at} \quad t &= 14 \to I = 200 \\
\text{at} \quad t &= 20 \to I = 0.
\end{aligned}
$$

For the right hand side vertical surface a different linear function is adopted,

$$
\begin{aligned}
\text{at} \quad t &= 4 \to I = 0 \\
\text{at} \quad t &= 10 \to I = 100 \\
\text{at} \quad t &= 15 \to I = 600 \\
\text{at} \quad t &= 20 \to I = 0.
\end{aligned}
$$

The surrounding temperature is described by another linear equation representing

$$
\begin{aligned}
\text{at} \quad t &= 4 \to u_f = 22\,°\text{C} \\
\text{at} \quad t &= 15 \to u_f = 34\,°\text{C} \\
\text{at} \quad t &= 20 \to u_f = 26\,°\text{C}.
\end{aligned}
$$

The thermal properties of the section are

$$
\begin{aligned}
k_0 &= 1.4\ \text{WmK} & &\text{thermal conductivity} \\
\varrho &= 2400\ \text{kg/m}^3 & &\text{mass density} \\
c &= 960\ \text{J/kgK} & &\text{specific heat.}
\end{aligned}
$$

Absorption of external radiation is defined by $(r\,I)$, where the r — absorptivity coefficients are,

top surface	0.7
bottom surface	0.5
left surface	0.5
right surface	0.5.

The heat transfer coefficients $h_0\ [\text{W/m}^2\,\text{K}]$ are,

top surface	23
bottom surface	7
left surface	15
right surface	9
internal surface	2.

Inside the cavity it is assumed that the heat is transferred from a point on one surface to the opposite surface according to the following equation

$$
q = h_0 (u_A - u_B)
$$

where A and B are points opposite each other on the two surfaces (the boundary conditions of the type defined by equation 3.3).

The section is modelled using 82 continuous quadratic boundary elements in space, with 164 nodal points and 164 continuous quadratic internal cells with 246 internal points. The variation on time was assumed to be linear within each time step. The solution was started at $t = 8$ with initial uniform temperature of the section equal to 20 °C and finished at $t = 20$ hours, using a time step $\Delta t_F = 2$ hours.

Isothermals for time 14, are shown in Fig. 3.14. Figure 3.15 shows the results on the central line for different times.

It is interesting to notice that due to the low conductivity of the concrete and its mass the temperature does not vary appreciably in its centre part but the variation is more marked on the right hand side where the thickness of the section is reduced.

The problem was run also with non linear conductivity but the results as expected did not show any appreciable difference to the linear ones and consequently are not presented here.

This application shows how the technique can be used to model practical problems. Although the results were not compared with the other solution they appear to be accurate for design purposes.

Notice that the results for isothermals at 14 hours agree reasonably well with these requested in Example 4, Chap. 2 of Vol. 1 in this series. Small differences where attributed to the different ways of dealing with the holes. As a matter of fact the example was used as a beach mash for the counter coals.

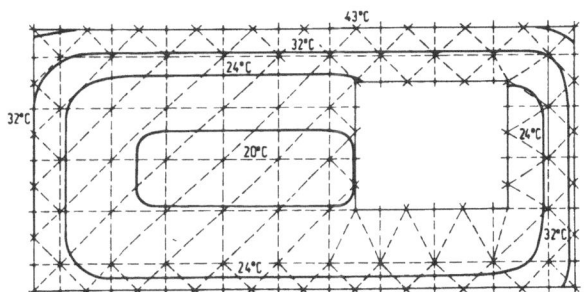

Fig. 3.14. Isothermals for time $t = 14$ hours

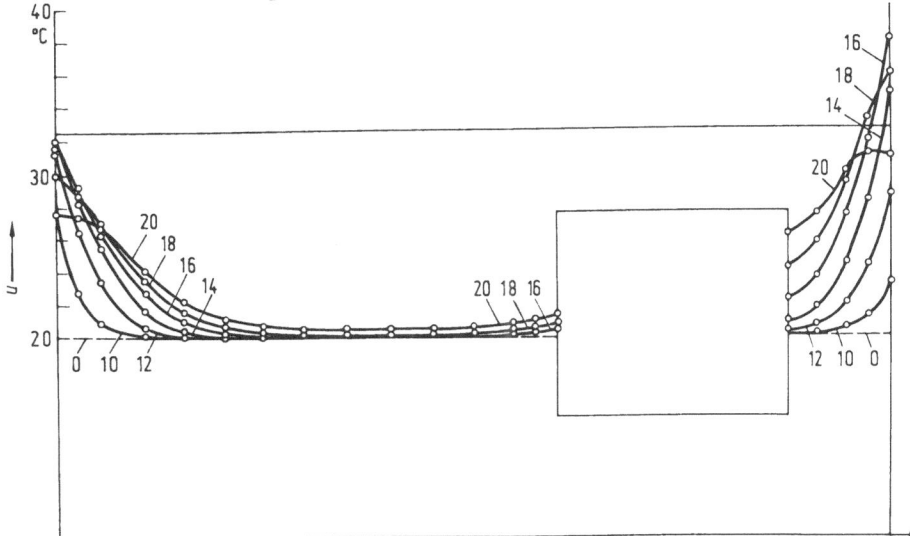

Fig. 3.15. Temperatures on central line for different times

3.8 Conclusions

This chapter demonstrates how the Kirchoff's transform can be extended to time dependent potential problems for which the conductivity is function of temperature or potential. The transformation effectively reduces the space dependent non linearities to the boundary. Cells in this case are only required for the computation of initial conditions. A time and space dependent fundamental solution has been applied here which gives extremely accurate and stable results even for large time steps. The resulting integral equation also incorporates mixed and radiation boundary conditions in addition to natural and essential boundary conditions. The non linearities were solved using a very efficient-Brown's-method.

References

1 Wrobel, L., Brebbia, C., Time dependent potential problems. Chapter 6 in *Progress in boundary element methods* 1, Pentech Press, London, Wiley, New York, 1981
2 Bialecki, R., Nowak, A., Boundary value problems, in heat conduction with non linear material and non linear boundary conditions. Appl. Math. Modelling 5, p. 417–421, 1981
3 Brebbia, C.A., Telles, J., Wrobel, L., *Boundary element methods – Theory and application.* Springer-Verlag, New York, 1982
4 Skerget, P., Brebbia, C.A., Non linear potential problems. Chapter 1 in *Progress in boundary element methods* 2, Pentech Press, London, Springer-Verlag, New York, 1983
5 Brebbia, C.A., Walkers, S., *Boundary element techniques in engineering.* Butterworths, 1980
6 Wrobel, L.C., Potential and viscous flow problems using the boundary element method. Ph.D. Thesis, Department of Civil Engineering of the University of Southampton, 1981
7 Onishi, K., Kuroki, T., Ohura, Y., Obata, K., Ito, T., Boundary element method in transient heat transfer problems. Fukuoka University, Fukuoka
8 Abramowitz, M., Stegun, I.A., *Handbook of mathematical functions.* Dover Publications, New York, 1965
9 Danson, D.J., BEASY – A boundary element analysis system. In: *Boundary element methods in engineering.* Proceedings of the fourth international seminar, Southampton, 1982

Chapter 4

Further Developments on the Solution of the Transient Scalar Wave Equation

by W. J. Mansur and C. A. Brebbia

4.1 Introduction

Since the last century integral equations have been used by mathematicians and physicists but only in the sixties they started being employed to obtain solutions of engineering problems. The numerical techniques that employ integral equations became known by engineers as boundary element methods.

At present the boundary element method (B.E.M.) can efficiently be applied to solve many practical problems such as steady state and transient heat transfer, elastostatics, plasticity, plate bending, etc. (for a comprehensive review see [1−3]).

Most of the research carried out on boundary elements has been concerned with solutions of elliptic and parabolic type differential equations. Quite a number of investigations have been carried out showing that the boundary element method is an efficient technique for these types of problem. However the same amount of effort has not been directed towards solving hyperbolic differential equations. This is a developing research area with a great deal to be accomplished in both analytical formulation and implementation of general numerical procedures.

In this chapter a review of the applications of the B.E.M. to wave equation problems is initially presented. Later in Sects. 4.2 to 4.9, one of the new procedures [4−6] that has been developed to solve two-dimensional transient scalar wave equation problems will be discussed.

Direct solution of hyperbolic differential equations using time-stepping techniques was first carried out by Friedman and Shaw [7] and later on by Shaw et al. [8−15].

The initial investigations carried out by these authors were concerned with the Kirchhoff's integral representation, and appear to have marked the shift to computer solutions of wave propagation problems using integral equations. They solved two-dimensional problems by considering them as three-dimensional cylindrical ones with arbitrary axes length. In this way the three-dimensional formulation could be employed, with the artificially introduced third spatial coordinate playing the role of a time like variable. With this procedure the time integration which is required in two-dimensional formulations is avoided at the expense of introducing an additional spatial dimension.

Further investigations related to Kirchhoff's integral equation were carried out by Mitzener [16], who presented a general numerical procedure to analyse transient scattering from a hard surface. Later on Groenenboom [17] using an approach similar to Mitzner's presented a general boundary element retarded potential technique to solve unsteady potential fluid flow problems in three dimensions.

So far, very few numerical schemes have been implemented to solve wave propagation problems using two-dimensional time dependent fundamental solutions.

Cole et al. [18] applied the well known two-dimensional time domain integral equation for the scalar wave equation [19] to solve transient elastodynamic anti-plane motions. In that work a time-stepping scheme was used to obtain numerical solutions for the problem of two welded half-planes excited by a concentrated source. Very accurate displacements at the common surface were obtained. Their formulation was however restricted to problems in which the boundary integral involving the potential (displacement) disappears, which implies that internal displacements could not be computed with their procedure. In spite of this their paper represents the first contribution towards finding a general formulation using a two-dimensional time-dependent fundamental solution.

Mansur and Brebbia [4, 5] have also applied the boundary element method to analyse transient problems governed by the two-dimensional scalar wave equation. Commencing with weighted residual considerations they initially derived the same integral equation obtained by Morse and Feshbach [19] using Green's theorem. Further transformations were than carried out to eliminate derivatives of Heaviside functions that appeared in the integral equation and a general approach amenable to numerical solutions was derived. Contributions due to initial conditions and source terms were also included. A time-stepping scheme similar to that proposed by Cole et al. was used to obtain time domain solutions. The numerical features of this approach were illustrated by three examples for all of which highly accurate results were obtained. The technique derived in [4, 5] was later on extended by Mansur [6] to solve transient two-dimensional elastodynamics.

4.2 The Boundary Initial Value Problem

The scalar wave equation can be written in terms of a potential u as:

$$\nabla^2 u\,(\mathbf{x},\,t) - \ddot{u}\,(\mathbf{x},\,t)/c^2 = -\,\gamma\,(\mathbf{x},\,t) \tag{4.1}$$

where c is the speed of wave propagation, t represents time, $\mathbf{x}$ is the position vector of a point with rectangular coordinates x_i $(i = 1, 2, 3)$ and $\gamma\,(\mathbf{x}, t)$ describes space and time dependence of the source density. Dots in Eq. (4.1) indicate time derivative, i.e. $\ddot{u} = \partial^2 u / \partial t^2$ and ∇^2 is the Laplace operator. The region Ω in which two-dimensional solutions of Eq. (4.1) are sought will be considered to be regular in the sense defined by Kellog [20], i.e. the Γ boundary of Ω can be composed of several closed regular surfaces which may have corners or edges [21].

In order to find the particular solution of Eq. (4.1) corresponding to the specific problem which needs to be solved it is necessary to specify initial conditions

$$u\,(\mathbf{x}, 0) = u_0\,(\mathbf{x})$$

$$v\,(\mathbf{x}, 0) = \left.\frac{\partial u\,(\mathbf{x}, t)}{\partial t}\right|_{t=0} = v_0\,(\mathbf{x}) \tag{4.2}$$

for every point **x** in Ω, at $t = 0$; and boundary conditions

$$u(\mathbf{x}, t) = \bar{u}(\mathbf{x}, t), \qquad\qquad\qquad \mathbf{x} \in \Gamma_1,$$

$$p(\mathbf{x}, t) = \mathbf{n} \cdot [\nabla u(\mathbf{x}, t)] = \frac{\partial u(\mathbf{x}, t)}{\partial n} = \bar{p}(\mathbf{x}, t), \quad \mathbf{x} \in \Gamma_2, \tag{4.3}$$

where $\Gamma = \Gamma_1 + \Gamma_2$, ∇u stands for grad u, '$\cdot$' represents the dot product and n is the coordinate in the direction parallel to the unit outward vector **n**, normal to Γ.

4.3 Dirac Delta and Heaviside Functions

When studying integral equations it is convenient to employ the Dirac delta function [22]. In one dimension the Dirac delta is defined by

$$\delta(x - a) = 0 \quad \text{when } x \neq a \text{ and}$$

$$\int_{-\infty}^{+\infty} \delta(x - a) f(x) \, dx = f(a). \tag{4.4}$$

The derivatives of the Dirac delta are functions such that,

$$\delta^{(k)}(x - a) = 0 \quad \text{when } x \neq a \text{ and}$$

$$\int_{-\infty}^{+\infty} \delta^{(k)}(x - a) f(x) \, dx = (-1)^k f^{(k)}(a), \tag{4.5}$$

where $\delta^{(k)}(x - a)$ and $f^{(k)}(a)$ stand for $\dfrac{\partial^k}{\partial x^k} \delta(x - a)$ and $\dfrac{\partial^k}{\partial x^k} f(x)\Big|_{x=a}$ respectively.

When a two- or three-dimensional domain Ω is considered the Dirac delta can be defined as follows

$$\delta(q - s) = 0 \quad \text{when } s \neq q \text{ and}$$

$$\int_{\Omega} \delta(q - s) f(q) \, d\Omega(q) = f(s), \tag{4.6}$$

where s and q represent two points within Ω.

Two-dimensional Green's functions corresponding to Eq. (4.1) can be conveniently represented using the Heaviside function (see Fig. 4.1) given, by

$$H(x - a) = \begin{cases} 1 & \text{if } x > a, \\ 0 & \text{if } x < a. \end{cases} \tag{4.7}$$

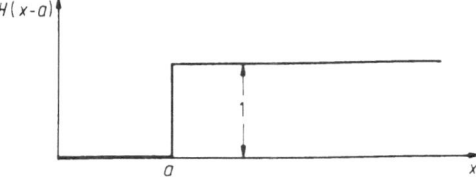

Fig. 4.1.   The Heaviside function

The Dirac delta and Heaviside functions can be related to each other as follows

$$\frac{d}{dx} H\,(x-a) = \delta\,(x-a)\,. \tag{4.8}$$

In the discussion just carried out, definitions and also certain basic properties of the Dirac delta and the Heaviside functions were presented. Additional properties to the ones previously described will be introduced where required. For a rigorous and detailed discussion on this subject attention should be directed to [23, 24].

4.4 Fundamental Solution in Three Dimensions

The Green's function (fundamental solution) for the scalar wave equation is the solution of Eq. (4.1) for an unbounded domain [19, 25] and a particular concentrated source, i.e.

$$\gamma = 4\,\pi\,\delta\,(q-s)\,\delta\,(t-\tau)\,. \tag{4.9}$$

Equation (4.1), in this case, can then be written as

$$\nabla^2 u^* - \ddot{u}^*/c^2 = -\,4\,\pi\,\delta\,(q-s)\,\delta\,(t-\tau)\,. \tag{4.10}$$

Thus u^* is the effect of a source represented by an impulse at $t=\tau$ located at $q=s$, whilst q and s are referred to in the literature as observation (field) and source points respectively.

The fundamental solution represented by Eq. (4.10) has the following properties [19, 25]

(i) causality

$$u^*\,(q,t;s,\tau) = 0 \quad \text{whenever } c\,(t-\tau) < |\,q - s\,|\,, \tag{4.11}$$

(ii) reciprocity

$$u^*\,(q,t;s,\tau) = u^*\,(s,-\tau;q,-t)\,, \tag{4.12}$$

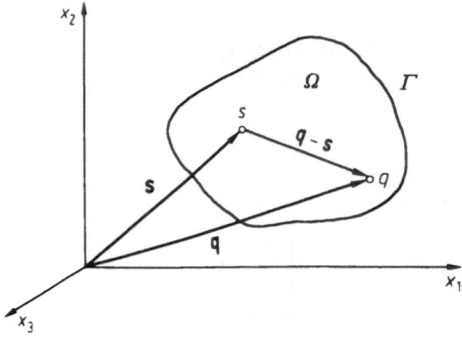

Fig. 4.2.   Definition of the vector $q - s$

(iii) time translation

$$u^* (q, t + t_1; s, \tau + t_1) = u^* (q, t; s, \tau) . \tag{4.13}$$

In three dimensions the solution of Eq. (4.10) is given by [19, 25]

$$u^* (q, t; s, \tau) = \frac{\delta [(r/c) - (t - \tau)]}{r} = \frac{c}{r} \delta [r - c (t - \tau)] \tag{4.14}$$

where $r = r (q, s) = |\, \mathbf{q} - \mathbf{s}\,|$, as shown in Fig. 4.2. In [25] substitution of u^* given by Eq. (4.14) into Eq. (4.1) is carried out in order to illustrate that the first is a solution of the second. A rigorous derivation of expression (4.14) can be found in [19].

4.5 Kirchhoff Integral Representation

When t is replaced by τ, Eq. (4.1) takes the following form

$$\nabla^2 u (q, \tau) - \frac{1}{c^2} \frac{\partial^2 u (q, \tau)}{\partial \tau^2} = - \gamma (q, \tau) . \tag{4.15}$$

From the reciprocity property Eq. (4.10) can be written as [19]

$$\nabla^2 u^* (q, t; s, \tau) - \frac{1}{c^2} \frac{\partial^2 u^* (q, t; s, \tau)}{\partial \tau^2} = - 4 \pi \, \delta (q - s) \, \delta (t - \tau) . \tag{4.16}$$

It is now convenient to introduce a notation which will be employed later. In future source and field points when over the Γ boundary will be denoted respectively by S and Q.

In order to deduce a singular boundary integral equation for the problem it is necessary to consider two distribution of potentials u^* and u that satisfy respectively Eqs. (4.15) and (4.16). In addition, u^* and u are assumed to be distributed respectively over the regions $\Omega + \Gamma$ and $\Omega^* + \Gamma^*$ (see Figs. 4.3 and 4.4) which have the same physical properties and are such that Ω^* contains $\Omega + \Gamma$. Only fundamental solutions concerning the infinite space are used in this work, therefore

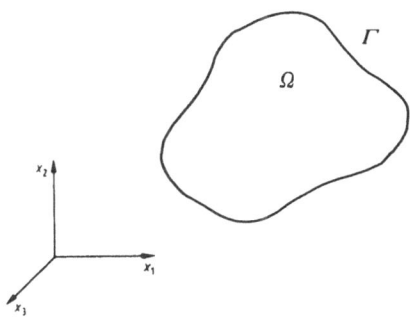

Fig. 4.3.   Three-dimensional region $\Omega + \Gamma$

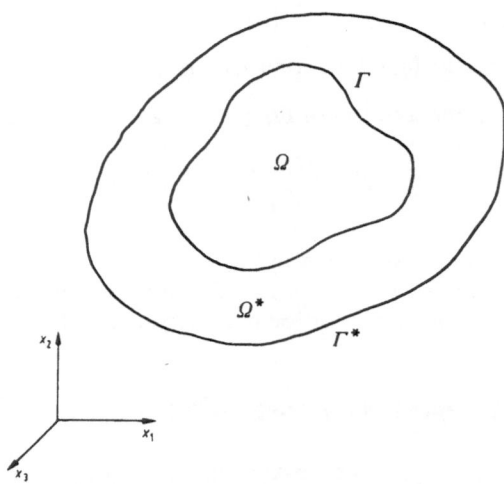

Fig. 4.4.   Region $\Omega^* + \Gamma^*$ containing $\Omega + \Gamma$

Γ^* must be placed at infinity. It is important to recognize that a procedure similar to the one described in this chapter can also be used when the fundamental solutions employed do not relate to the infinite space [26, 27].

A weighted residual statement for the problem under consideration can be written as [1–3]

$$\int_0^{t^+}\!\!\int_\Omega \left(\nabla^2 u - \frac{1}{c^2}\frac{\partial^2 u}{\partial \tau^2} + \gamma\right) u^* \, d\Omega \; d\tau = \int_0^{t^+}\!\!\int_{\Gamma_2} (p - \bar{p}) \, u^* d\Gamma \; d\tau - \int_0^{t^+}\!\!\int_{\Gamma_1} (u - \bar{u}) \, p^* \, d\Gamma \; d\tau \tag{4.17}$$

where $p^* = \partial u^*/\partial n$. t^+ in Eq. (4.17) represents $t + \varepsilon$, ε being arbitrarily small. Applying this procedure avoids terminating the integration exactly at the peak of a Dirac delta function. In Eq. (4.17) space integration and derivatives refer to the coordinates of the field points q or Q. Applying the divergence theorem twice to the term of Eq. (4.17) that contains the Laplacian operator $(\nabla^2 u)$ and integrating by parts twice with respect to τ the term that contains the time derivative $\partial^2 u/\partial \tau^2$, the following expression is obtained [4–6]

$$\int_0^{t^+}\!\!\int_\Gamma (u^* p - u p^*) \, d\Gamma \; d\tau + \int_0^{t^+}\!\!\int_\Omega \left(\nabla^2 u^* - \frac{1}{c^2}\frac{\partial^2 u^*}{\partial \tau^2}\right) u \, d\Omega \; d\tau$$

$$+ \int_0^{t^+}\!\!\int_\Omega u^* \, \gamma \, d\Omega \; d\tau + \frac{1}{c^2}\int_\Omega \left[\frac{\partial u^*}{\partial \tau} u - \frac{\partial u}{\partial \tau} u^*\right]_0^{t^+} d\Omega = 0. \tag{4.18}$$

Bearing in mind Eq. (4.16) and that due to the causality property

$$\left|\frac{\partial u^*}{\partial \tau} u\right|^{\tau = t^+} = \left|\frac{\partial u}{\partial \tau} u^*\right|^{\tau = t^+} = 0 \tag{4.19}$$

Eq. (4.18) can be written as

$$\int_0^{t^+}\int_\Gamma (u^* p - u p^*)\, d\Gamma\, d\tau - \int_0^{t^+}\int_\Omega 4\pi\,\delta\,(q-s)\,\delta\,(t-\tau)\,u\, d\Omega\, d\tau$$

$$+ \int_0^{t^+}\int_\Omega u^*\,\gamma\, d\Omega\, d\tau - \frac{1}{c^2}\int_\Omega (v_0^*\, u_0 - v_0\, u_0^*)\, d\Omega = 0, \qquad (4.20)$$

where

$$v_0^* = \left.\left|\frac{\partial u^*}{\partial \tau}\right.\right|_{\tau=0} \qquad (4.21)$$

$$u_0^* = \left| u^* \right|_{\tau=0}.$$

When the Dirac delta properties are applied to the second term on the left-hand side of Eq. (4.20) the following integral equation is obtained,

$$u\,(s,t) = \frac{1}{4\pi}\left(\int_0^{t^+}\int_\Gamma u^*\,(Q,t;s,\tau)\,p\,(Q,\tau)\, d\Gamma\,(Q)\, d\tau\right.$$

$$- \int_0^{t^+}\int_\Gamma p^*\,(Q,t;s,\tau)\,u\,(Q,\tau)\, d\Gamma\,(Q)\, d\tau$$

$$- \frac{1}{c^2}\int_\Omega v_0^*\,(q,t;s)\,u_0\,(q)\, d\Omega\,(q)$$

$$+ \frac{1}{c^2}\int_\Omega u_0^*\,(q,t;s)\,v_0\,(q)\, d\Omega\,(q)$$

$$\left.+ \int_0^{t^+}\int_\Omega u^*\,(q,t;s,\tau)\,\gamma\,(q,\tau)\, d\Omega\,(q)\, d\tau\right). \qquad (4.22)$$

The properties of the Dirac delta function can be used to eliminate the time integrations in Eq. (4.22) [17, 19, 25, 28]. Taking into consideration u^* given by Eq. (4.14) the following operations can be carried out for the first term on the right-hand side of Eq. (4.22)

$$\int_0^{t^+}\int_\Gamma u^*\,p\, d\Gamma\, d\tau = \int_\Gamma \frac{1}{r}\frac{1}{r}\int_0^{t^+}\delta\,[(r/c)-(t-\tau)]\,p\,(Q,\tau)\, d\tau\, d\Gamma$$

$$= \int_\Gamma \frac{1}{r}\frac{1}{r}\int_0^{t^+}\delta\,(\tau-t_r)\,p\,(Q,\tau)\, d\tau\, d\Gamma = \int_\Gamma \frac{1}{r}\,p\,(Q,t_r)\, d\Gamma \qquad (4.23)$$

where t_r stands for 'retarded time', equal to $[t - r/c]$.

The fundamental traction can be computed from

$$p^*\,(Q,t;s,\tau) = \frac{\partial}{\partial n}\,[u^*\,(Q,t;s,\tau)] = \frac{\partial r}{\partial n}\frac{\partial u^*}{\partial r}. \qquad (4.24)$$

The derivatives indicated in Eq. (4.24) refers to boundary points Q. Using formula (4.14), p^* can be written as

$$p^* = \frac{\partial r}{\partial n} \left(-\frac{1}{r^2} \delta \left[(r/c) - (t - \tau) \right] + \frac{1}{r} \frac{\partial}{\partial r} \delta \left[(r/c) - (t - \tau) \right] \right). \qquad (4.25)$$

Equation (4.25) can also be written as

$$p^* = \frac{\partial r}{\partial n} \left(-\frac{1}{r^2} \delta \left(\tau - t_r \right) + \frac{1}{c\,r} \frac{\partial}{\partial \tau} \left[\delta \left(\tau - t_r \right) \right] \right). \qquad (4.26)$$

In view of expression (4.26) the second term on the right-hand side of Eq. (4.22) can be written in the following way

$$\int_0^{t^+} \int_\Gamma p^* \, u \, d\Gamma \, d\tau = \int_\Gamma \frac{\partial r}{\partial n} \int_0^{t^+} \left(-\frac{1}{r^2} \delta \left(\tau - t_r \right) + \frac{1}{c\,r} \frac{\partial}{\partial \tau} \left[\delta \left(\tau - t_r \right) \right] \right) u \, d\tau \, d\Gamma. \qquad (4.27)$$

Taking expression (4.5) into consideration the following equation is then obtained

$$\int_0^{t^+} \int_\Gamma p^* \, u \, d\Gamma \, d\tau = -\int_\Gamma \frac{\partial r}{\partial n} \left(\frac{1}{r^2} u \left(Q, t_r \right) + \frac{1}{c\,r} \left[\frac{\partial u \left(Q, \tau \right)}{\partial \tau} \right]_{\tau = t_r} \right) d\Gamma. \qquad (4.28)$$

The integral involving source density in Eq. (4.22) can be operated as follows,

$$\int_0^{t^+} \int_\Omega u^* \, \gamma \, d\Omega \, d\tau = \int_\Omega \frac{1}{r} \int_0^{t^+} \gamma \left(q, \tau \right) \delta \left(\tau - t_r \right) d\tau \, d\Omega = \int_\Omega \frac{1}{r} \gamma \left(q, t_r \right) d\Omega. \qquad (4.29)$$

Dirac delta properties can also be applied to the terms that involve initial conditions [25] in Eq. (4.22). The final integral equation then obtained has the following form

$$u \left(s, t \right) = \frac{1}{4 \pi} \int_\Gamma \frac{1}{r \left(s, Q \right)} \, p \left(Q, t_r \right) d\Gamma \left(Q \right)$$

$$+ \frac{1}{4 \pi} \int_\Gamma \frac{\partial r \left(s, Q \right)}{\partial n \left(Q \right)} \left(\frac{1}{r^2 \left(s, Q \right)} u \left(Q, t_r \right) + \frac{1}{c\,r \left(s, Q \right)} \left[\frac{\partial u \left(Q, \tau \right)}{\partial \tau} \right]_{\tau = t_r} \right) d\Gamma \left(Q \right)$$

$$+ t \, N_0 + \frac{\partial}{\partial t} \left(t \, M_0 \right) + \frac{1}{4 \pi} \int_\Omega \frac{1}{r \left(s, q \right)} \gamma \left(q, t_r \right) d\Omega \left(q \right) \qquad (4.30)$$

where M_0 and N_0 are respectively the mean value of u_0 and v_0 over a spherical surface with centre at s and with a variable radius $c\,t$. It should be noticed that when $t_r < 0$ the terms on the right-hand side of expression (4.30) give no contribution to $u \left(s, t \right)$.

Equation (4.30) is known as the Kirchhoff's integral representation [29] and can be considered as the mathematical representation of Huygens' principle [19, 30].

The singular integrands of the integrals referring to initial conditions in Eq. (4.22) have been eliminated. However computation of source density contributions requires integrations of a singular function (γ/r) to be performed. This is not much of a problem and can easily be done numerically using the ordinary concept of integration.

Kirchhoff's integral representation can be used to compute u at internal points in terms of u, $\partial u/\partial n$ and $\partial u/\partial \tau$ on the Γ boundary and in terms of source density and initial conditions. However, in a well-posed problem u and p are not known over the entire Γ boundary. As a result Eq. (4.30) alone does not represent the complete solution of the boundary-initial value problem described in Sect. 4.2. A boundary integral equation from which boundary unknowns can be computed, can be derived by taking Eq. (4.30) to the Γ boundary. The integral equation obtained, unlike Kirchhoff's representation, has boundary integrals of singular functions which must be computed in the Cauchy principal value sense. The analytical manipulations required will be described next.

When the Γ boundary is assumed to satisfy the Liapunov smoothness condition [21], the domain Ω can be augmented by a small hemisphere of radius ε, whose centre is at a boundary point S as depicted in Fig. 4.5; Γ_ε displayed in this figure is the boundary of the hemisphere. In this situation, when initial conditions and source density are not considered, Eq. (4.30) can be written as

$$u\,(S,\,t) = \frac{1}{4\,\pi} \int_{\Gamma-\Gamma_\varepsilon} \frac{1}{r\,(S,\,Q)}\, p\,(Q,\,t_r)\, d\Gamma\,(Q)$$

$$+ \frac{1}{4\,\pi} \int_{\Gamma-\Gamma_\varepsilon} \frac{\partial r\,(S,Q)}{\partial n\,(Q)} \left(\frac{1}{r^2\,(S,Q)}\, u\,(Q,t_r) + \frac{1}{c\,r\,(S,Q)} \left[\frac{\partial u\,(Q,\tau)}{\partial \tau} \right]_{\tau=t_r} \right) d\Gamma\,(Q)$$

$$+ \frac{1}{4\,\pi}\,(S_p + S_u + S_v) \tag{4.31}$$

where

$$S_p = \int_{\Gamma_\varepsilon} \frac{1}{r\,(S,\,Q)}\, p\,(Q,\,t_r)\, d\Gamma_\varepsilon\,(Q) \tag{4.32}$$

$$S_u = \int_{\Gamma_\varepsilon} \frac{\partial r\,(S,\,Q)}{\partial n\,(Q)}\, \frac{1}{r^2\,(S,\,Q)}\, u\,(Q,\,t_r)\, d\Gamma_\varepsilon\,(Q) \tag{4.33}$$

$$S_v = \int_{\Gamma_\varepsilon} \frac{\partial r\,(S,\,Q)}{\partial n\,(Q)}\, \frac{1}{c\,r\,(S,\,Q)} \left[\frac{\partial u\,(Q,\,\tau)}{\partial \tau} \right]_{\tau=t_r} d\Gamma_\varepsilon\,(Q)\,. \tag{4.34}$$

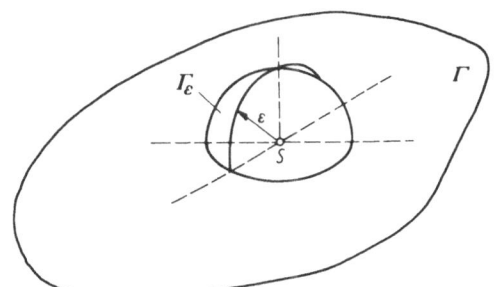

Fig. 4.5. Domain augmented by a hemisphere of radius ε whose centre is at a boundary point S

When $\varepsilon \to 0$; $\Gamma - \Gamma_\varepsilon \to \Gamma$, and as shown in [6, 18]

$$S_p = 0 , \quad \frac{1}{4\pi} S_u = \tfrac{1}{2} u (S, t) , \quad S_v = 0 . \tag{4.35}$$

Therefore for boundary points located on smooth parts of the Γ boundary the following boundary integral equation can be written,

$$\tfrac{1}{2} u (S, t) = \frac{1}{4\pi} \int_\Gamma \frac{1}{r(S, Q)} p (Q, t_r) \, d\Gamma (Q)$$

$$+ \frac{1}{4\pi} \int_\Gamma \frac{\partial r (S, Q)}{\partial n (Q)} \left(\frac{1}{r^2 (S, Q)} u (Q, t_r) + \frac{1}{cr (S, Q)} \left[\frac{\partial u (Q, \tau)}{\partial \tau} \right]_{\tau = t_r} \right) d\Gamma (Q)$$

$$+ t N_0 + \frac{\partial}{\partial t} (t M_0) + \frac{1}{4\pi} \int_\Omega \frac{1}{r (S, q)} \gamma (q, t_r) \, d\Omega (q) . \tag{4.36}$$

It should be noticed that the integrals outlined in Eq. (4.36) are to be computed in the Cauchy principal value sense.

It is important to point out that at points s located outside $\Omega + \Gamma$ the potential is equal to zero. The integral equation corresponding to this situation can be obtained by making the left-hand side of Eq. (4.30) equal to zero, i.e., $u (s, t) = 0$.

Occasionally a physical phenomenon can be best represented by a concentrated source given as

$$\gamma (q, t) = f (t) \delta (q - q_c) \tag{4.37}$$

where q_c gives the position of the source. The last integral on the right-hand side of Eq. (4.36) then becomes

$$\frac{1}{r_c} f (t_c) \tag{4.38}$$

where $r_c = | S - q_c|$ and $t_c = t - r_c/c$.

The numerical implementation of Eq. (4.36) is discussed in [17]. A special feature of the three-dimensional analysis is that no time integration is required. The same does not apply for the two-dimensional case as will be shown in the next section.

4.6 Two-Dimensional Boundary Integral Equation

A two-dimensional problem can be considered as being a three-dimensional one in which u is a function of two rectangular coordinates only, i.e.

$$u (\mathbf{x}, t) = u (x_1, x_2, t) . \tag{4.39}$$

Expression (4.39) implies that tractions, source density and initial conditions are also independent of x_3. In this case the domain in which the problem is studied can be considered to be a cylinder whose axis has infinite length and is parallel to the x_3-direction. Then, the two-dimensional domain Ω and the Γ boundary are

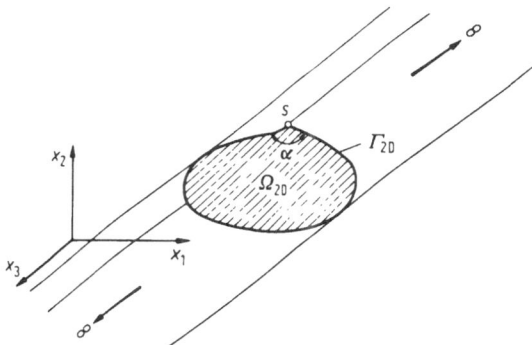

Fig. 4.6. Two-dimensional domain with a Kellog type Γ boundary

defined by the intersection of the cylinder with the (x_1, x_2) plane as depicted in
Fig. 4.6. Therefore, for this particular three-dimensional situation the first term on
the right-hand side of Eq. (4.22) can be operated as outlined below

$$\int_0^{t^+} \int_{\Gamma_{3D}} u_{3D}^* \, p \, d\Gamma \, d\tau = \int_0^{t^+} \int_{\Gamma_{2D}} p \int_{-\infty}^{+\infty} u_{3D}^* \, dx_3 \, d\Gamma_{2D} \, d\tau = \int_0^{t^+} \int_{\Gamma_{2D}} p \, u_{2D}^* \, d\Gamma_{2D} \, d\tau \quad (4.40)$$

where u_{2D}^* is the two-dimensional fundamental solution given by

$$u_{2D}^* = \int_{-\infty}^{+\infty} u_{3D}^* \, dx_3 . \quad (4.41)$$

The subscripts symbols $2D$ and $3D$ used in Eqs. (4.40) and (4.41) indicate respectively two and three dimensions and will be used hereafter only when confusion is
likely.

Transformations similar to the ones shown in expression (4.40) can be carried
out on the other integrals in Eq. (4.22). When the resulting expression is taken to
the Γ boundary the following integral equation is obtained

$$c(S) \, u(S, t) = \frac{1}{4\pi} \left(\int_0^{t^+} \int_{\Gamma} u^*(Q, t; S, \tau) \, p(Q, \tau) \, d\Gamma(Q) \, d\tau \right.$$

$$- \int_0^{t^+} \int_{\Gamma} p^*(Q, t; S, \tau) \, u(Q, \tau) \, d\Gamma(Q) \, d\tau$$

$$- \frac{1}{c^2} \int_{\Omega} v_0^*(q, t; S) \, u_0(q) \, d\Omega(q)$$

$$+ \frac{1}{c^2} \int_{\Omega} u_0^*(q, t; S) \, v_0(q) \, d\Omega(q)$$

$$\left. + \int_0^{t^+} \int_{\Omega} u^*(q, t; S, \tau) \, \gamma(q, \tau) \, d\Omega(q) \, d\tau \right) \quad (4.42)$$

where $u^* = u_{2D}^*$ is given by expression (4.41) and

$$p^* = p_{2D}^* = \int_{-\infty}^{+\infty} p_{3D}^* \, dx_3 = \frac{\partial}{\partial n} (u_{2D}^*) \tag{4.43}$$

$$v_0^* = v_{02D}^* = \left| \frac{\partial u_{2D}^*}{\partial \tau} \right|_{\tau=0} \tag{4.44}$$

$$u_0^* = u_{02D}^* = | u_{2D}^* |_{\tau=0} . \tag{4.45}$$

It should be understood that as it is clear that Eq. (4.42) refers to two dimensions, the subscript symbol $2D$ was not used in that case.

In the three-dimensional analysis, only Liapunov boundaries were considered, hence $c(S)$ in that situation was equal to $\frac{1}{2}$. However, in the two-dimensional formulation a generalization was introduced, namely that the Γ boundary can be of Kellog type. The parameter $c(S)$ in this case, as shown in [6], is represented by

$$c(S) = \frac{\alpha}{2\pi} \tag{4.46}$$

where α is the internal angle depicted in Fig. 4.6. In a similar manner to the three-dimensional case, two-dimensional integral equations that apply to points located inside and outside $\Omega + \Gamma$ can be obtained by considering $c(S)$ in Eq. (4.42) to be respectively equal to one and zero.

The methodology used here to obtain a two-dimensional boundary integral equation for the scalar wave equation is called the method of descent [19, 25].

The two-dimensional fundamental solution evolved from carrying out the integration indicated in expression (4.41) (for further details see [6]) is

$$u^* (q, t; s, \tau) = \frac{2c}{\sqrt{c^2 (t-\tau)^2 - r^2}} H [c (t-\tau) - r] . \tag{4.47}$$

The integral equation for the two-dimensional scalar wave equation was first obtained by Volterra [31]. A comparison between Volterra's and Kirchhoff's formulas displays a significant difference between two- and three-dimensional waves. Kirchhoff's formula demonstrates that at a time t, only the signal emitted at a point s at a time $(t - |q - s|/c)$ affects a point q. Volterra's formula, however, implies that in two dimensions a point q is affected at an instant t, by signals emitted at a point s, at all times previous to $(t - |q - s|/c)$. A more comprehensive discussion of this interesting discrepancy of behaviour of two- and three-dimensional waves can be found in [19, 25].

In addition to being of great benefit to the more complete understanding of wave propagation phenomena, Volterra's formula can also be used to obtain analytical solutions. However it has to undergo further transformations before it can be used in a numerical analysis.

4.7 Additional Transformations to Volterra's Integral Representation

The objective of the operations carried out in (i) and (ii) that follow is to remove the time and space derivatives of the Heaviside function that appear in Volterra's integral equation.

(i) The second term on the right-hand side of Eq. (4.42) can be operated as follows,

$$\int_0^{t^+}\int_r p^* \, u \, d\Gamma \, d\tau = \int_0^{t^+}\int_r u \, \frac{\partial u^*}{\partial n} \, d\Gamma \, d\tau = \int_0^{t^+}\int_r \frac{\partial r}{\partial n} \, u \, \frac{\partial u^*}{\partial r} \, d\Gamma \, d\tau. \qquad (4.48)$$

Substituting formula (4.47) into expression (4.48), the following expression is obtained

$$\int_{r_0}^{t^+}\int \frac{\partial r}{\partial n} \, u \, \frac{\partial u^*}{\partial r} \, d\tau \, d\Gamma = \int_{r_0}^{t^+}\int \frac{\partial r}{\partial n} \, u \, \frac{2cr}{\sqrt{[c^2(t-\tau)^2 - r^2]^3}} \, H\,[c\,(t-\tau) - r]\, d\tau \, d\Gamma$$

$$+ \int_r \frac{\partial r}{\partial n} \int_0^{t^+} u \, \frac{2c}{\sqrt{c^2(t-\tau)^2 - r^2}} \, \frac{\partial}{\partial r}\,(H\,[c\,(t-\tau) - r])\, d\tau \, d\Gamma. \qquad (4.49)$$

Further operations must now be performed on the second term on the right-hand side of Eq. (4.49). The following relationship will be used

$$\frac{\partial}{\partial r}\,(H\,[c\,(t-\tau) - r]) = \frac{\partial}{\partial (c\,\tau)}\,(H\,[c\,(t-\tau) - r])$$

$$= \frac{\partial}{\partial r}\,(1 - H\,[r - c\,(t-\tau)]) = -\,\delta\,[r - c\,(t-\tau)]. \qquad (4.50)$$

Therefore, using the notation,

$$L = L\,(r, t, \tau) = 2\,[c^2(t-\tau)^2 - r^2]^{-1/2}, \qquad (4.51)$$

$$L_0 = L_0\,(r, t, 0) = 2\,(c^2 t^2 - r^2)^{-1/2} \qquad (4.52)$$

and bearing in mind expressions (4.4) and (4.50) the following transformations can be carried out

$$\int_r \frac{\partial r}{\partial n} \int_0^{t^+} u \, \frac{2c}{\sqrt{c^2(t-\tau)^2 - r^2}} \, \frac{\partial}{\partial r}\,(H\,[c\,(t-\tau) - r])\, d\tau \, d\Gamma$$

$$= \int_r \frac{\partial r}{\partial n} \int_0^{t^+} u \, c \, L \, \frac{\partial}{\partial r}\,(H\,[c\,(t-\tau) - r])\, d\tau \, d\Gamma$$

$$= -\int_r \frac{\partial r}{\partial n} \int_0^{ct^+} u \, L \, \delta\,[c\,\tau - (c\,t - r)]\, d\,(c\,\tau)\, d\Gamma$$

$$= -\int_r \frac{\partial r}{\partial n}\,[u\,L]_{c\tau = ct - r}\, d\Gamma$$

$$= -\int_r \frac{\partial r}{\partial n} \int_0^{ct - r} \frac{\partial}{\partial (c\,\tau)}\,(u\,L)\, d\,(c\,\tau)\, d\Gamma - \int_r \frac{\partial r}{\partial n}\, u_0 \, L_0 \, H\,[c\,t - r]\, d\Gamma$$

$$= -\int_0^{t^+}\int_r \frac{\partial r}{\partial n}\left[c^2(t-\tau)\,u\,\frac{L^3}{4} + (\partial u/\partial \tau)\,L\right] H\,[c\,(t-\tau) - r]\, d\Gamma \, d\tau$$

$$-\int_r \frac{\partial r}{\partial n}\, u_0 \, L_0 \, H\,[c\,t - r]\, d\Gamma. \qquad (4.53)$$

Taking expressions (4.48), (4.49), (4.51), (4.52) and (4.53) into consideration the following expression can be derived

$$\int_0^{t^+}\int_r p^* \, u \, d\Gamma \, d\tau = -\int_0^{t^+}\int_r \frac{\partial r}{\partial n}\left(u \, \frac{2c\,[c\,(t-\tau)-r]}{\sqrt{[c^2\,(t-\tau)^2-r^2]^3}} + \frac{2\,(\partial u/\partial \tau)}{\sqrt{c^2(t-\tau)^2-r^2}}\right)$$

$$\cdot H\,[c\,(t-\tau)-r]\,d\Gamma \, d\tau - \int_r \frac{\partial r}{\partial n}\,\frac{2\,u_0}{\sqrt{c^2\,t^2-r^2}}\,H\,[c\,t-r]\,d\Gamma. \qquad (4.54)$$

(ii) The following property of the Heaviside function

$$\frac{\partial}{\partial \tau}\,H\,[c\,(t-\tau)-r] = c\,\frac{\partial}{\partial r}\,H\,[c\,(t-\tau)-r] \qquad (4.55)$$

is required in the transformations regarding the third term of the right-hand side of Eq. (4.42), given by

$$\int_\Omega v_0^* \, u_0 \, d\Omega. \qquad (4.56)$$

Taking account of expression (4.47) it is possible to write

$$\int_\Omega v_0^* \, u_0 \, d\Omega = \int_\Omega \left[\frac{\partial u^*}{\partial \tau}\right]_{\tau=0} u_0 \, d\Omega$$

$$= \int_\Omega u_0 \, \frac{2c^3 t}{\sqrt{[c^2\,t^2-r^2]^3}}\,H\,[c\,t-r]\,d\Omega$$

$$+ \int_\Omega u_0 \, \frac{2c^2}{\sqrt{c^2\,t^2-r^2}}\,\frac{\partial}{\partial r}\,(H\,[c\,t-r])\,d\Omega. \qquad (4.57)$$

A further investigation concerning the second term on the right-hand side of expression (4.57) is now required. If this term is called I_2, and a system of polar coordinates is adopted (see Fig. 4.7) whose origin is at the source point s, I_2 can be written as

$$I_2 = \int_{\theta_1}^{\theta_2}\int_{r=0}^{r=r_\Gamma(\theta)} r \, u_0 \, \frac{2c^2}{\sqrt{c^2\,t^2-r^2}}\,\frac{\partial}{\partial r}\,(H\,[c\,t-r])\,dr\,d\theta \qquad (4.58)$$

where $\theta_1 = 0$, $\theta_2 = 2\,\pi$, and

$$r_\Gamma(\theta) = r\,(s,\,Q) = |\,Q-s\,| \qquad (4.59)$$

defines the Γ boundary in polar coordinates (see Fig. 4.7).

If expression (4.58) is integrated by parts with respect to r, the following expression is obtained,

$$I_2 = \int_{\theta_1}^{\theta_2}\left[r\,u_0\,\frac{2c^2}{\sqrt{c^2\,t^2-r^2}}\,H\,[c\,t-r]\right]_0^{r_\Gamma(\theta)}d\theta$$

$$- \int_{\theta_1}^{\theta_2}\int_{r=0}^{r=r_\Gamma(\theta)}\frac{\partial}{\partial r}\,(r\,u_0\,c^2\,L_0)\,H\,[c\,t-r]\,dr\,d\theta. \qquad (4.60)$$

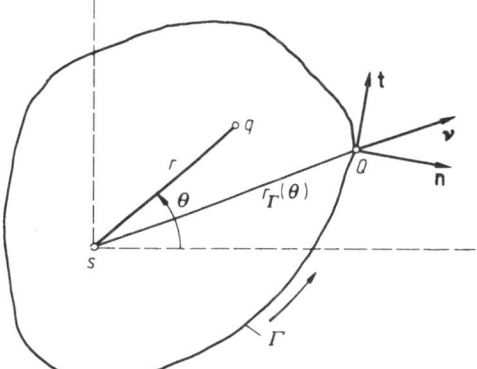

Fig. 4.7.   System of polar coordinates

Further manipulations give

$$I_2 = \int_{\theta_1}^{\theta_2} u_0 \, [r_\Gamma(\theta)] \, \frac{2c^2}{\sqrt{c^2 \, t^2 - r_\Gamma^2(\theta)}} \, H \, [c \, t - r_\Gamma(\theta)] \, r_\Gamma(\theta) \, d\theta$$

$$- \int_\Omega \frac{1}{r} u_0 \, c^2 \, L_0 \, H \, [c \, t - r] \, d\Omega - \int_\Omega \frac{\partial u_0}{\partial r} \, c^2 \, L_0 \, H \, [c \, t - r] \, d\Omega$$

$$- \int_\Omega u_0 \, c^2 \, r \, \frac{L_0^3}{4} \, H \, [c \, t - r] \, d\Omega \, . \tag{4.61}$$

The first integral on the right-hand side of expression (4.61) can be transformed by applying the following formula [6]

$$\int_{\theta_1}^{\theta_2} f \, [r_\Gamma(\theta)] \, r_\Gamma(\theta) \, d\theta = \int_\Gamma f \, [r(s, Q)] \, \frac{\partial r \, (s, Q)}{\partial n \, (Q)} \, d\Gamma(Q) \tag{4.62}$$

and so it is possible to write

$$\int_{\theta_1}^{\theta_2} u_0 \, [r_\Gamma(\theta)] \, \frac{2 \, c^2}{\sqrt{c^2 \, t^2 - r_\Gamma^2(\theta)}} \, H \, [c \, t - r_\Gamma(\theta)] \, r_\Gamma(\theta) \, d\theta$$

$$= \int_\Gamma \frac{\partial r \, (s, Q)}{\partial n \, (Q)} \, \frac{2 \, c^2 \, u_0 \, (Q)}{\sqrt{c^2 \, t^2 - r^2 \, (s, Q)}} \, H \, [c \, t - r \, (s, Q)] \, d\Gamma(Q) \, . \tag{4.63}$$

Taking expressions (4.52), (4.57), (4.61) and (4.63) into consideration, the following relationship can be stated

$$\int_\Omega v_0^* \, u_0 \, d\Omega = - \int_\Omega \left(u_0 \left[\frac{2c^2 \, (r - c \, t)}{\sqrt{[c^2 \, t^2 - r^2]^3}} + \frac{2 \, c^2}{r \sqrt{c^2 \, t^2 - r^2}} \right] + \frac{\partial u_0}{\partial r} \, \frac{2 \, c^2}{\sqrt{c^2 \, t^2 - r^2}} \right)$$

$$\cdot H \, [c \, t - r] \, d\Omega + \int_\Gamma \frac{\partial r}{\partial n} \, u_0 \, \frac{2 \, c^2}{\sqrt{c^2 \, t^2 - r^2}} \, H \, [c \, t - r] \, d\Gamma \, . \tag{4.64}$$

The last terms on the right-hand side of expressions (4.54) and (4.64) will cancel out within Eq. (4.42) to produce the final integral statement which for points located on the Γ boundary is written as

$$4 \pi c (S) \, u (S, t) = \int_0^{t^+} \int_\Gamma u^* p \, d\Gamma \, d\tau + \int_0^{t^+} \int_\Gamma \frac{\partial r}{\partial n} \left(B^* u + u^* \frac{v}{c} \right) d\Gamma \, d\tau$$

$$+ \frac{1}{c^2} \int_\Omega u_0^* v_0 \, d\Omega + \frac{1}{c} \int_\Omega \left(- B_0^* u_0 + u_0^* \frac{\partial u_0}{\partial r} + u_0^* \frac{u_0}{r} \right) d\Omega$$

$$+ \int_0^{t^+} \int_\Omega \gamma u^* \, d\Omega \, d\tau \tag{4.65}$$

where u^* and u_0^* are given respectively by expressions (4.47) and (4.21),

$$B^* = B^* (Q, t; S, \tau) = \frac{2 c \, [c \, (t - \tau) - r]}{\sqrt{[c^2 \, (t - \tau)^2 - r^2]^3}} \, H \, [c \, (t - \tau) - r], \tag{4.66}$$

$$B_0^* = B_0^* (Q, t; S) = B^* (Q, t; S, 0) \tag{4.67}$$

and v indicates velocity as given by

$$v = \frac{\partial u}{\partial \tau}. \tag{4.68}$$

It should be recognized that Eq. (4.65) can also be used for points inside the domain Ω. As stated previously, $c \, (s)$ must be regarded as equal to 1 in this situation.

Two distinct types of singularity can occur in the integrands of Eq. (4.65). The first type of singularity occurs in the integral of the initial conditions when $r = 0$ and in the boundary integrals when r and $c \, (t - \tau)$ are simultaneously null. The second type of singularity occurs at points located at the front of the wave represented by the Green's function, that is, in the boundary and source density integrals when $r = c \, (t - \tau)$, and in the integrals of the initial conditions when $r = c \, t$. Nevertheless numerical integration of Eq. (4.65) does not present any notable difficulty as it will be discussed next.

4.8 Numerical Implementation

Time and space interpolation functions, similar to the ones used in finite elements, can be employed to transform the integral Eq. (4.65) into a system of algebraic equations whose solution supplies the boundary unknowns u and p. The potential $u \, (s, t)$ at internal points can then be calculated by using Eq. (4.65) with $c \, (s) = 1$. This procedure is standard in boundary element formulations [1, 2]; but a discussion about this subject is necessary in order to clarify certain factors which only appear in the problem under consideration.

The usual time marching schemes consider each time step as a new problem and consequently at the end of each time interval, values of displacements and velocities are calculated for a sufficient number of internal points; this is in order to use them as pseudo-initial conditions for the next step, i.e. the integral Eq. (4.65) is applied from 0 to Δt, Δt to $2\Delta t$, etc. However in the numerical applications carried out in this section the time integration process is always considered to start at the time '0' and so values of displacements and velocities do not need to be calculated at intermediate steps. With this procedure the domain discretization is restricted to regions where source density and initial conditions do not disappear. Domain integrations at a time step 'j' are then avoided at the cost of having to compute time integrations for all time steps previous to 'j'. This technique is especially suitable for infinite and semi-infinite domains. A comparison of the performance of both techniques for transient heat transfer problems can be found in [32].

4.8.1 Boundary Integrals

In order to implement a numerical scheme to solve Eq. (4.65), it is necessary to consider a set of discrete points (nodes) Q_j, $j = 1, \ldots, J$ on the Γ boundary, and also a set of values of time t_n, $n = 1, \ldots, N$. $u(Q,t)$, $v(Q,t)$ and $p(Q,t)$ can be approximated using a set of interpolation functions as indicated below

$$u(Q,t) = \sum_{j=1}^{J} \sum_{m=1}^{N} \phi^m(t)\, \eta_j(Q)\, u_j^m$$

$$v(Q,t) = \sum_{j=1}^{J} \sum_{m=1}^{N} \frac{d\phi^m(t)}{dt}\, \eta_j(Q)\, u_j^m \tag{4.69}$$

$$p(Q,t) = \sum_{j=1}^{J} \sum_{m=1}^{N} \theta^m(t)\, v_j(Q)\, p_j^m$$

where m and j refer to time and space respectively. $\phi^m(t)$, $\eta_j(Q)$, $\theta^m(t)$ and $v_j(Q)$ are chosen such that

$$\eta_j(Q_i) = \delta_{ij}$$
$$v_j(Q_i) = \delta_{ij}$$
$$\phi^m(t_n) = \delta_{mn} \tag{4.70}$$
$$\theta^m(t_n) = \delta_{mn}$$

where δ_{ij} is the Kronecker delta. Therefore

$$u_j^m = u(Q_j, t_m)$$
$$p_j^m = p(Q_j, t_m)\,. \tag{4.71}$$

If Eq. (4.65) is written for every node i and for every value of time t_n, and u, v and p are replaced by their approximations given by expression (4.69), the following system of algebraic equations is then obtained

$$c(S_i)\, u_i^n + \frac{1}{4\pi} \sum_{m=1}^{N} \sum_{j=1}^{J} H_{ij}^{nm}\, u_j^m = \frac{1}{4\pi} \left\{ \sum_{m=1}^{N} \sum_{j=1}^{J} G_{ij}^{nm}\, p_j^m + F_i^n + S_i^n \right\} \tag{4.72}$$

where

$$H_{ij}^{nm} = -\int_{\Gamma} \frac{\partial r(S_i, Q)}{\partial n(Q)}\, \eta_j(Q) \int_0^{t_n} \left[\phi^m(\tau)\, B^*(Q, t_n; S_i, \tau) \right.$$

$$\left. + \frac{1}{c}\, \frac{d\phi^m(\tau)}{d\tau}\, u^*(Q, t_n; S_i, \tau) \right] d\tau\, d\Gamma(Q) \tag{4.73}$$

$$G_{ij}^{nm} = \int_{\Gamma} v_j(Q) \int_0^{t_n} \theta^m(\tau)\, u^*(Q, t_n; S_i, \tau)\, d\tau\, d\Gamma(Q) \tag{4.74}$$

$$F_i^n = \frac{1}{c^2} \int_{\Omega} u_0^*(q, t_n; S_i)\, v_0(q)\, d\Omega(q)$$

$$+ \frac{1}{c} \int_{\Omega} u_0^*(q, t_n; S_i)\, \frac{\partial u_0(q)}{\partial r(S_i, q)}\, d\Omega(q)$$

$$+ t_n \int_{\Omega} \frac{1}{r(S_i, q)}\, B_0^*(q, t_n; S_i)\, u_0(q)\, d\Omega(q) \tag{4.75}$$

$$S_i^n = \int_0^{t_n} \int_{\Omega} u^*(q, t_n; S_i, \tau)\, \gamma(q, \tau)\, d\Omega(q)\, d\tau. \tag{4.76}$$

It should be recognized that the third term on the right-hand side of Eq. (4.75) is the sum of the first and third terms of the integrand of the fourth integral on the right-hand side of Eq. (4.65).

When

$$t_{m+1} - t_m = \Delta t = \text{const}, \tag{4.77}$$

$\phi^m(t)$ can be assigned the time translation property, i.e.

$$\phi^m(t) = \phi^{m+l}(t + l\,\Delta t). \tag{4.78}$$

Hence

$$H_{ij}^{nm} = H_{ij}^{(n+l)(m+l)}$$
$$G_{ij}^{nm} = G_{ij}^{(n+l)(m+l)}. \tag{4.79}$$

If expression (4.79) is taken into consideration, a large number of redundant operations can be avoided in the numerical analysis.

It is important to recognize that due to the causality property the integrals indicated in expressions (4.73) to (4.76) do not always extend over the hole domain and boundary. This results in quite a number of coefficients H_{ij}^{nm}, G_{ij}^{nm}, F_i^n and S_i^n being equal to null (for a more comprehensive discussion, see [6, 18]).

A time-stepping scheme in which Eq. (4.72) is successively solved for $n = 1, \dots, N$ can be used to calculate unknowns u_j^N and q_j^N at the time t_N. The actual numerical implementation of such a scheme requires of course, the specification of the type of interpolation functions to be used.

In the numerical applications presented in Sect. 4.9 it was decided to always use $\phi^m(\tau)$ linear [see expression (4.80) and Fig. 4.8] and to try a linear or constant [see expression (4.81) and Fig. 4.9] time interpolation function $\theta^m(\tau)$.

$$\phi^m(\tau) = \theta^m(\tau) = \begin{cases} \dfrac{1}{\Delta t}(\tau - t_{m-1}) & \text{if } t_{m-1} < \tau < t_m, \\[2mm] \dfrac{1}{\Delta t}(t_{m+1} - \tau) & \text{if } t_m < \tau < t_{m+1}, \\[2mm] 0 & \text{otherwise}. \end{cases} \tag{4.80}$$

$$\theta^m(\tau) = \begin{cases} 1 & \text{if } t_{m-1} < \tau < t_m, \\ 0 & \text{otherwise}. \end{cases} \tag{4.81}$$

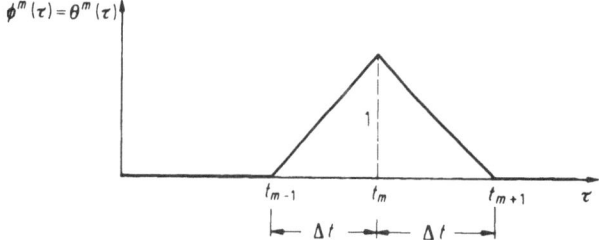

Fig. 4.8. Linear time interpolation functions for u and p on the Γ boundary

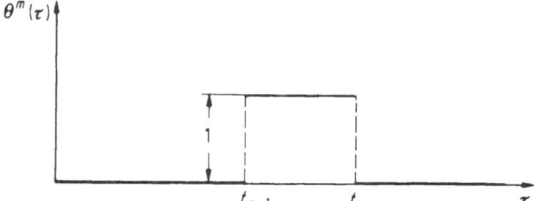

Fig. 4.9. Constant time interpolation for p

Substitution of expressions (4.80) and (4.81) into expressions (4.73) and (4.74) leads to expressions which can be integrated analytically on time. A detailed investigation concerning this subject can be found in [6].

In order of perform numerically the integrations over Γ indicated in expressions (4.73) and (4.74) the Γ boundary must be replaced by an approximated one. Linear discretization is used in this work, that is, Γ is represented by a series of straight line segments, e_k (elements), each one joining two consecutives nodes of Γ. l_k and $\mathbf{n}_k$ are the length of e_k and the unit outward vector normal to e_k respectively (see Fig. 4.10). Then the integrals over Γ given by expressions (4.73) and (4.74) can be replaced by a sum of integrals over elements.

When two elements e_p and e_q with a common node j are considered, and the interpolation functions $\eta_j(Q)$ and $v_j(Q)$ are linear, the use of natural coordinates gives (see Fig. 4.11)

$$\eta_j(\xi) = v_j(\xi) = \begin{cases} \frac{1}{2}(\xi_p + 1), & Q \in e_p, \\ -\frac{1}{2}(\xi_q - 1), & Q \in e_q \\ 0 & \text{otherwise}. \end{cases} \tag{4.82}$$

When Eqs. (4.73) and (4.74) are integrated analytically on time, and formula (4.82) is substituted into the resulting expressions, Cauchy principal values of integrals must be calculated in order to obtain the coefficients (H_{ii}^{nn}). The function being integrated has a singularity of the type $1/r$ [6], however when linear discretization is used these integrals disappear due to the orthogonality of Γ_k and $\mathbf{n}_k$ (see Fig. 4.10) which makes $\partial r/\partial n = 0$. This problem deserves special attention when interpolation functions of order higher than linear are used to approximate the geometry of the Γ boundary. The coefficients (G_{ii}^{nn}) contain integrals which

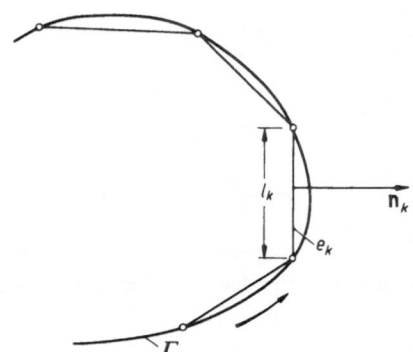

Fig. 4.10.   Linear discretization of the Γ boundary

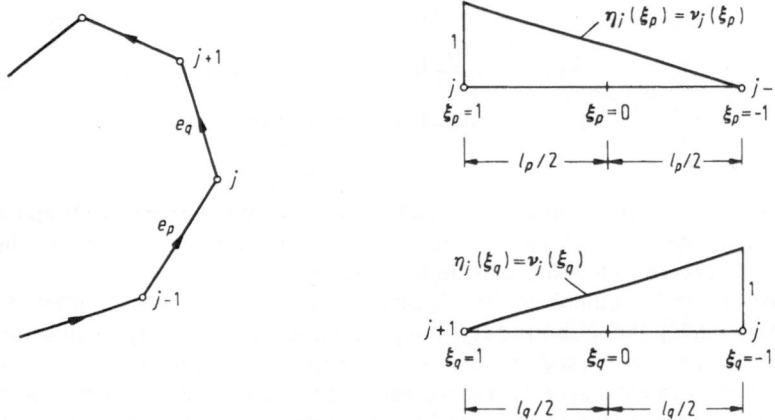

Fig. 4.11.   Linear space interpolation functions for u and p on the Γ boundary

have a singularity of the type $\ln r$. These integrals can be computed in the ordinary sense using Gaussian quadrature. However, a greater precision can be obtained if these integrals are carried out analytically rather than numerically [6].

The rest of the coefficients (H_{ij}^{nm}) and (σ_{ij}^{nm}) can be calculated using standard Gauss quadrature formulae.

Another situation to be examined is that in which $\eta_j(Q)$ and $v_j(Q)$ are constant, i.e.

$$\eta_j(Q) = v_j(Q) = \begin{cases} 1 & \text{when } Q \in e_j, \\ 0 & \text{otherwise}. \end{cases} \tag{4.83}$$

In this case a node j can be considered as belonging to a set of discrete points Q_j on the Γ boundary, $j = 1, \ldots, J$ where each Q_j is placed at the middle of an element e_j as shown in Fig. 4.12. It should be recognized that in this case $c(s_j)$ is always equal to $\frac{1}{2}$.

Because of the causality property a situation exists in which it is necessary to carry out numerical integrations of functions which are null over part of an

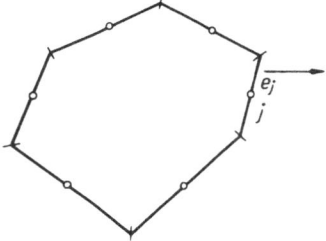

Fig. 4.12. Position of nodes when constant interpolation functions η_j and ν_j are used

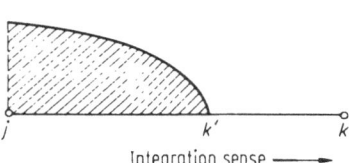

Integration sense ────▶

Fig. 4.13. Integration over part of an element

element. In this case it became clear that greater precision could be obtained if such integrations were performed from j to k' instead of from j to k as depicted in Fig. 4.13.

The fundamental solution of the problem under consideration [see Eq. (4.47)] suggests that the number of Gauss points can be gradually reduced as $(t - \tau)$ gets bigger. This procedure was used in the numerical applications discussed in Sect. 4.9, in order to save computer time.

4.8.2 Domain Integrals

The domain contributions due to initial conditions can be calculated from expression (4.75), which can be written as

$$F_i^n = \frac{1}{c^2} \int_\Omega u_{0i}^{*n} v_0 \, d\Omega \, (q) + \frac{1}{c} \int_\Omega u_{0i}^{*n} \frac{\partial u_0}{\partial r} \, d\Omega \, (q) + t_n \int_\Omega \frac{1}{r} B_{0i}^{*n} u_0 \, d\Omega \, (q) \qquad (4.84)$$

where $u_{0i}^{*n} = u_0^* (q, t_n; S_i)$ and $B_{0i}^{*n} = B_0^* (q, t_n; S_i)$.

In order to carry out the integrations indicated in expression (4.84) the domain Ω is divided into L triangular subdomains, O_l (cells), as shown in Fig. 4.14. Then the expression (4.84) can be written as

$$F_i^n = \sum_{l=1}^{L} \left(\frac{1}{c^2} \int_{O_l} u_{0i}^{*n} v_0 \, d\Omega \, (q) + \frac{1}{c} \int_{O_l} u_{0i}^{*n} \frac{\partial u_0}{\partial r} \, d\Omega \, (q) + t_n \int_{O_l} \frac{1}{r} B_{0i}^{*n} u_0 \, d\Omega \, (q) \right). \qquad (4.85)$$

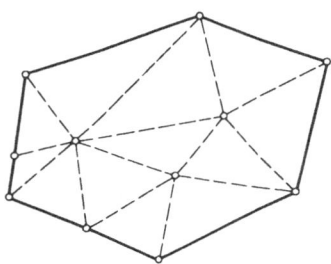

Fig. 4.14. Discretization of the domain Ω into triangular cells

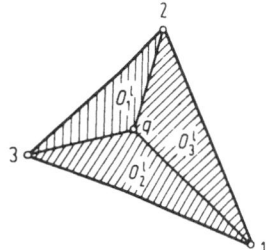

Fig. 4.15. Areas for the definition of triangular coordinates

The position of a point q inside a cell can be more conveniently defined by triangular coordinates [33], i.e.

$$\mu_1 = \frac{A_1}{A}$$

$$\mu_2 = \frac{A_2}{A}$$

$$\mu_3 = \frac{A_3}{A}$$

$$\mu_1 + \mu_2 + \mu_3 = 1$$

(4.86)

where A_1, A_2 and A_3 are respectively the areas of the triangles O_1^i, O_2^i and O_3^i depicted in Fig. 4.15, and $A = A_1 + A_2 + A_3$ is the area of the cell O_l.

When u_0 and v_0 are linearly interpolated inside each cell the following expression can be written

$$u_0 = \sum_{\alpha=1}^{3} u_{0\alpha} \mu_\alpha$$

$$v_0 = \sum_{\alpha=1}^{3} v_{0\alpha} \mu_\alpha$$

(4.87)

where $u_{0\alpha}$ and $v_{0\alpha}$ are respectively initial displacement and initial velocity at a node α of the cell O_l. $\partial u_0/\partial r$ is also required and can be calculated from expression (4.87), giving

$$\frac{\partial u_0}{\partial r} = \sum_{\alpha=1}^{3} u_{0\alpha} \frac{\partial \mu_\alpha}{\partial r}.$$

(4.88)

Triangular coordinates can be related to rectangular coordinates in the following way

$$\mu_\alpha = \frac{A_\alpha^0}{A} + \frac{1}{2A} (b_\alpha x_1 + a_\alpha x_2)$$

(4.89)

where

$$a_\alpha = x_1^\gamma - x_1^\beta$$
$$b_\alpha = x_2^\beta - x_2^\gamma$$
$$2 A_\alpha^0 = x_1^\beta x_2^\gamma - x_1^\gamma x_2^\beta$$
$$A = \tfrac{1}{2} (b_1 a_2 - b_2 a_1).$$

(4.90)

In expression (4.90) $\alpha = 1, 2, 3$ for $\beta = 2, 3, 1$ and $\gamma = 3, 1, 2$.

Considering a system of polar coordinates (r, θ) with origin at the source point S_i as depicted in Fig. 4.16, expression (4.89) becomes

$$\mu_\alpha = C_\alpha + D_\alpha (\theta) r$$

(4.91)

where

$$C_\alpha = A_\alpha^0/A$$

$$D_\alpha = \frac{1}{2A} (b_\alpha \cos \theta + a_\alpha \sin \theta).$$

(4.92)

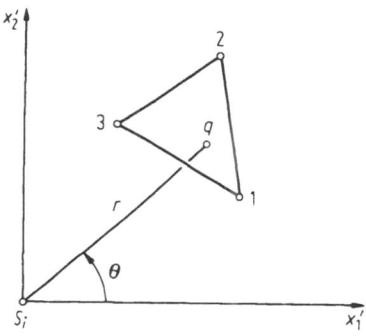

Fig. 4.16. Polar coordinates based at the source point S_i

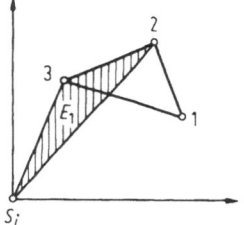 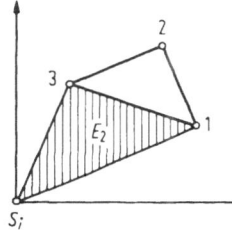 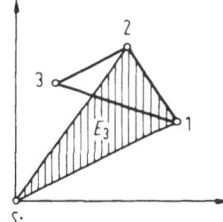

Fig. 4.17. Domains used to integrate over a cell

Taking formulas (4.87), (4.88) and (4.91) into consideration, u_0, v_0 and $\partial u_0/\partial r$ can be expressed as

$$u_0 = \sum_{\alpha=1}^{3} u_{0\alpha} \left[C_\alpha + D_\alpha(\theta) \, r \right]$$

$$v_0 = \sum_{\alpha=1}^{3} v_{0\alpha} \left[C_\alpha + D_\alpha(\theta) \, r \right] \tag{4.93}$$

$$\frac{\partial u_0}{\partial r} = \sum_{\alpha=1}^{3} u_{0\alpha} D_\alpha(\theta) \, .$$

Integration over a cell can now be performed using polar coordinates. A convenient procedure [6] is that in which the integral over a cell is obtained as a sum of three integrals over the domains E_1, E_2 and E_3 depicted in Fig. 4.17. Then, using formula (4.93) the first integral on the right-hand side of expression (4.85) can be written as

$$\int_{O_i} u_{0i}^{*n} v_0 \, d\Omega = \sum_{\alpha=1}^{3} v_{0\alpha} \sum_{t=1}^{3} \int_{\theta_u}^{\theta_v} \int_{0}^{g_t^i(\theta)} u_{0i}^{*n} \left[C_\alpha + D_\alpha(\theta) \, r \right] r \, dr \, d\theta . \tag{4.94}$$

In expression (4.94), $t = 1, 2, 3$ for $u = 2, 3, 1$ and $v = 1, 3, 2$,

$$g_t^i(\theta) = \begin{cases} r_t^i(\theta) & \text{when } r_t^i(\theta) < c \, t_n \, , \\ c \, t_n & \text{when } r_t^i(\theta) > c \, t_n \, , \end{cases} \tag{4.95}$$

and $r_t^i(\theta)$, θ_t, θ_u and θ_r are shown in Fig. 4.18.

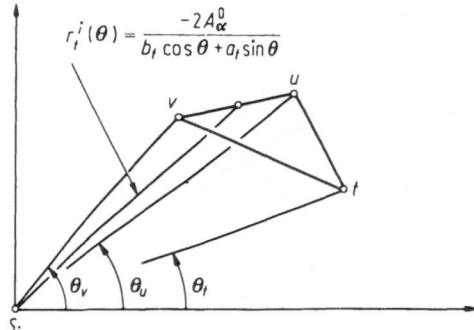

Fig. 4.18. Definitions with cell integration purpose

In [6] expression (4.94) is integrated analytically with respect to r. Subsequently the change of coordinates indicated in expression (4.96) is carried out and one dimensional Gaussian quadrature is used to integrate numerically with respect to θ. In that work a similar procedure regarding the other integrals that appear in expression (4.85) is also discussed.

$$\theta = \frac{\xi}{2} \, (\theta_v - \theta_u) + \tfrac{1}{2} \, (\theta_v + \theta_u) \tag{4.96}$$

4.8.3 Double Nodes

A very common situation in wave propagation problems concerns p being discontinuous on the boundary. A convenient way of analysing these types of problem is that in which two distinct values of tractions, p^r and p^l, and two values of displacements u^r and u^l are considered on the neighbourhood of each point where a discontinuity can occur (see Fig. 4.19). So, for each of these points two extra boundary unknowns are introduced in the analysis. When, p^r and p^l, or $u^r (u^l)$ and $p^l (p^r)$ are prescribed the continuity condition for displacements, namely

$$u^r = u^l \tag{4.97}$$

gives the extra equation required. When constant elements are used, this problem is naturally considered by the discontinuous nature of these elements. However,

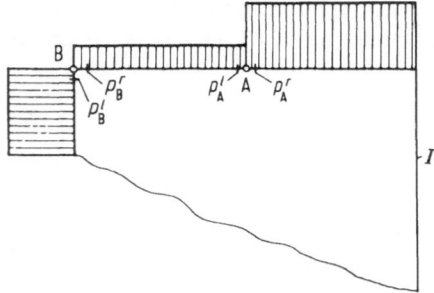

Fig. 4.19. Discontinuous p on the Γ boundary

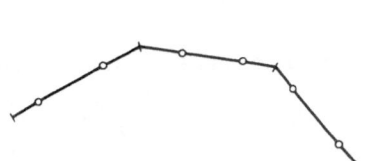

Fig. 4.20. Discontinuous linear elements

when linear or higher order elements are used special considerations are required. The system of equations given by expression (4.72) can still be used and the condition (4.97) can be introduced using "double nodes", i.e. two different nodes being placed at points where p can be discontinuous. An extensive study on this subject can be found in [34−36].

A more involved situation is that in which p^l and p^r are different from each other in the neighbourhood of a point where the potential is prescribed. The approach to be followed in this case can be found in [35−37].

In quite a number of situations it is not possible to determine a priori when and where tractions are discontinuous. In this case the mean value of the unknowns is to be expected from the numerical analysis.

Another method of dealing with discontinuities is by using discontinuous elements [38, 39]. The discontinuity is then avoided because as shown in Fig. 4.20 the nodes of the discontinuous elements are placed inside them, rather than on their extremities. It should be recognized that this procedure can also be used when time discontinuities occur in a problem.

4.9 Examples

If it is desired to find boundary unknowns at a time t_n, it is convenient to write Eq. (4.72) in the following way (summation convention does not apply)

$$C(S_i) u_i^n + \frac{1}{4\pi} \sum_{j=1}^{J} H_{ij}^{nn} u_j^n = \frac{1}{4\pi} \left\{ \sum_{j=1}^{J} G_{ij}^{nn} p_j^n - \sum_{m=1}^{n-1} \sum_{j=1}^{J} H_{ij}^{nm} u_j^m \right.$$

$$\left. + \sum_{m=1}^{n-1} \sum_{j=1}^{J} G_{ij}^{nm} p_j^m + F_i^n + S_i^n \right\}. \qquad (4.98)$$

Equation (4.98) can also be written as

$$\mathbf{H}\,\mathbf{u} = \mathbf{G}\,\mathbf{p} + \mathbf{B} \qquad (4.99)$$

where $\mathbf{H}$ and $\mathbf{G}$ are square matrices of order $(J \times J)$ and $\mathbf{u}$, $\mathbf{p}$ and $\mathbf{B}$ are vectors.

If the boundary conditions at the time t_n are considered and the system of equations that arises is reordered expression (4.99) can be written as

$$\mathbf{A}\,\mathbf{y} = \mathbf{C} \qquad (4.100)$$

where the vector $\mathbf{y}$ is formed by unknowns u_j^n and p_j^n at boundary nodes.

Within the examples analysed in this chapter the boundary conditions at boundary nodes are always of the same type, i.e. a node at which u (or p) is initially prescribed will only have prescribed u (or p) until the end of the transient analysis. Consequently, due to the time translation property [see expression (4.79)], $\mathbf{A}$ requires to be inverted only once. Gauss elimination is used in this work to obtain the inverse of $\mathbf{A}$.

In the examples discussed here the numerical integrations mentioned previously in Sect. 4.8 were carried out using a maximum of ten Gauss points.

The choice of domain discretization to be used when solving a problem is fairly simple because u_0, v_0 and γ are known functions. However boundary discretization and time division depend on what the problem under consideration is like. For this reason, in many problems, more than one numerical analysis has to be carried out in which the boundary discretization and the time division are successively refined. The quantity of work required is considerably reduced as experience is gained in the method adopted. The observation of certain physical characteristics of the problem can also be of great help. For instance when studying wave propagation care should be taken on the choice of time intervals and boundary discretization in order to avoid contradicting the causality property too far, that is, in a time interval, waves should not be allowed to travel between nodes far from each other. There are also certain precautions which must be taken when choosing the parameter β given by

$$\beta = \frac{c \, \Delta t}{l_j} \, . \tag{4.101}$$

It is quite commonly regarded that there exist strict rules concerning the choice of a similar parameter in finite differences and finite elements; which if not followed can result in a completely invalid analysis. In boundary elements conclusive analytical studies regarding the choice of β have not yet been completed, consequently the discussion based on numerical experiments presented in the examples can be very helpful.

The numerical procedure discussed previously in this chapter was converted into FORTRAN and implemented on an ICL 2970 computer. The computer code was used to analyse a number of examples which will be presented next.

4.9.1 One–Dimensional Rod under a Heaviside Type Forcing Function

The results obtained from using the two-dimensional boundary element computer code were compared with the analytical results for a one-dimensional rod under a Heaviside type forcing function. The boundary element solution considered a rectangular domain with sides of length a and b ($b = a/2$) as depicted in Fig. 4.21. The u displacements were assumed to be zero at $x_1 = a$ and their normal derivative p were also taken as null at $x_2 = 0$ and $x_2 = b$ for any time 't'. At $x_1 = 0$ and $t = 0$ a load Ep was suddenly applied and kept constant until the end of the analysis (E is the Young's modulus). Due to the topology and boundary conditions the problem is actually one-dimensional and its analytical solution can be found elsewhere [40].

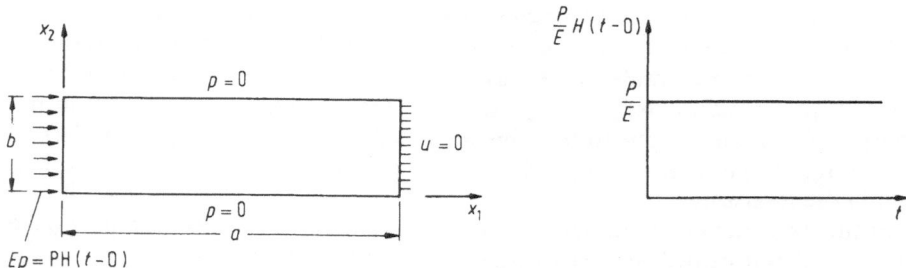

Fig. 4.21. Boundary conditions and geometry definitions for one-dimensional rod under a Heaviside type forcing function

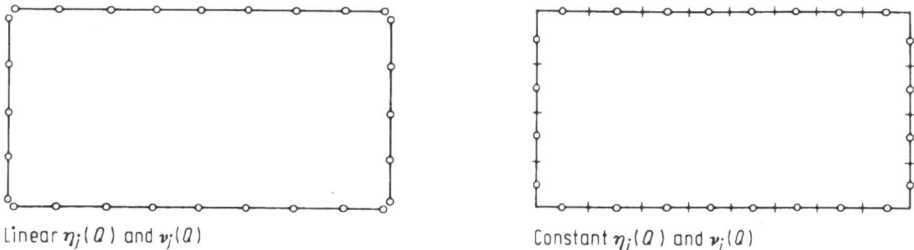

Linear $\eta_j(Q)$ and $v_j(Q)$ Constant $\eta_j(Q)$ and $v_j(Q)$

Fig. 4.22.   Boundary discretization for one-dimensional rod

Table 4.1.   Combination of interpolation functions

Combination	Interpolation function		
	$\eta_j(Q)$ and $v_j(Q)$	$\phi^m(t)$	$\theta^m(t)$
1	Linear	Linear	Linear
2	Linear	Linear	Constant
3	Constant	Linear	Constant

Three different combinations of interpolation functions were used in the analysis as given in Table 4.1.

The boundary was discretized into twenty four constant and linear elements as shown in Fig. 4.22, double nodes were used at the corners for the latter model.

Combination 1 was tried with $\beta = 0.6$ and gave good results for the displacements u (the degree of accuracy was the same as combination 2). The numerical values of p, however, oscillated around the analytical solution, displaying the onset

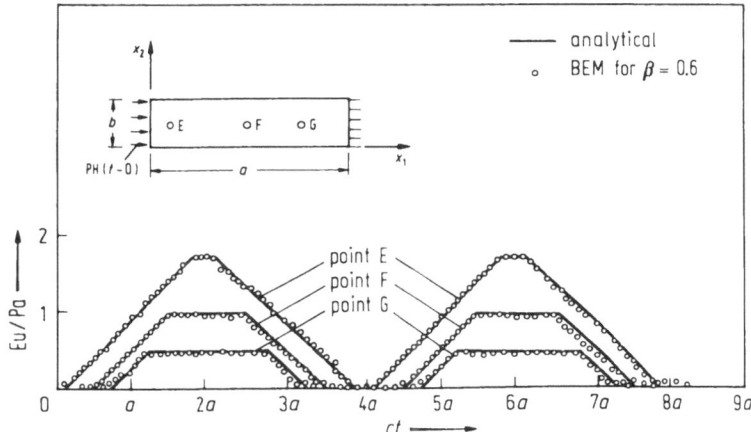

Fig. 4.23.   Displacements at internal points $E(a/8, b/2)$, $F(a/2, b/2)$ and $G(3a/4, b/2)$ for one-dimensional rod under a Heaviside type forcing function. $\eta_j(Q)$, $v_j(Q)$, $\phi^m(t)$ are linear and $\theta^m(t)$ is constant

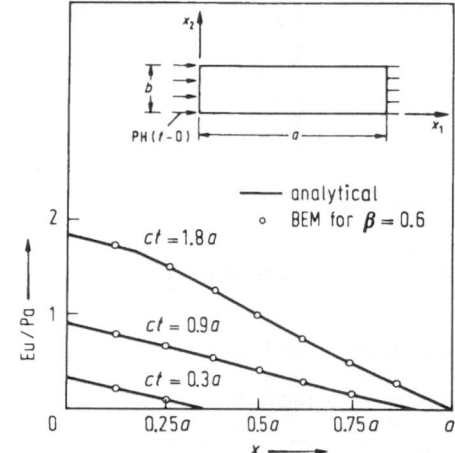

Fig. 4.24. Displacements along boundary $y = 0$ at times $t = 0.3\,a/c$, $t = 0.9\,a/c$, $t = 1.8\,a/c$ for one-dimensional rod under a Heaviside type forcing function. $\eta_j(Q)$, $v_j(Q)$, $\phi^m(t)$ are linear, and $\theta^m(t)$ is constant

of instability. This unstable behaviour of p can be avoided in this particular analysis by replacing the jump of the forcing function $PH(t-0)$ by a steep slope, however more applications must be carried out before one can find out in which cases this procedure leads to acceptable results. Therefore it was decided not to use combination 1 for problems in which the u function is prescribed on parts of the boundary.

Combinations 2 and 3 were then compared and it was found that for the same number of boundary elements and the same time division, better results were obtained for linear $\eta_j(Q)$ and $v_j(Q)$ (combination 2) than for constant $\eta_j(Q)$ and

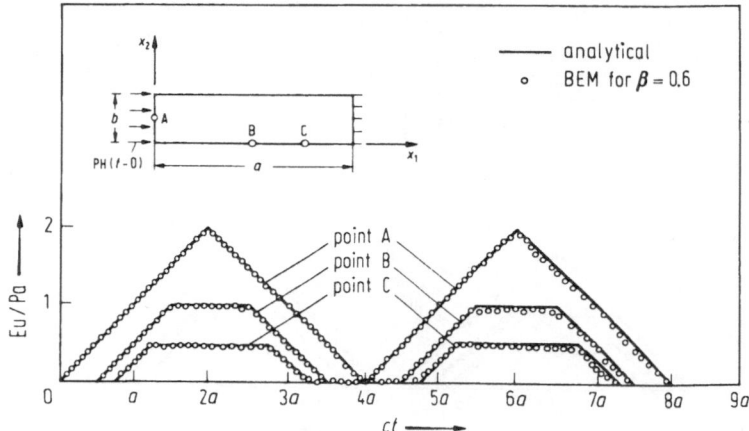

Fig. 4.25. Displacements at boundary points A$(0, b/2)$, B$(a/2, 0)$ and C$(3a/4, 0)$ for one-dimensional rod under a Heaviside type forcing function. $\eta_j(Q)$, $v_j(Q)$, $\phi^m(t)$ are linear and $\theta^m(t)$ is constant

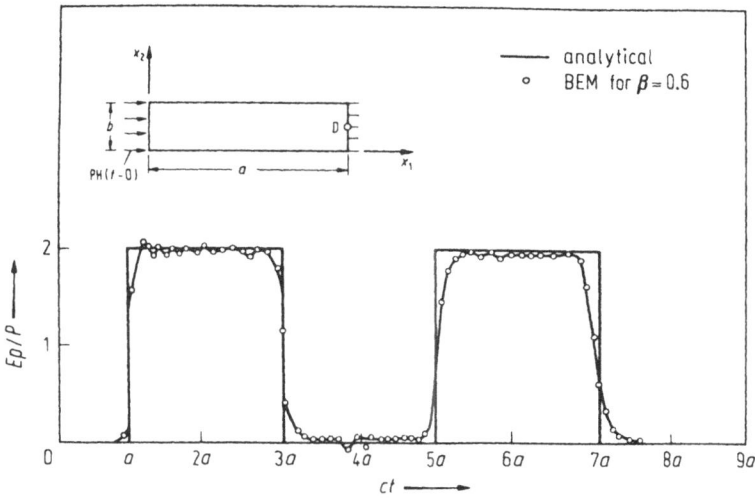

Fig. 4.26. Normal derivative of displacement at point $D(a, b/2)$ for one-dimensional rod ander a Heaviside type forcing function. $\eta_j(Q), v_j(Q), \phi^m(t)$ are linear and $\theta^m(t)$ is constant

$v_j(Q)$ (combination 3). As the computing time is much the same for both cases it was concluded that combination 2 is more efficient than combination 3. Therefore all the boundary element method (B.E.M.) results presented from now on are based on combination 2.

Figures 4.23–4.25 show B.E.M. and analytical displacement results at internal and boundary points. The degree of accuracy of B.E.M. results is quite good. In Fig. 4.26 the normal derivative of the u displacement at point $(a, b/2)$ versus $c\,t$ is presented. Except for the presence of a comparatively small amount of noise, boundary elements and analytical solutions are in good agreement.

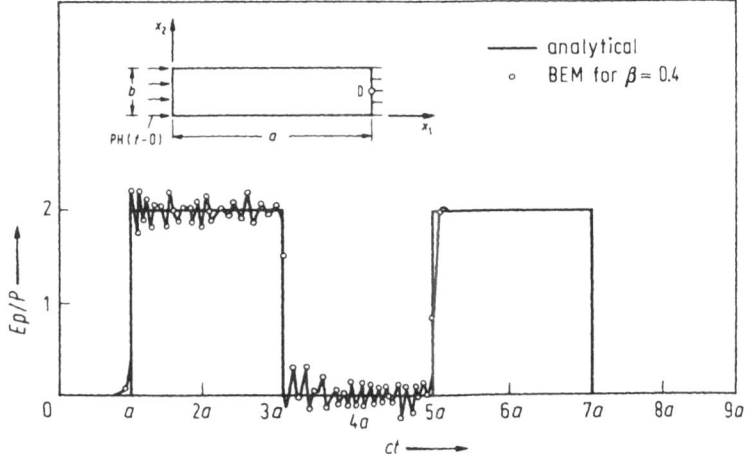

Fig. 4.27. Normal derivative of displacement at point $D(a, b/2)$ for one-dimensional rod under a Heaviside type forcing function. $\eta_j(Q), v_j(Q), \phi^m(t)$ are linear and $\theta^m(t)$ is constant

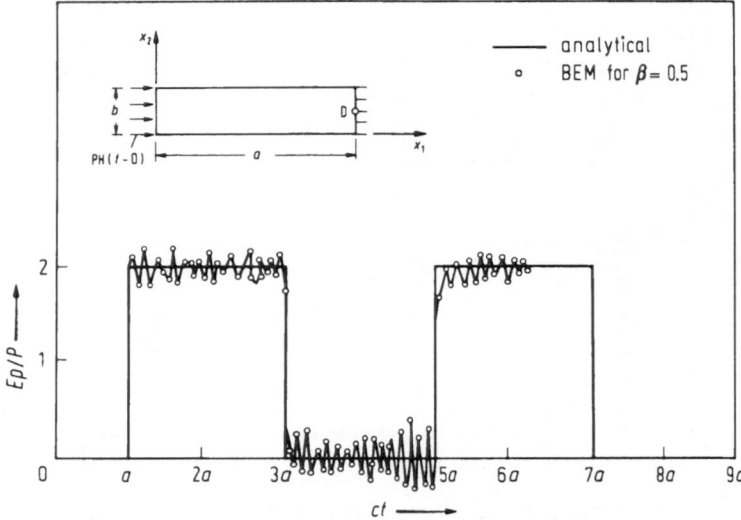

Fig. 4.28. Normal derivative of displacement at point $D(a, b/2)$ for one-dimensional rod under a Heaviside type forcing function. $\eta_j(Q)$, $v_j(Q)$, $\phi^m(t)$ are linear and $\theta^m(t)$ is constant

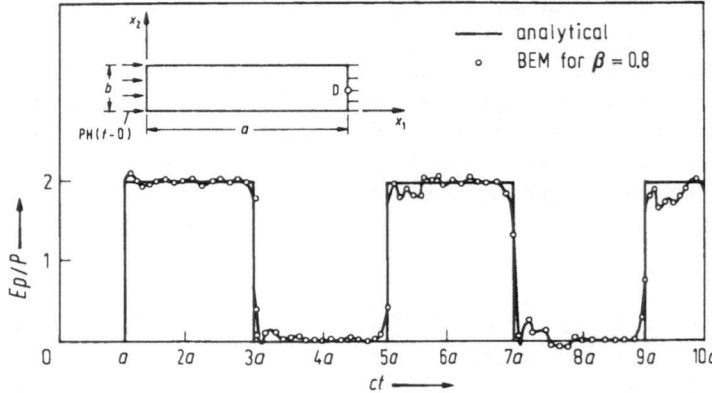

Fig. 4.29. Normal derivative of displacement at point $D(a, b/2)$ for one-dimensional rod under a Heaviside type forcing function. $\eta_j(Q)$, $v_j(Q)$, $\phi^m(t)$ are linear and $\theta^m(t)$ is constant

Considerable care must be taken with the choice of β in order to avoid noise, which although usually not critical for displacements, can often be excessive for tractions. In order to study the effect of varying the parameter β on the level of noise four other values of β were investigated; 0.4, 0.5, 0.8 and 1.0 in addition to $\beta = 0.6$. The results for p at point $(a, b/2)$ are plotted in Figs. 4.27−4.30. It is apparent that excessive noise occurred for $\beta < 0.6$. The value $\beta = 0.6$ was considered the optimum for this problem.

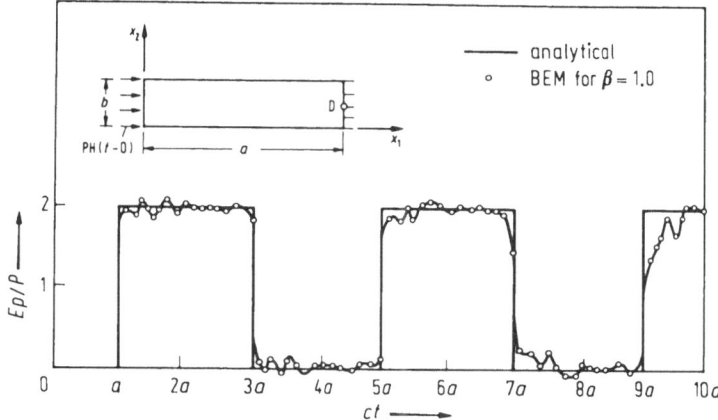

Fig. 4.30.   Normal derivative of displacement at point $D(a, b/2)$ for one-dimensional rod under a Heaviside type forcing function. $\eta_j(Q)$, $v_j(Q)$, $\phi^m(t)$ are linear and $\theta^m(t)$ is constant

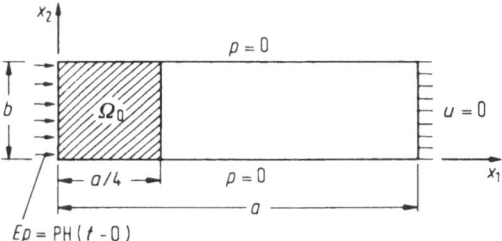

Fig. 4.31.   Geometry definitions, boundary and initial conditions for one-dimensional rod

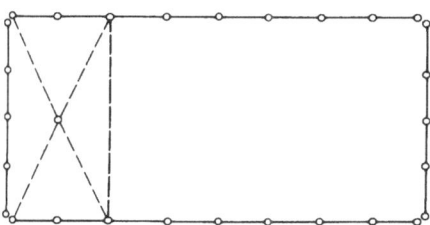

Fig. 4.32.   Domain and boundary discretization for one-dimensional rod under prescribed initial conditions

4.9.2 One-Dimensional Rod under Prescribed Initial Velocity and Displacement

For this problem the geometry and boundary conditions were indentical to the previous case and, in addition, over the domain Ω_0 depicted in Fig. 4.31 the following initial conditions were prescribed

$$u_0 (x_1, x_2) = \frac{P}{E} \left(\frac{a}{4} - x_1 \right)$$

$$v_0 (x_1, x_2) = \frac{P}{E} \frac{c}{.}$$

(4.102)

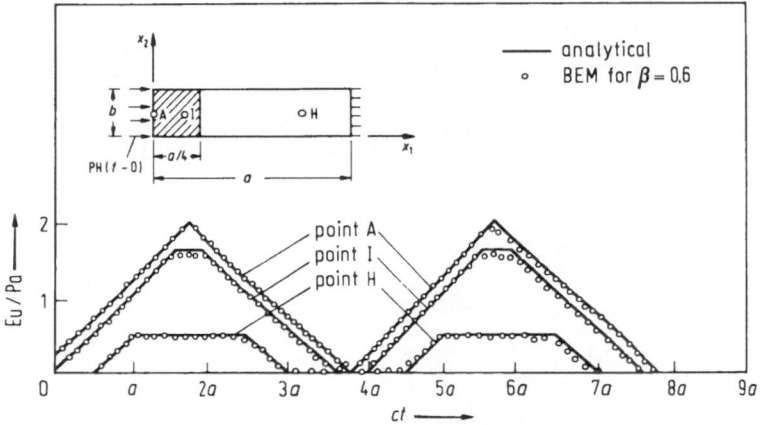

Fig. 4.33. Displacements at boundary point $A(0, b/2)$ and internal points $I(3a/16, b/2)$, $H(3a/4, b/2)$ for one-dimensional rod under prescribed initial conditions. $\eta_j(Q)$, $v_j(Q)$, $\phi^m(t)$ are linear and $\theta^m(t)$ is constant

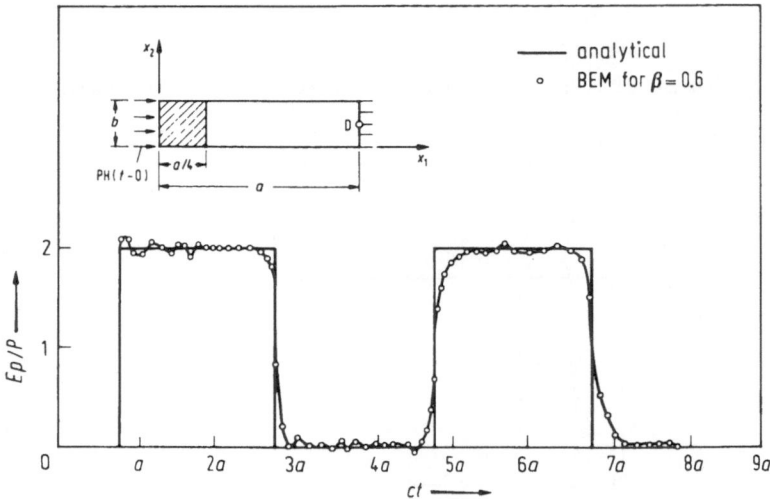

Fig. 4.34. Normal derivative of displacement at point $D(a, b/2)$ for one-dimensional rod under prescribed initial conditions. $\eta_j(Q)$, $v_j(Q)$, $\phi^m(t)$ are linear and $\theta^m(t)$ is constant

The analytical solution for this problem is the same as for the previous one but with the time t dephased by $a/4\,c$, i.e.

$$u'(q, t) = u\left(q, t - \frac{a}{4c}\right)$$

$$p'(q, t) = p\left(q, t - \frac{a}{4c}\right)$$

(4.103)

where u' and p' refer to the problem studied in Sect. 4.9.1.

Twenty four linear elements were used to discretize the boundary and Ω_0 was subdivided into four triangular cells as depicted in Fig. 4.32. The time steps were such that $\beta = 0.6$.

Displacements at points $(0, b/2)$, $(3\,a/16, b/2)$, $(3\,a/4, b/2)$ and traction at point $(a, b/2)$ are presented in Figs. 4.33 and 4.34 respectively. The accuracy of the results is similar to that obtained in the previous problem.

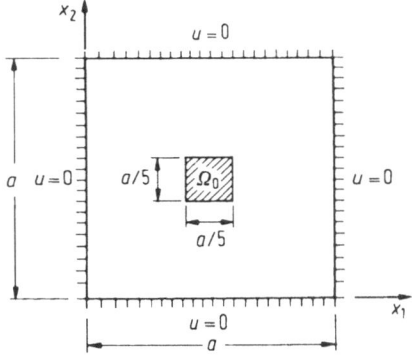

Fig. 4.35. Geometry definition, boundary and initial conditions for membrane analysis

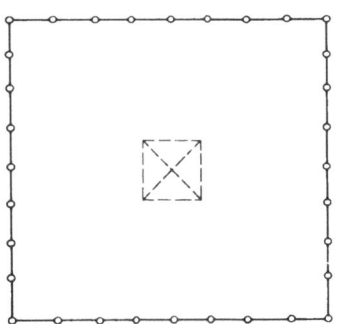

Fig. 4.36. Membrane discretized into 32 elements and four cells

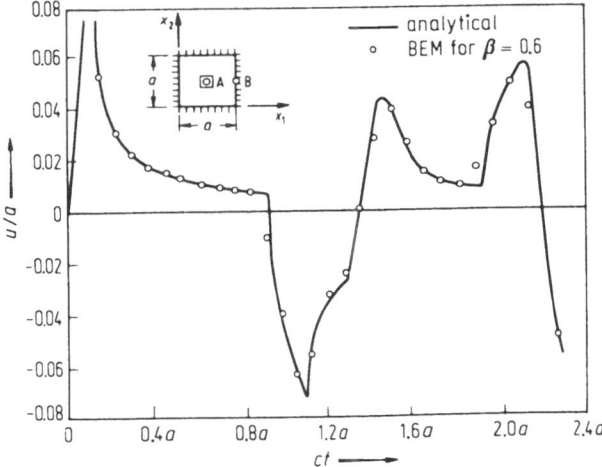

Fig. 4.37. Displacement at point A $(a/2, a/2)$. 32 boundary elements

4.9.3 Square Membrane under Prescribed Initial Velocity

The subject of this investigation is the transverse motion of a square membrane with initial velocity $v_0 = c$ prescribed over the domain Ω_0 depicted in Fig. 4.35 and zero displacements prescribed over all the boundary.

The boundary was discretized into thirty two elements and Ω_0 was divided into four cells as shown in Fig. 4.36. Analytical [6] and boundary element method results for displacements at point $(a/2, a/2)$ and the normal derivative of displacements at point $(a, a/2)$ were compared.

The values of u and p for $\beta = 0.6$ are plotted in Figs. 4.37 and 4.38 respectively. Although the agreement for displacements is reasonable, it was found that a more

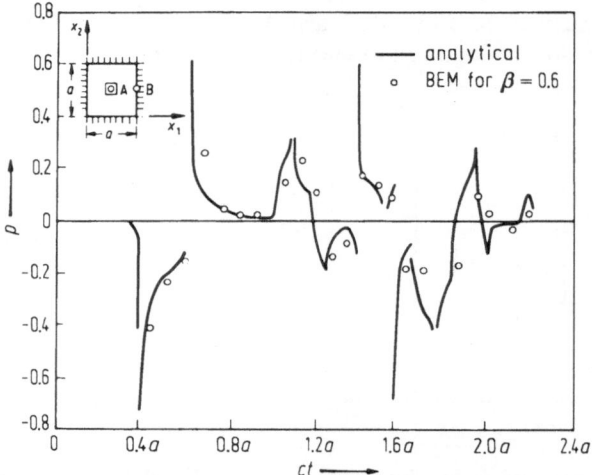

Fig. 4.38.   Normal derivative of displacement at point $B\,(a, a/2)$.   32 boundary elements

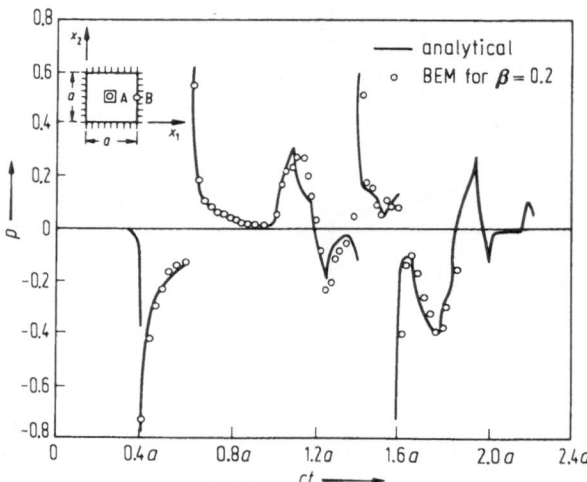

Fig. 4.39.   Normal derivative of displacement at point $B\,(a, a/2)$.   32 boundary elements

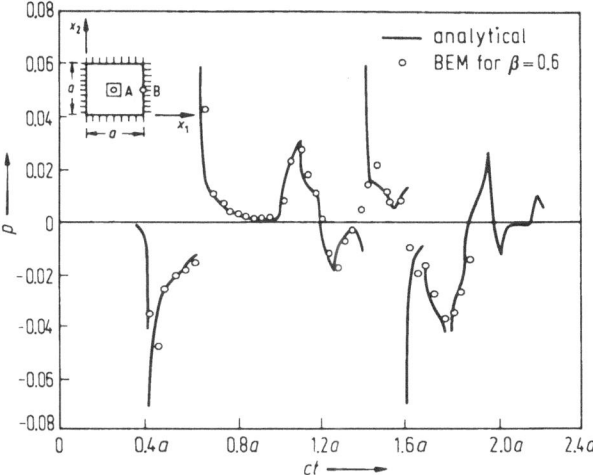

Fig. 4.40. Normal derivative of displacement at point B $(a, a/2)$. 64 boundary elements

refined time division was needed to represent p more accurately. Another boundary element analysis was then carried out, with $\beta = 0.2$ and the results obtained for p, plotted in Fig. 4.39 show a better agreement. A final analysis was performed, in which the boundary was discretized into sixty four rather than thirty two elements, and the value of β was taken as 0.6. The results (see Fig. 4.40) were only slightly better than those for the previous case, apparently because unlike the rod analysis, $\beta < 0.6$ did not introduce any great amount of noise into the numerical results.

Acknowledgement. This work has been partially supported by NATO grant 282/84, Double Jump Program.

References

1 Brebbia, C.A., *The Boundary Element Method for Engineers*, Pentech Press, London 1978
2 Brebbia, C.A. and Walker, S., *The Boundary Element Techniques in Engineering*, Newes-Butterworths, London, 1980
3 Brebbia, C.A., Telles, J.C., and Wrobel, L.C., *Boundary Element Techniques – Theory and Application in Engineering*, Springer-Verlag, Berlin and New York, 1984
4 Mansur, W.J. and Brebbia, CA., Formulation of the Boundary Element Method for Transient Problems Governed by the Scalar Wave Equation. Appl. Math. Modelling **6**, 307–311, 1982
5 Mansur, W.J. and Brebbia, C.A., Numerical Implementation of the Boundary Element Method for Two-Dimensional Transient Scalar Wave Propagation Problems. Appl. Math. Modelling **6**, 299–306, 1982
6 Mansur, W.J., A Time-steeping Technique to Solve Wave Propagation Problems Using the Boundary Element Method. Ph.D. Thesis, Southampton University, 1983
7 Friedman, M.B. and Shaw, R., Diffraction of Pulses by Cylindrical Obstacles of Arbitrary Cross Section. Trans. ASME, J. Appl. Mech., Series E, **29**, 40–46, 1962

8 Shaw, R.P., Diffraction of Acoustic Pulses by Obstacles of Arbitrary Shape with a Robin Boundary Condition. Part A, J.A.S.A. **41**, 855–859, 1967

9 Shaw, R.P., Scattering of Plane Acoustic Pulses by an Infinite Plane with a General First Order Boundary Condition. J. Appl. Mech. **34**, 770–772, 1967

10 Shaw, R.P., Retarded Potential Approach to the Scattering of Elastic Pulses by Rigid Obstacles of Arbitrary Shape. J.A.S.A. **44**, 745–748, 1968

11 Shaw, R.P., Diffraction of Pulses by Obstacles of Arbitrary Shape with an Impedance Boundary Condition. J.A.S.A. **44**, 1962–1968, 1968

12 Shaw, R.P., Singularities in Acoustic Pulse Scattering by Free Surface Obstacles with Sharp Corners. J. Appl. Mech. **38**, 526–528, 1971

13 Shaw, R.P. and English, J.A., Transient Acoustic Scattering by a Free Sphere. J. Sound Vibration **20**, 321–331, 1972

14 Shaw, R.P., Transient Scattering by a Circular Cylinder. J. Sound Vibration **42**, 295–304, 1975

15 Shaw, R.P., An Outer Boundary Integral Equation Applied to Transient Wave Scattering in an Inhomogeneous Medium. J. Appl. Mech. **42**, 147–152, 1975

16 Mitzner, R.M., Numerical Solution for Transient Scattering from a Hard Surface of Arbitrary Shape – Retarded Potential Technique. J. Acoust. Soc. Amer. **42**, 391–397, 1967

17 Groenenboom, P.H.L., Wave Propagation Phenomena. Chapter 2 in *Progress in Boundary Element Methods* (C.A. Brebbia ed.). Pentech Press, London; Springer-Verlag, NY, 1983

18 Cole, D.M., Kosloff, D.D., and Minster, J.B., A Numerical Boundary Integral Equation Method for Elastodynamics. I. Bull. Seis. Soc. America **68** (5), 1331–1357, 1978

19 Morse, P.M. and Feshbach, H., *Methods of Theoretical Physics*. McGraw-Hill, New York, Toronto and London, 1953

20 Kellog, O.D., *Foundations of Potential Theory*. Springer, Berlin, 1929

21 Jason, M.A. and Symn, G.T., *Integral Equation Methods in Potential Theory and Elastostatics*. Academic Press, London, 1977

22 Dirac, P.A.M., *The Principles of Quantum Mechanics*. Second edition, Clarendon, Oxford, 1935

23 Jones, D.S., *Generalised Functions*. McGraw-Hill, London, 1966

24 Lighthill, M.J., *Fourier Analysis and Generalised Functions*. Cambridge University Press, 1958

25 Eringen, A.C. and Suhubi, E.S., *Elastodynamics*. Vols. I and II, Academic Press, New York, San Francisco and London, 1975

26 Nakaguma, R.K., Three Dimensional Elastostatics Using the Boundary Element Method. Ph.D. Thesis, University of Southampton, 1979

27 Telles, J.C.F., On the Application of the Boundary Element Method to Inelastic Problems. Ph.D. Thesis, University of Southampton, 1981

28 Love, A.E.H., *A Treatise on the Mathematical Theory of Elasticity*. Dover, New York, 1944

29 Kirchhoff, G., Zur Theorie der Lichtstrahlen. Ann. Physik **18**, 663–695, 1883

30 Baker, B.B. and Copson, E.T., The Mathematical Theory of Huygens' Principle. Oxford Univ. Press, London, 1939

31 Volterra, V., Sur les Vibrations des Corps Elastiques Isotropes. Acta Math. **18**, 161–332, 1894

32 Wrobel, L.C., Potential and Viscous Flow Problems Using the Boundary Element Method. Ph.D. Thesis, University of Southampton, 1981

33 Brebbia, C.A. and Connor, J.J., Fundamentals of Finite Elements Techniques for Structural Engineers. Butterworths, London, 1973

34 Halbritter, A.L., Telles, J.C.F., and Mansur, W.J., Application of the Boundary Element Method to Field Problems. Proceedings of the Conference, on Structural Analysis, Design and Construction in Nuclear Power Plants (in Portuguese), **39**, pp. 707–724, Porto Alegre, Brasil, 1978

35 Chaudonneret, M., On the Discontinuity of the Stress Vector in the Boundary Integral Equation Method for Elastic Analysis. Proceedings of the First International Seminar on Recent Advances in Boundary Element Methods (C.A. Brebbia ed.), Southampton, 185–194, 1978

36 Mansur, W.J., Halbritter, A.L., and Telles, J.C.F., Boundary Element Method – Formulation for Two-Dimensional Elasticity. Commemorative Annals of the 15th Anniversary of COPPE-UFRJ (COPPE-UFRJ ed. in Portuguese), 1–22, Rio de Janeiro, Brasil, 1979
37 Venturini, W.S., Application of the Boundary Element Formulation to Solve Geomechanical Problems. Ph.D. Thesis, University of Southampton, 1982
38 Patterson, C. and Sheikh, M.A., Non-Conforming Boundary Elements for Stress Analysis. Proceedings of the Second International Conference on Boundary Element Methods (C.A. Brebbia ed.). Springer-Verlag, Berlin, Heidelberg and New York, 1981
39 Patterson, C. and Elseba, N.A.S., A Regular Boundary Method Using Non-Conforming Elements for Potential Problems in Three Dimensions. Proceedings of the Fourth International Seminar on Boundary Element Methods in Engineering (C.A. Brebbia ed.), Springer-Verlag, Berlin, Heidelberg and New York, 1982
40 Miles, J.W., *Modern Mathematics for the Engineer* (E.F. Beckenbach ed.), 82–84, McGraw-Hill, London, 1961

Chapter 5

Transient Elastodynamics

by W. J. Mansur and C. A. Brebbia

5.1 Introduction

Cruse and Rizzo [1–3] were the first researchers to use the boundary integral method to solve elastodynamic problems. In their approach the Laplace transform was proposed to remove the time dependence of the problem. The resulting space dependent system of equations obtained in this way is then solved for various values of the Laplace parameter using boundary elements and a numerical algorithm of inversion [4] is then employed to find the time domain solution.

As an extension of Cruse's work Manolis et al. [5, 6] used the Fourier's transform rather than the Laplace's transform and concluded that their formulation gives better numerical results than that employed by Cruse and Rizzo.

Time stepping techniques have also been used to solve transient two-dimensional elastodynamic problems [7–9]. The approach employed by Niwa et al. [7] consists in solving two-dimensional problems using the simpler three-dimensional fundamental solution. Two-dimensional problems are considered to be cylinders with arbitrary length in the direction of the third axes. In this manner the three-dimensional representation of elastodynamics, can be used to solve the two-dimensional problem, with the third spatial coordinate playing the role of a time related variable.

A boundary element scheme that properly solves the two-dimensional transient elastodynamics problem using a two-dimensional time dependent fundamental solution was first derived by Mansur and Brebbia [8]. The numerical applications carried out by these authors have produced very encouraging results.

An original new procedure for the boundary element solution of eigenvalues and transient dynamic problems has been presented by Nardini and Brebbia [10]. The technique points the formulation of a mass matrix in function of the boundary nodes only. This allows for elastodynamics problems to be treated in a similar way as in finite elements or finite differences, i.e. the problem is reduced to a set of time-dependent differential equations expressed in matrix form. The free vibrations problem can then be reduced to the solution of an algebraic eigenvalue problem. The main advantage of the new approach is that the boundary integrals need to be computed only once as they are frequency independent. Hence the procedure is extremely economic for free vibrations when compared to the ones previously presented. The technique also allows for the general elastodynamics case to be solved in time rather than in the transform domain employing the simple

fundamental solution of elastostatics. The results obtained are highly accurate. For
further details of this novel approach the reader is referred to Chap. 7 of this book.

The present chapter is primarily concerned with the use of time dependent
fundamental solutions for solving elastodynamic problems, including those with
infinite or semi-infinite domains. Three- and two-dimensional integral representations are derived, but most of the ensuing discussions concentrate on the latter case
which is where the most of the mathematical complications occur. Some numerical
examples are presented to illustrate the accuracy of the technique versus other
methods of numerical analysis. It is important to point out that the results were
obtained integrating on time always starting from the initial time rather than with
the usual scheme which updates the initial conditions at the end of each time step.
In this way it is possible to avoid carrying out domain integrations and all terms
need to be defined only on the boundary. This technique is evidently of great importance in problems with infinite or semi-infinite domains.

5.2 Basic Theory

Throughout this chapter the Cartesian tensor notation is used. This notation
permits expressions to be written in a compact form and it is very useful when
considering equations related to mathematical physics. Such notation makes use of
subscript indices $(1, 2, 3)$ to represent (x, y, z). In this chapter the summation
convention will be employed, i.e., a repeated index (subscript or superscript) in a
term implies summation with respect to that index over its range. Hence in three
dimensions,

$$a_{ij}\, x_j = a_{i1}\, x_1 + a_{i2}\, x_2 + a_{i3}\, x_3 \; . \tag{5.1}$$

In addition, the Kronecker delta symbol δ_{ij} as defined by the following
expression will be used.

$$\delta_{ij} = \begin{cases} 1 & \text{when } i = j \\ 0 & \text{when } i \neq j. \end{cases} \tag{5.2}$$

Another useful convention refers to partial differentiation of functions. The
following representation is used,

$$\frac{\partial f_{ij}}{\partial x_k} = f_{ij,k} \; . \tag{5.3}$$

Within this chapter unless otherwise stated indices are assumed to have
respectively a range of three or two for three- or two-dimensional analysis.

Consider an infinitesimal parallelepiped surrounding a point within a body. If
one isolates such a parallelepiped the remainder of the body can be replaced by
the components of the stress tensor σ_{ij} (force per unit area) as depicted in Fig. 5.1.
The sign convention for stresses is such that if σ_{ij} is positive the vector representing
σ_{ij} (stress vector) points in the positive or negative x_j-direction if the outer normal
to the surface element under consideration points respectively in the positive or
negative x_i-direction. Therefore, the components of the stress tensor illustrated in
Fig. 5.1 are positive.

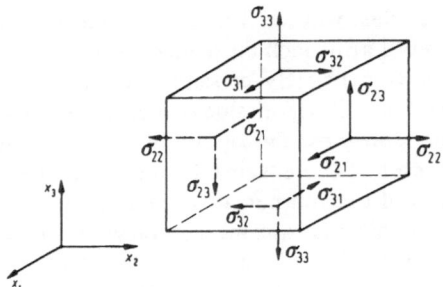

Fig. 5.1. Positive stresses

Once the components of the stress tensor are known, surface forces p_i (force per unit area) acting across any surface in the body, including its boundary, can be computed from

$$p_i = \sigma_{ji}\, n_j \tag{5.4}$$

where n_j stands for the components of the unit vector **n** normal to the surface at the point under consideraction. p_i must be interpreted according to the sense of the vector **n**. It is apparent that the surface over which p_i is being computed can be considered to divide the body into two others. p_i stands for the forces exerted by the body for which **n** is inwards over the body for which **n** is outwards.

Dynamic equilibrium of forces acting on the parallelepiped shown in Fig. 5.1 requires that

$$\sigma_{ij,i} + b_j = \varrho\,\ddot{u}_j \tag{5.5}$$

where b_j stands for the components of the body forces (force per unit volume) and ϱ is the density of the body (mass per unit volume). Time derivatives are indicated by dots, i.e., $\partial^2 u_i/\partial t^2 = \ddot{u}_i$. Equation (5.5) will be referred to hereafter as the stress equations of motion.

Furthermore, if there are no body moments present, dynamic equilibrium of moments requires that

$$\sigma_{ij} = \sigma_{ji}\,. \tag{5.6}$$

Let **x** represent the position vector of a point within a body in its undeformed configuration. Under the action of loads this point moves into a new position described by the coordinates x'_i. The displacement components u_i are given by

$$u_i(\mathbf{x}, t) = x'_i(\mathbf{x}, t) - x_i\,. \tag{5.7}$$

If the u_i displacement components are such that their first derivatives are so small that the squares and products of the partial derivatives of u_i are negligible, then strains can be computed using Cauchy's infinitesimal strain tensor,

$$\varepsilon_{ij} = \tfrac{1}{2}(u_{i,j} + u_{j,i})\,. \tag{5.8}$$

Consider a point P' in the neighbourhood of a point P within a body. Let the coordinates of P and P' be represented by x_i and $x_i + dx_i$ respectively. The relative displacement of P' with respect to P is given by

$$du_i = u_{i,j}\, dx_j\,. \tag{5.9}$$

In the above expression the time variation of the displacement field has not been included, therefore it is valid for the static case. However the discussion now under consideration also applies to elastodynamics if one considers the displacement field corresponding to a fixed instant. Equation (5.9) can also be written as [11]

$$du_i = \tfrac{1}{2}(u_{i,j} + u_{j,i})\, dx_j + \tfrac{1}{2}(u_{i,j} - u_{j,i})\, dx_j \tag{5.10}$$

or

$$du_i = \tfrac{1}{2}\, \varepsilon_{ij}\, dx_j - \tfrac{1}{2}\, \omega_{ij}\, dx_j \tag{5.11}$$

where

$$\omega_{ij} = \tfrac{1}{2}(u_{j,i} - u_{i,j}). \tag{5.12}$$

The tensor ω_{ij} is called the infinitesimal rotation tensor. From expressions (5.8) and (5.12) it is easy to see that the tensors ε_{ij} and ω_{ij} are respectively symmetric and antisymmetric, i.e.

$$\begin{aligned} \varepsilon_{ij} &= \varepsilon_{ji} \\ \omega_{ij} &= -\omega_{ji} . \end{aligned} \tag{5.13}$$

In addition to the stress equations of motion, Hooke's Law relating strain and stress must also be considered when formulating the elastodynamic problem. For isotropic elastic materials in which there is no change in temperature, Hooke's law can be stated in the form

$$\sigma_{ij} = \lambda \varepsilon_{mm}\, \delta_{ij} + 2G\, \varepsilon_{ij} \tag{5.14}$$

or inversely

$$\varepsilon_{ij} = \frac{1}{2G}\left(\sigma_{ij} - \frac{v}{1+v}\, \sigma_{kk}\, \delta_{ij} \right) \tag{5.15}$$

where λ and G are the Lamé's constants and v is the Poisson ratio. λ and G can be computed from v and the elasticity (Young's) modulus E as follows

$$\begin{aligned} \lambda &= \frac{E\,v}{(1-2v)(1+v)} \\ G &= \frac{E}{2(1+v)}. \end{aligned} \tag{5.16}$$

Equations (5.5), (5.8) and (5.14) represent a set of 15 equations for the 15 unknowns σ_{ij}, ε_{ij} and u_i. σ_{ij} can be eliminated by substituting Eq. (5.14) into (5.5). Then, using Eq. (5.8) one obtains Navier's equations which are outlined below

$$G u_{j,kk} + (\lambda + G)\, u_{k,kj} + b_j = \varrho \ddot{u}_j . \tag{5.17}$$

Equations (5.17) are also referred to as the displacement equations of motion and constitute a linear system of hyperbolic differential equations for the dependent variable u_i.

The waves that occur in an elastodynamic state for which the rotations are null ($\omega_{ij} = 0$) are called dilatational waves and propagate with a speed c_d given by

$$c_d = \sqrt{(\lambda + 2G)/\varrho} . \tag{5.18}$$

When the dilatation is null instead, i.e.

$$e = u_{i,i} = 0 \tag{5.19}$$

the waves in an elastic body are called equivoluminal waves and propagate with a speed c_s given by

$$c_s = \sqrt{G/\varrho}. \tag{5.20}$$

Each of the body waves mentioned above can be identified by several distinct physical characteristics. For this reason dilatational waves are also known as primary, irrotational, compressional or longitudinal waves. The corresponding names for equivoluminal waves are secondary, shear, transverse and distortional. (For a fuller discussion of wave theory the reader is referred to Chap. 7 of "Progress in Boundary Element Methods", Vol. 1, i.e. "Elastodynamics" by J. Dominguez and E. Alarcon, published by Pentech Press, London, Halstead Press, N.Y., 1981.)

The constants c_d and c_s can be used to write Navier's equation for elastodynamics (i.e. the equilibrium equation in terms of displacements) by substituting (5.18) and (5.20) into (5.17). This gives

$$(c_d^2 - c_s^2)\, u_{k,kj} + c_s^2\, u_{j,kk} + f_j = \ddot{u}_j \tag{5.21}$$

where

$$f_j = \frac{b_j}{\varrho}.$$

Consider that the body initially at rest has part of its domain (or boundary) disturbed. As time elapses this disturbance propagates setting in motion points of the body that initially were at rest. The moving surface which separates the disturbed from the undisturbed part of the body is called the *wave front*. Wavefronts are also referred to as surfaces of discontinuity because stresses, strains and velocities $\{\partial u_i/\partial t\}$ can be discontinuous there. It should however be realised that discontinuities do not in reality exist in the physical problem but are mathematical idealizations of physical quantities that vary rapidly in a small interval of space and time. Wave fronts do not need necessarily to be considered as moving into an undisturbed region of a body. It is quite common to find situations in which a region is already disturbed before the wave front of an additional disturbance arrives.

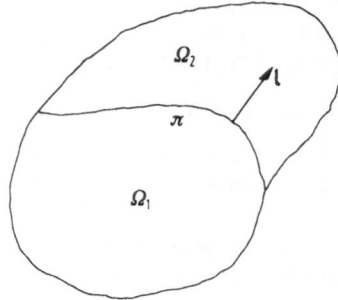

Fig. 5.2.   Surface of discontinuity

Consider a surface of discontinuity Π (Fig. 5.2) moving through Ω; Π moves normal to itself with a speed c, from region Ω_1 to Ω_2 as shown in the figure. Let ℓ_i be the components of the unit vector normal to Π pointing out from the region 1 to the region 2. The jump conditions for displacements in Ω in the neighbourhood of Π are given by

$$[u_i] = (u_i)_2 - (u_i)_1 = 0. \tag{5.22}$$

This indicates that displacements are continuous functions of space and time, stresses and velocities, however can be discontinuous. In the neighbourhood of Π the kinematical conditions applies for velocities, i.e.

$$[\dot{u}_i] = -c\,\ell_j\,[u_{i,j}] \tag{5.23}$$

as well as the dynamical condition for stresses, i.e.

$$[\sigma_{ij}\,\ell_j] = -\varrho\,c\,[\dot{u}_i]. \tag{5.24}$$

5.3 The Initial Value Problem of Elastodynamics

When solving an isotropic elastodynamic problem, it is necessary to determine displacement components $u_i(\mathbf{x}, t)$ that satisfy,

  (i) Equation (5.21) for $t \geq t_0$ at all points inside the domain Ω.
 (ii) Initial conditions,

$$u_i(\mathbf{x}, t_0) = u_{0i}(\mathbf{x})$$

$$\dot{u}_i(\mathbf{x}, t_0) = \left[\frac{\partial}{\partial t}\,u_i(\mathbf{x}, t)\right]_{t=t_0} = v_{0i}(\mathbf{x}) \tag{5.25}$$

     prescribed all over Ω including its boundary Γ.
(iii) Boundary conditions,

$$u_i(\mathbf{x}, t) = \bar{u}_i(\mathbf{x}, t) \qquad\qquad \text{on } \Gamma_1$$

$$p_i(\mathbf{x}, t) = \sigma_{ij}\,n_j = \bar{p}_i(\mathbf{x}, t) \quad \text{on } \Gamma_2 \tag{5.26}$$

specified over the boundary Γ ($\Gamma = \Gamma_1 + \Gamma_2$). Γ may be the union of several closed surfaces with a piecewise continuous exterior unit normal.

From Eqs. (5.8) and (5.14) the stresses can also be written as,

$$\sigma_{ij} = \varrho(c_d^2 - 2c_s^2)\,u_{m,m}\,\delta_{ij} + \varrho c_s^2(u_{i,j} + u_{j,i}). \tag{5.27}$$

Hence, using Eq. (5.4), the second of the conditions given by Eq. (5.26) can be described in terms of displacement components as

$$\varrho(c_d^2 - 2c_s^2)\,u_{m,m}\,n_i + \varrho c_s^2(u_{i,j} + u_{j,i})\,n_j = \bar{p}_i. \tag{5.28}$$

5.4 One-Dimensional Motions

Let us consider the case of simple waves to better understand some of the above concepts. If the displacement is a function of one space variable only,

$$u_i = u_i(x_1, t) \tag{5.29}$$

and body forces are null ($b_i = 0$) Eq. (5.21) reduces to the following three uncoupled one-dimensional wave equations [12]

$$\frac{\partial^2 u_1}{\partial x_1^2} = \frac{1}{c_d^2} \ddot{u}_1, \quad \frac{\partial^2 u_j}{\partial^2 x_1} = \frac{1}{c_s^2} \ddot{u}_j \quad (j = 2, 3). \tag{5.30}$$

u_1, u_2 and u_3 represent displacement waves travelling in the infinite strip shown in Fig. 5.3. Solutions of Eqs. (5.30) can also be regarded as representing waves in one-dimensional bodies, such as strings, rods, etc. The dilatational component of the displacement u_1 is directed along the direction of propagation x_1, whereas the equivoluminal components of the displacements u_2 and u_3 are directed along directions perpendicular to x_1. As $c_d > c_s$ the dilatational disturbance travels faster than the equivoluminal one. If the plane that contains x_1 and x_2 in Fig. 5.3 is the horizontal one u_1, u_2 and u_3 can be identified respectively with the P, SH and SV waves of seismology.

   Boundary conditions must be specified on two planes parallel to each other. If the planes $x_1 = 0$ and $x_1 = 1$ are chosen the boundary conditions can be of type (i), (ii) or (iii) as described below.

(i) Displacement boundary conditions

$$u_i(0, t) = \bar{u}_i^1(t)$$

$$u_i(\ell, t) = \bar{u}_i^2(t). \tag{5.31}$$

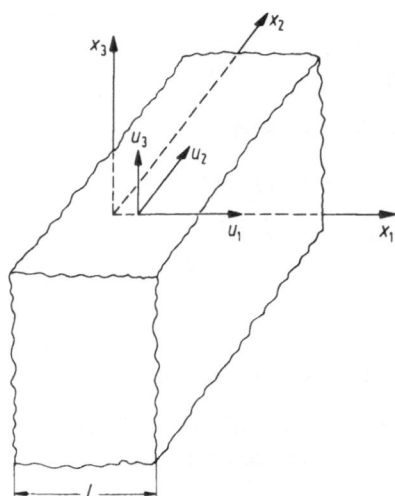

Fig. 5.3.   Infinite strip of width

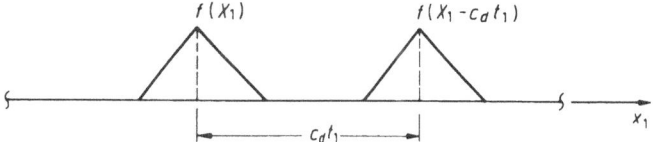

Fig. 5.4.   Propagation of one-dimensional waves

(ii) Traction boundary conditions

$$p_i(0, t) = \bar{p}_i^1(t)$$
$$p_i(\ell, t) = \bar{p}_i^2(t).$$

(5.32)

(iii) Mixed boundary conditions

$$u_i(0, t) = \bar{u}_i(t)$$
$$p_i(\ell, t) = \bar{p}_i(t).$$

(5.33)

In addition initial conditions

$$u_i(x_1, 0) = u_{0i}(x_1)$$
$$\dot{u}_i(x_1, 0) = v_{0i}(x_1).$$

(5.34)

must also be prescribed.

Analytical solutions for the one-dimensional wave equation are not difficult to find. The general solution of an equation such as the first of those given by expression (5.30) was first derived by D'Alembert, and reads,

$$u_1 = f(x_1 - c_d t) + g(x_1 + c_d t).$$

(5.35)

Equation (5.35) has a very simply physical interpretation; it can be regarded as being composed of two one-dimensional waves $f(x_1 - c_d t)$ and $g(x_1 + c_d t)$ propagating in the positive and negative x_1 direction respectively. A consideration for instance of contributions due to $f(x_1 - c_d t)$ only, results in the conclusion that at $t = 0$, $u_1 = f(x_1)$. At a time instant $t = t_1$ the shape of the wave given by $u_1 = f(x_1 - c_d t_1)$ is that which is obtained by displacing the initial shape by a distance $c_d t_1$ in the positive x_1 direction as illustrated in Fig. 5.4.

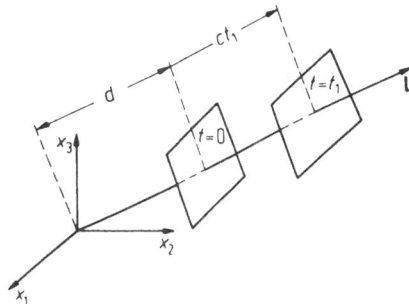

Fig. 5.5.   Propagation of plane waves

A plane displacement wave propagating in the direction of a unit vector ℓ can be represented by

$$u_i(\mathbf{x}, t) = u_i(x_i \ell_i - c t) \tag{5.36}$$

where $x_i \ell_i = d + ct$ defines planes normal to ℓ over which u_i is constant. The argument of u_i, $x_i \ell_i - ct = d$ is called the *phase of the wave*. Figure 5.5 shows two planes of constant phase, L_0 and L_1, that correspond respectively to $t = 0$ and $t = t_1$. It should be noticed that u_i over L_0 is equal to u_i over L_1, therefore, plane waves have the same characteristics of propagation exhibited by D'Alembert solution for one-dimensional cases.

Waves like those represented by Eq. (5.36) only obey Navier's equations if,

(i) $\ell_i u_i = \pm \sqrt{u_i u_i}$ and $c = c_d$

(ii) $\ell_i u_i = 0$ and $c = c_s$. $\tag{5.37}$

It can be demonstrated [12–15] that waves defined by Eq. (5.36) and which consequently obey conditions (i) and (ii) in expression (5.37) are in fact equivoluminal and dilatational displacement waves respectively. Hence a complete analogy with the one-dimensional case previously studied can be forthcoming if one considers that the coordinate axis x_1 is parallel to the direction of propagation defined by the unit vector ℓ.

5.5 Plane Motions

If the displacement is a function of two rectangular coordinates only, i.e.

$$u_i(\mathbf{x}, t) = u_i(x_1, x_2, t) \tag{5.38}$$

the problem is termed elastodynamic in the plane [12] or complete plane strain [16]. In view of Eq. (5.38), $u_{3,3} = 0$ and all other derivatives of the displacement components are functions of x_1 and x_2 only. Therefore the Navier's equations take the following form,

$$(c_d^2 - c_s^2)u_{k,kj} + c_s^2 u_{j,kk} + f_j = \ddot{u}_j \tag{5.39}$$

$$c_s^2 u_{3,kk} + f_3 = \ddot{u}_3 \tag{5.40}$$

where j and k can be 1 or 2.

The domains in which complete plane strain problems are studied are infinite cylinders whose axes are parallel to the x_3-direction. The mathematical problem of solving the differential equations (5.39) and (5.40) can then be considered as two-dimensional. The domain Ω and the boundary Γ in this case are defined by the intersection of the infinite cylinder with the (x_1, x_2) plane. Of course the physical problem is three-dimensional because displacements and stresses in the x_3-direction do not equal null. Equation (5.27) in this case is written as

$$\sigma_{ij} = \varrho(c_d^2 - 2c_s^2) u_{m,m} \delta_{ij} + \varrho c_s^2 (u_{i,j} + u_{j,i})$$

$$\sigma_{33} = \lambda u_{k,k} \tag{5.41}$$

$$\sigma_{i3} = G u_{3,i}.$$

Equations (5.39)–(5.41) show that Eqs. (5.39) and (5.40) can be solved independently. For this reason, complete plane strain can also be seen as resulting from the
superposition of the plane strain and antiplane motions governed respectively by
Eqs. (5.39) and (5.40). These motions are described in a) and b) below.

a) Antiplane Motion

This motion is governed by the scalar wave equation [Eq. (5.40)]. The boundary
conditions in this case are given by

$$
\begin{aligned}
u_3 &= \bar{u}_3(x_1, x_2, t) && \text{on } \Gamma_1' \\
p_3 &= \sigma_{i3}\, n_i = G u_{3,i}\, n_i = \bar{p}_3(x_1, x_2, t) && \text{on } \Gamma_2'
\end{aligned}
\tag{5.42}
$$

where $\Gamma_1' + \Gamma_2' = \Gamma$. The initial conditions for the antiplane motion are written as

$$
\begin{aligned}
u_3(x_1, x_2, 0) &= u_{03}(x_1, x_2) \\
\dot{u}_3(x_1, x_2, 0) &= v_{03}(x_1, x_2)
\end{aligned}
\quad \text{in } \Omega.
\tag{5.43}
$$

In this problem the normal stress σ_{33} is null, therefore only the shear stresses
$\sigma_{13} = \sigma_{31}$ and $\sigma_{23} = \sigma_{32}$ are present in the analysis. In addition the vector representing the displacement u_3 is perpendicular to the direction of propagation of the
displacement waves. For these reasons this motion is also called shear antiplane or
horizontally polarized shear motion [14].

b) Plane Strain Motion

Plane strain motions are governed by Eq. (5.39), which is of the same form as
Navier's equations for three dimensions. The only difference is that in the present
situation the indices range from 1 to 2, rather than from 1 to 3. The boundary
conditions for this problem are given by

$$
\begin{aligned}
u_i &= \bar{u}_i(x_1, x_2, t) && \text{on } \Gamma_1 \\
p_i &= \sigma_{ij}\, n_j = \bar{p}_i(x_1, x_2, t) && \text{on } \Gamma_2
\end{aligned}
\tag{5.44}
$$

where $\Gamma = \Gamma_1 + \Gamma_2$. The initial conditions for plane strain read

$$
\begin{aligned}
u_i(x_1, x_2, 0) &= u_{0i}(x_1, x_2) \\
v_i(x_1, x_2, 0) &= v_{0i}(x_1, x_2)
\end{aligned}
\quad \text{in } \Omega.
\tag{5.45}
$$

In a plane strain problem

$$
u_3 = \varepsilon_{13} = \varepsilon_{31} = \varepsilon_{23} = \varepsilon_{32} = 0.
\tag{5.46}
$$

However, the stress σ_{33} is not null and can be computed from the second of
Eqs. (5.41).

When the domain of the problem being analysed does not extend to infinity in
the x_3-direction a plane strain condition can not be assumed to exist. In this case a
three-dimensional analysis must be carried out, however when the dimensions of
the body in the x_3-direction are small, a condition known as plane stress can be
assumed. This situation occurs when analysing thin plates acted on by forces

parallel to its midplane. The plane stress hypothesis assumes that

$$\sigma_{33} = \sigma_{31} = \sigma_{13} = \sigma_{32} = \sigma_{23} = 0. \tag{5.47}$$

In this case the same equations of plane strain can be used provided that the constants v and E are replaced by fictitious ones, $\bar{v}$ and $\bar{E}$, given by

$$\bar{v} = v/(1 + v)$$
$$\bar{E} = E(1 + 2v)/(1 + v)^2 \tag{5.48}$$

which implies that

$$\bar{G} = G$$
$$\bar{\lambda} = 2\lambda\, G/(\lambda + 2G). \tag{5.49}$$

It is important to state that since ε_{33} is not necessarily null, u_i depends on x_3 and the problem is not really two-dimensional. However, plane stress can be considered a good assumption when the plate being studied is sufficiently thin [11].

5.6 Fundamental Solutions for Transient Elastodynamics

An integral representation for elastodynamics can be obtained following a procedure similar to that described in [9, 17, 19] for the scalar wave equation, where a weighted residual procedure was employed. In this chapter instead Graffi's elastodynamic reciprocal theorem will be employed.

The fundamental singular solution of elastodynamics which is used in this chapter is the function u_{ik}^* which satisfies the following equations

$$\sigma_{ijk,j}^* - \varrho \ddot{u}_{ik}^* = - \delta_{ik}\, \delta(q - s)\, \delta(t - \tau) \tag{5.50}$$

in an unbounded domain Ω^*, which is free from any imposed initial condition. The body forces in Eqs. (5.50) correspond to a concentrated force in the x_i-direction which is an impulse at $t = \tau$ located at $q = s$.

In three dimensions, the solution of Eqs. (5.50) can be written as follows [12]

$$u_{ik}^*(q, t; s, \tau) = \frac{t'}{4\pi\varrho\, r^2} \left\{ \left(\frac{3 r_i\, r_k}{r^3} - \frac{\delta_{ik}}{r} \right) \left[H\left(t' - \frac{r}{c_d}\right) - H\left(t' - \frac{r}{c_s}\right) \right] \right.$$
$$\left. + \frac{r_i\, r_k}{r^2} \left[\frac{1}{c_d} \delta\left(t' - \frac{r}{c_d}\right) - \frac{1}{c_s} \delta\left(t' - \frac{r}{c_s}\right) \right] + \frac{\delta_{ik}}{c_s} \delta\left(t' - \frac{r}{c_s}\right) \right\} \tag{5.51}$$

where

$$t' = t - \tau$$
$$r = |q - s| = (r_i\, r_i)^{1/2} \tag{5.52}$$
$$r_i = x_i(q) - x_i(s).$$

It should be recognized that equivalent forms for u_{ik}^* other than that described by expression (5.51) are frequently to be found in the literature.

Equations (5.4), (5.27) and (5.51) can be used to obtain the fundamental traction given by

$$p_{ik}^* = \sigma_{ijk}^* n_j = \sigma_{ikj}^* n_j = \varrho \left[(c_d^2 - 2c_s^2) \, u_{im,m}^* \, \delta_{jk} + c_s^2 (u_{ik,j}^* + u_{ij,k}^*) \right] n_j \qquad (5.53)$$

where

$$\sigma_{ijk}^* (q, t; s, \tau) = \frac{1}{4\pi} \left\{ - 6 \, \frac{c_s^2 \, t'}{r^2} \left[5 \, \frac{r_i \, r_j \, r_k}{r^5} - \frac{\delta_{ij} \, r_k + \delta_{ik} \, r_j + \delta_{jk} \, r_i}{r^3} \right] \right.$$

$$\times \left[H \left(t' - \frac{r}{c_d} \right) - H \left(t' - \frac{r}{c_s} \right) \right] + 2 \left[6 \, \frac{r_i \, r_j \, r_k}{r^5} - \frac{\delta_{ij} \, r_k + \delta_{ik} \, r_j + \delta_{jk} \, r_i}{r^3} \right]$$

$$\times \left[\delta \left(t' - \frac{r}{c_s} \right) - \frac{c_s^2}{c_d^2} \, \delta \left(t' - \frac{r}{c_d} \right) \right]$$

$$+ 2 \, \frac{r_i \, r_j \, r_k}{r^4 \, c_s} \left[\dot{\delta} \left(t' - \frac{r}{c_s} \right) - \frac{c_s^3}{c_d^3} \, \dot{\delta} \left(t' - \frac{r}{c_d} \right) \right]$$

$$- \frac{r_i \, \delta_{jk}}{r^3} \left(1 - 2 \, \frac{c_s^2}{c_d^2} \right) \left[\delta \left(t' - \frac{r}{c_d} \right) + \frac{r}{c_d} \, \dot{\delta} \left(t' - \frac{r}{c_d} \right) \right]$$

$$- \frac{\delta_{ik} \, r_j + \delta_{ij} \, r_k}{r^3} \left[\delta \left(t' - \frac{r}{c_s} \right) + \frac{r}{c_s} \, \dot{\delta} \left(t' - \frac{r}{c_s} \right) \right] \right\}. \qquad (5.54)$$

The two-dimensional fundamental solution of elastodynamics can be obtained by employing the method of descent. A procedure similar to that given in [19] for the scalar wave equation can be followed giving

$$u_{ik}^* (q, t; s, \tau) = \frac{1}{2\pi \varrho \, c_s} \left\{ \left[\frac{\delta_{ik}}{\sqrt{c_s^2 (t')^2 - r^2}} + \frac{\delta_{ik}}{r^2} \sqrt{c_s^2 (t')^2 - r^2} - \frac{r_{,i} \, r_{,k}}{r^2} \frac{2 c_s^2 (t')^2 - r^2}{\sqrt{c_s^2 (t')^2 - r^2}} \right] \right.$$

$$\times H(c_s \, t' - r) - \frac{c_s}{c_d} \left[\frac{\delta_{ik}}{r^2} \sqrt{c_d^2 (t')^2 - r^2} - \frac{r_{,i} \, r_{,k}}{r^2} \frac{2 c_d^2 (t')^2 - r^2}{\sqrt{c_d^2 (t')^2 - r^2}} \right] H(c_d \, t' - r) \right\} \qquad (5.55)$$

where

$$r_{,i} = \frac{\partial r}{\partial x_i (q)} = - \frac{\partial r}{\partial x_i (s)} = \frac{r_i}{r}. \qquad (5.56)$$

Equations (5.53) and (5.55) can be applied to derive an expression for the two-dimensional fundamental traction, which is given by

$$p_{ik}^* (q, t; s, \tau) = \frac{1}{2\pi \varrho \, c_s} \left\{ A_{ik} \left[\frac{r}{\sqrt{[c_s^2 (t')^2 - r^2]^3}} \, H(c_s \, t' - r) \right. \right.$$

$$+ \left. \left. \frac{1}{\sqrt{c_s^2 (t')^2 - r^2}} \frac{\partial}{\partial (c_s \, \tau)} \, H(c_s \, t' - r) \right] \right.$$

$$+ \left[B_{ik} \frac{2c_s^2(t')^2 - r^2}{\sqrt{c_s^2(t')^2 - r^2}} + D_{ik} \frac{r^3}{\sqrt{[c_s^2(t')^2 - r^2]^3}} \right]$$

$$\times H(c_s\, t' - r) + D_{ik} \frac{2c_s^2(t')^2 - r^2}{\sqrt{c_s^2(t')^2 - r^2}} \frac{\partial}{\partial(c_s\,\tau)} H(c_s\, t' - r)$$

$$- \frac{c_s}{c_d} \left(\left[B_{ik} \frac{2c_d^2(t')^2 - r^2}{\sqrt{c_d^2(t')^2 - r^2}} + D_{ik} \frac{r^3}{\sqrt{[c_d^2(t')^2 - r^2]^3}} \right] \right.$$

$$\left. \times H(c_d\, t' - r) + D_{ik} \frac{2c_d^2(t')^2 - r^2}{\sqrt{c_d^2(t')^2 - r^2}} \frac{\partial}{\partial(c_d\,\tau)} H(c_d\, t' - r) \right) \Big\} \tag{5.57}$$

where

$$A_{ik} = G \left(2\theta\, n_k\, r_{,i} + \delta_{ik} \frac{\partial r}{\partial n} + n_i\, r_{,k} \right)$$

$$B_{ik} = -\frac{2G}{r^3} \left(\delta_{ik} \frac{\partial r}{\partial n} + n_i\, r_{,k} + n_k\, r_{,i} - 4 \frac{\partial r}{\partial n}\, r_{,i}\, r_{,k} \right) \tag{5.58}$$

$$D_{ik} = -\frac{2G}{r^2} \left(\theta\, n_k\, r_{,i} + \frac{\partial r}{\partial n}\, r_{,i}\, r_{,k} \right)$$

$$\theta = \lambda/2\,G = (c_d^2 - 2c_s^2)/2c_s^2 .$$

The fundamental solutions studied in this section have the following properties [12]:

(i) causality

$$u_{ik}^*(q, t; s, \tau) = 0 \quad \text{whenever } c_s(t - \tau) < |\mathbf{q} - \mathbf{s}| \tag{5.59}$$

(ii) reciprocity

$$u_{ik}^*(q, t; s, \tau) = u_{ik}^*(s, -\tau; q, -t) \tag{5.60}$$

(iii) time translation

$$u_{ik}^*(q, t + t_1; s, \tau + t_1) = u_{ik}^*(q, t; s, \tau). \tag{5.61}$$

The symmetry of the tensors given by Eqs. (5.51) and (5.55) implies that [12] the k-component of the displacement at q due to the i-component of the concentrated force at s is equal to the i-component of the displacement at q due to the k-component of the concentrated force at s, i.e.

$$u_{ik}^*(q, t; s, \tau) = u_{ki}^*(q, t; s, \tau). \tag{5.62}$$

5.7 Time Domain Elastodynamic Boundary Integral Representation

The reciprocal theorem for elastodynamics, to be derived in this section, effectively relates two elastodynamic states whose displacement fields will be denoted by u_k

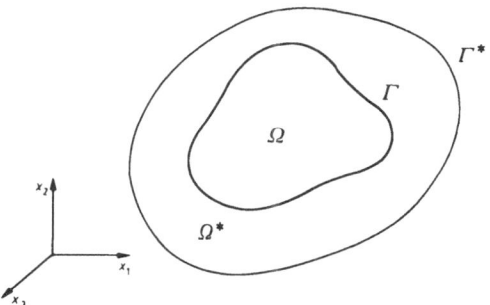

Fig. 5.6.   Region $\Omega^* + \Gamma^*$ containing $\Omega + \Gamma$

and u_k^*. These are defined over regions $\Omega + \Gamma$ and $\Omega^* + \Gamma^*$ respectively so that Ω^* contains $\Omega + \Gamma$ as depicted in Fig. 5.6. The bodies enclosed by Γ and Γ^* have the same physical properties, and u_k and u_k^* satisfy the elastodynamic equilibrium equations, i.e.

$$\sigma_{kj,j} + \beta_k = 0 \quad \text{in } \Omega$$
$$\sigma_{kj,j}^* + \beta_k^* = 0 \quad \text{in } \Omega^* \tag{5.63}$$

where

$$\beta_k = b_k - \varrho \, \frac{\partial^2 u_k}{\partial \tau^2}$$

$$\beta_k^* = b_k^* - \varrho \, \frac{\partial^2 u_k^*}{\partial \tau^2} \, . \tag{5.64}$$

Using Hooke's law the following integral statement can easily be inferred

$$\int_\Omega \sigma_{ij}^* \, \varepsilon_{ij} \, d\Omega = \int_\Omega \sigma_{ij} \, \varepsilon_{ij}^* \, d\Omega \, . \tag{5.65}$$

If the divergence theorem is applied to both sides of Eq. (5.65) and Eqs. (5.15) and (5.63) are used, the following statement is inferred:

$$\int_\Omega \beta_k^* \, u_k \, d\Omega + \int_\Gamma p_k^* \, u_k \, d\Gamma = \int_\Omega \beta_k \, u_k^* \, d\Omega + \int_\Gamma p_k \, u_k^* \, d\Gamma \, . \tag{5.66}$$

Equation (5.66) corresponds to Betti's second reciprocal work theorem for two distinct elastostatic states with body forces β and β^*.

When Eq. (5.66) is integrated from 0 to t and expression (5.64) is taken into consideration, the following equation is obtained:

$$\int_0^t \int_\Omega b_k^* \, u_k \, d\Omega \, d\tau - \varrho \int_0^t \int_\Omega \frac{\partial^2 u_k^*}{\partial \tau^2} \, u_k \, d\Omega \, d\tau + \int_0^t \int_\Gamma p_k^* \, u_k \, d\Gamma \, d\tau$$

$$= \int_0^t \int_\Omega b_k \, u_k^* \, d\Omega \, d\tau - \varrho \int_0^t \int_\Omega \frac{\partial^2 u_k}{\partial \tau^2} \, u_k^* \, d\Omega \, d\tau + \int_0^t \int_\Gamma p_k \, u_k^* \, d\Gamma \, d\tau \, . \tag{5.67}$$

Integrating by parts with respect to time gives

$$\int_0^t \frac{\partial^2 u_k}{\partial \tau^2}\, u_k^* \, d\tau = v_k(q, t)\, u_k^*(q, t) - v_{0k}\, u_{0k}^* - \int_0^t v_k(q, \tau)\, v_k^*(q, \tau)\, d\tau$$

$$\int_0^t \frac{\partial^2 u_k^*}{\partial \tau^2}\, u_k \, d\tau = v_k^*(q, t)\, u_k(q, t) - v_{0k}^*\, u_{0k} - \int_0^t v_k^*(q, \tau)\, v_k(q, \tau)\, d\tau$$

(5.68)

where

$$v_k(q, t) = \left| \frac{\partial u_k}{\partial \tau} \right|_{\tau = t} \tag{5.69}$$

and

$$v_{0k}(q) = v_k(q, 0). \tag{5.70}$$

When expression (5.68) is substituted into Eq. (5.67), the reciprocal theorem of elastodynamics is obtained, i.e.

$$\int_0^t \int_\Omega b_k^* u_k \, d\Omega \, d\tau - \varrho \int_\Omega v_k^* u_k \, d\Omega + \varrho \int_\Omega v_{0k}^* u_{0k} \, d\Omega + \int_0^t \int_\Gamma p_k^* u_k \, d\Gamma \, d\tau$$

$$= \int_0^t \int_\Omega b_k u_k^* \, d\Omega \, d\tau - \varrho \int_\Omega v_k u_k^* \, d\Omega + \varrho \int_\Omega v_{0k} u_{0k}^* \, d\Omega + \int_0^t \int_\Gamma p_k u_k^* \, d\Gamma \, d\tau. \quad (5.71)$$

If one of the elastodynamic states is taken at a time $t' = t - \tau$ the reciprocal theorem given by Eq. (5.71) can be cast into Graffi's theorem, in the form in which it is presented in [12] and [15].

In order to obtain a boundary integral equation for the problem being studied, one of the elastodynamic states in expression (5.71) will be considered to be that governed by Eq. (5.50). In this case, due to the reciprocity property

$$\frac{\partial^2 u_{ik}^*}{\partial t^2} = \frac{\partial^2 u_{ik}^*}{\partial \tau^2}, \tag{5.72}$$

and as a result of the causality property

$$\int_\Omega v_{ik}^*(q, t; s, t^+)\, u_k \, d\Omega = \int_\Omega v_k \, u_{ik}^*(q, t; s, t^+)\, d\Omega = 0. \tag{5.73}$$

Then, if the time integration limits indicated in Eq. (5.71) are taken to be zero and t^+ ($t^+ = t + \varepsilon, \varepsilon \to 0$) the following equation is obtained

$$\int_0^{t^+} \int_\Omega u_k \, \delta_{ik} \, \delta(q - s)\, \delta(t - \tau)\, d\Omega \, d\tau + \int_0^{t^+} \int_\Gamma p_{ik}^* u_k \, d\Gamma \, d\tau$$

$$= \int_0^{t^+} \int_\Gamma p_k \, u_{ik}^* \, d\Gamma \, d\tau + \varrho \int_\Omega v_{0k} \, u_{0ik}^* \, d\Omega - \varrho \int_\Omega v_{0ik}^* \, u_{0k} \, d\Omega + \int_0^{t^+} \int_\Omega b_k \, u_{ik}^* \, d\Omega \, d\tau. \quad (5.74)$$

Taking account of the Dirac delta properties

$$\int_0^{t^+} \int_\Omega u_k \, \delta_{ik} \, \delta(q - s)\, \delta(t - \tau)\, d\Omega\,(q)\, d\tau = u_i(s, t), \tag{5.75}$$

the following integral statement is then obtained

$$u_i(s,t) = \int_0^{t^+} \int_\Gamma u_{ik}^*(Q,t;s,\tau)\, p_k(Q,\tau)\, d\Gamma(Q)\, d\tau - \int_0^{t^+} \int_\Gamma p_{ik}^*(Q,t;s,\tau)\, u_k(Q,\tau)\, d\Gamma(Q)\, d\tau$$

$$+ \varrho \int_\Omega u_{0ik}^*(q,t;s)\, v_{0k}(q)\, d\Omega(q) - \varrho \int_\Omega v_{0ik}^*(q,t;s)\, u_{0k}(q)\, d\Omega(q)$$

$$+ \int_0^{t^+} \int_\Omega u_{ik}^*(q,t;s,\tau)\, b_k(q,\tau)\, d\Omega(q)\, d\tau. \tag{5.76}$$

Equation (5.76) gives the u_i-component of the displacement, at an internal point s, as a function of boundary tractions and displacements, initial conditions and body forces. When $s \to S$ a procedure similar to that discussed in [18, 19] can be followed giving

$$c_{ik}(S)\, u_k(S,t) = \int_0^{t^+} \int_\Gamma u_{ik}^*(Q,t;S,\tau)\, p_k(Q,\tau)\, d\Gamma(Q)\, d\tau$$

$$- \int_0^{t^+} \int_\Gamma p_{ik}^*(Q,t;S,\tau)\, u_k(Q,\tau)\, d\Gamma(Q)\, d\tau$$

$$+ \varrho \int_\Omega u_{0ik}^*(q,t;S)\, v_{0k}(q)\, d\Omega(q)$$

$$- \varrho \int_\Omega v_{0ik}^*(q,t;S)\, u_{0k}(q)\, d\Omega(q)$$

$$+ \int_0^{t^+} \int_\Omega u_{ik}^*(q,t;S,\tau)\, b_k(q,\tau)\, d\Omega(q)\, d\tau \tag{5.77}$$

where

$$c_{ik}(S) = \tfrac{1}{2}\delta_{ik} \tag{5.78}$$

whenever the Γ boundary is smooth. It should be recognized that the integrals indicated in Eq. (5.77) are to be calculated in the Cauchy principal value sense.

Equation (5.77) can also be used when the source point is outside $\Omega + \Gamma$. In this case c_{ij} must be regarded as being equal to zero.

Additional information on how Eq. (5.77) can be obtained from Eq. (5.76), for both, three and two dimensions, can be found in [7, 12, 18]. In these references, a discussion concerning expression (5.78) is also considered.

In order to implement a numerical time-stepping algorithm to solve the three-dimensional boundary integral equation analytical integrations must be performed first, to eliminate the Dirac-delta functions and its derivatives that appear in Eqs. (5.51) and (5.54). This matter is discussed in references [7] and [20] where two-dimensional elastodynamic problems are analysed using three-dimensional fundamental solutions. In these papers the two-dimensional problem is considered to be a cylinder, whose axis has infinite length and is parallel to the x_3-direction. As this approach is essentially three-dimensional, an extra integration with respect to the coordinate x_3 is required.

In the present investigation, two-dimensional elastodynamic problems are analysed using a two-dimensional boundary integral equation, i.e., the fundamental solution considered is that given by Eq. (5.55). In order to implement a general two-dimensional numerical time-stepping algorithm, some additional transformations must first be carried out in order to eliminate the derivatives of Heaviside functions that appear in Eq. (5.57). This is discussed in the next section.

5.8 Additional Transformations to the Two-Dimensional Boundary Integral Equation of Elastodynamics

In the numerical analysis concerning two-dimensional elastodynamics, initial conditions and body forces will not be considered. Consequently when u_{ik}^* and p_{ik}^* given by expressions (5.55) and (5.57) respectively, are substituted into Eq. (5.77) the following expression is obtained

$$
\begin{aligned}
c_{ik}(S)\, u_k(S, t) = \frac{1}{2\pi \varrho\, c_s} \Bigg\{ & \int_0^{t^+} \int_\Gamma (A_{ik}\, L_2^3\, M_2 - B_{ik}\, L_2\, N_2 + D_{ik}\, L_2^3\, O_2)\, u_k\, H[c_s\, t' - r]\, d\Gamma\, d\tau \\
& -\frac{c_s}{c_d} \int_0^{t^+} \int_\Gamma (-B_{ik}\, L_1\, N_1 + D_{ik}\, L_1^3\, O_1)\, u_k\, H[c_d t' - r]\, d\Gamma\, d\tau \\
& +\frac{1}{c_s} \int_0^{t^+} \int_\Gamma (A_{ik}\, L_2 + D_{ik}\, L_2\, N_2)\, v_k\, H[c_s\, t' - r]\, d\Gamma\, d\tau \\
& -\frac{c_s}{c_d^2} \int_0^{t^+} \int_\Gamma D_{ik}\, L_1\, N_1\, v_k\, H[c_d t' - r]\, d\Gamma\, d\tau \\
& +\int_0^{t^+} \int_\Gamma (\delta_{ik}\, L_2 + F_{ik}\, L_2^{-1} + J_{ik}\, L_2\, N_2)\, p_k\, H[c_s\, t' - r]\, d\Gamma\, d\tau \\
& -\frac{c_s}{c_d} \int_0^{t^+} \int_\Gamma (F_{ik}\, L_1^{-1} + J_{ik}\, L_1\, N_1)\, p_k\, H[c_d t' - r]\, d\Gamma\, d\tau \Bigg\}
\end{aligned}
\tag{5.79}
$$

where A_{ik}, B_{ik} and D_{ik} are given by expression (5.58),

$$
F_{ik} = \frac{\delta_{ik}}{r^2},
$$
$$
J_{ik} = -\frac{r_{,i}\, r_{,k}}{r^2}
\tag{5.80}
$$

and

$$
\begin{aligned}
L_1 &= L_1\, (Q, t; S, \tau) = [c_d^2 (t')^2 - r^2]^{-1/2} \\
M_1 &= M_1\, (Q, t; S, \tau) = c_d\, t' - r \\
N_1 &= N_1\, (Q, t; S, \tau) = 2 c_d^2 (t')^2 - r^2 \\
O_1 &= O_1\, (Q, t; S, \tau) = 3 c_d\, t'\, r^2 - 2 c_d^3 (t')^3 - r^3 .
\end{aligned}
\tag{5.81}
$$

L_2, M_2, N_2 and O_2 can be respectively obtained from L_1, M_1, N_1 and O_1 replacing c_d by c_s in expression (5.81).

In items (i) and (ii) described below details are given of the modifications required to obtain Eq. (5.79) from expression (5.77).

(i) Applying the same procedure used in Sect. 4.7 of this book it is possible to write

$$\int_0^{t^+} \int_\Gamma A_{ik} u_k L_2 \frac{\partial}{\partial(c_s \tau)} H [c_s t' - r] d\Gamma \, d\tau = -\frac{1}{c_s} \int_\Gamma A_{ik} u_{0k} L_{02} H [c_s t - r] d\Gamma$$

$$-\frac{1}{c_s} \int_0^{t^+} \int_\Gamma A_{ik} v_k L_2 H [c_s t' - r] d\Gamma \, d\tau - \int_0^{t^+} \int_\Gamma A_{ik} u_k c_s t' L_2^3 H [c t' - r] d\Gamma \, d\tau \quad (5.82)$$

where

$$L_{02} = L_2(Q, t; S, 0). \quad (5.83)$$

The first term on the right-hand side of expression (5.82) was regarded as being equal to null because non zero initial conditions have not been considered in the elastodynamic formulation.

(ii) The remaining term in Eq. (5.77) that requires to be further manipulated is given by

$$I = \int_0^{t^+} \int_\Gamma D_{ik} u_k L_2 N_2 \frac{\partial}{\partial(c_s \tau)} H[c_s t' - r] d\Gamma \, d\tau$$

$$= \frac{1}{c_s} \int_\Gamma D_{ik} \int_0^{t^+} u_k L_2 N_2 \frac{\partial}{\partial \tau} H [c_s t' - r] d\tau \, d\Gamma . \quad (5.84)$$

Integration by parts with respect to time gives,

$$\int_0^{t^+} u_k L_2 N_2 \frac{\partial}{\partial \tau} H [c_s t' - r] d\tau = \left| u_k L_2 N_2 H [c_s t' - r] \right|_0^{t^+}$$

$$-\int_0^{t^+} v_k L_2 N_2 H [c_s t' - r] d\tau - \int_0^{t^+} u_k \frac{\partial}{\partial \tau} (L_2 N_2) H [c_s t' - r] d\tau. \quad (5.85)$$

In view of the causality property and the fact that

$$\frac{\partial}{\partial \tau} (L_2 N_2) = (- 2 c_s^4 (t')^3 + 3 c_s^2 t' r^2) L_2^3 \quad (5.86)$$

the following expression results

$$I = -\frac{1}{c_s} \int_\Gamma D_{ik} u_{0k} N_{02} L_{02} H [c_s t - r] d\Gamma$$

$$-\frac{1}{c_s} \int_0^{t^+} \int_\Gamma D_{ik} v_k N_2 L_2 H [c_s t' - r] d\Gamma \, d\tau$$

$$+ \int_0^{t^+} \int_\Gamma D_{ik} u_k [2 c_s^3 (t')^3 - 3 c_s t' r^2] L_2^3 H [c_s t' - r] d\Gamma \, d\tau \quad (5.87)$$

where

$$N_{02} = N_2(Q, t; S, 0). \quad (5.88)$$

The first term on the right-hand side of expression (5.87) was not included in Eq. (5.79) because u_{0k} was taken as being equal to zero.

The operations carried out in sub-sections (i) and (ii) above refer to terms in Eq. (5.79) that account for waves which propagate with speed c_s. The term in expression (5.77) given by

$$\int_0^{t^+} \int_\Gamma D_{ik} \, u_k \, L_1 \, N_1 \, \frac{\partial}{\partial(c_d \tau)} \, H \, [c_s \, t' - r] \, d\Gamma \, d\tau \qquad (5.89)$$

which refers to dilatational waves also has to undergo additional transformations. The final expression for this case can easily be obtained if c_s is replaced by c_d in Eq. (5.87).

An additional difficulty in the two-dimensional elastodynamic boundary element formulation is discussed in [20] and refers to the singularities that appear when $r \to 0$ and

$$c_s (t - \tau) \neq 0$$
$$c_d(t - \tau) \neq 0. \qquad (5.90)$$

These singularities, however, are only apparent ones and disappears if contributions from similar terms referring to equivoluminal and and dilational waves are calculated together in expression (5.79). The type of manipulations required will now be discussed by considering the integrals that involve B_{ik} in expression (5.79).

When considered alone $B_{ik} L_2 N_2$ and $B_{ik} L_1 N_1$ behave like $1/r^3$ when $r \to 0$. However these singularities can easily be eliminated from the integral equation if it is realized that

$$B_{ik} \left[- L_2 \, N_2 + \frac{c_s}{c_d} \, L_1 \, N_1 \right] = - B_{ik} \, r^4 \, \frac{[(c_d^4 - c_s^4) \, (t')^2 - (c_d^2 - c_s^2) \, r^2] \, L_1 \, L_2}{c_d(c_d \, N_2 \, L_1^{-l} + c_s \, N_1 \, L_2^{-l})}. \qquad (5.91)$$

Therefore, the only singularities present in the numerical analysis are those that occur when r and t' go to zero simultaneously.

5.9 Numerical Implementation for Two Dimensions

A time-stepping scheme to solve Eq. (5.79) will be discussed in this section. The procedure employed for two-dimensional transient elastodynamics is similar to that discussed in Chap. 4 of this book concerning the scalar wave equation.

After the boundary unknowns $u_i(S, t)$ and $p_i(S, t)$ have been obtained, internal displacements $u_i(s, t)$ can be calculated by applying the integral equation that results from Eq. (5.79) when S is replaced by s and $c_{ik}(s)$ is made to equal to δ_{ij}. In elasticity problems it is important to compute stresses as well. The scheme implemented in this section to calculate internal stresses is similar to the simplest one used in finite elements. Triangular cells are employed and stresses at their centroids are obtained by carrying out derivatives of displacements, which are linearly interpolated inside each cell as a function of the displacements at the cell nodes. Following this procedure one avoids performing analytical derivatives of the integral equation for internal displacements.

Interpolation functions are used to approximate u_k and p_k in Eq. (5.79). Analytical time integration can then be carried out, resulting in expressions which are considerably long and for this reason will not be presented here. A certain degree of care must be taken when integrating analytically with respect to time. If conveniently manipulated, the final expressions obtained will have no singularity at the fronts of the equivoluminal and dilatational waves represented by the Green's function. Convenient operations like those described by expression (5.91) must be also be carried out in order to remove apparent singularities that occur when $r \to 0$ and $t' \neq 0$. Consequently the only singularities which remain occur on the first time step, when $r \to 0$, and are of the same type as those for two-dimensional elastostatics, i.e., the integrands behave like $1/r$ and $\ln r$ on the boundary integrals involving u_k and p_k respectively.

The implementation of a numerical scheme to solve Eq. (5.79) requires the consideration of a set of discrete points $Q_j, j = 1, \ldots, J$, on the Γ boundary and a set of values of time t_n, $n = 1, \ldots, N$. $u_k(Q, t)$, $v_k(Q, t)$ and $p_k(Q, t)$ can be approximated as follows

$$u_k(Q, t) = \sum_{j=1}^{J} \sum_{m=1}^{N} \phi^m(t) \, \eta_j(Q) \, \bar{u}_{kj}^m$$

$$v_k(Q, t) = \sum_{j=1}^{J} \sum_{m=1}^{N} \frac{d\phi^m(t)}{dt} \, \eta_j(Q) \, \bar{u}_{kj}^m \qquad (5.92)$$

$$p_k(Q, t) = \sum_{j=1}^{J} \sum_{m=1}^{N} \theta^m(t) \, v_j(Q) \, \bar{p}_{kj}^m$$

where m and j refer to time and space respectively, $k = 1, 2$ relates to the x_k-direction and

$$\bar{u}_{kj}^n = u_k(Q_j, t_m)$$
$$\bar{p}_{kj}^m = p_k(Q_j, t_m). \qquad (5.93)$$

When Eq. (5.79) is written for every node l and also for all values of time t_n and u_k, v_k and p_k are replaced by their approximations as given by expression (5.92), the following system of algebraic equations is then obtained

$$\sum_{k=1}^{2} c_{ik}(S_l) \, \bar{u}_{kl}^n + \frac{1}{2\pi \varrho \, c_s} \sum_{k=1}^{2} \sum_{m=1}^{N} \sum_{j=1}^{J} \bar{H}_{iljk}^{nm} \, \bar{u}_{kj}^m = \frac{1}{2\pi \varrho \, c_s} \sum_{k=1}^{2} \sum_{m=1}^{N} \sum_{j=1}^{J} \bar{G}_{iljk}^{nm} \, \bar{p}_{kj}^m \qquad (5.94)$$

where

$$\bar{H}_{iljk}^{nm} = - \int_0^{t_n} \int_\Gamma \left(\left[(A_{ik} \, \bar{L}_2^3 \, \bar{M}_2 - B_{ik} \, \bar{L}_2 \, \bar{N}_2 + \bar{D}_{ik} \, \bar{L}_2^3 \, \bar{O}_2) \right. \right.$$

$$- \frac{c_s}{c_d} (- B_{ik} \, \bar{L}_1 \, \bar{N}_1 + D_{ik} \, \bar{L}_1^3 \, \bar{O}_1) \bigg] \phi^m(\tau) \, \eta_j(Q)$$

$$\qquad (5.95)$$

$$+ \left[\frac{1}{c_s} (A_{ik} \, \bar{L}_2 + D_{ik} \, \bar{L}_2 \, \bar{N}_2) - \frac{c_s}{c_d^2} D_{ik} \, \bar{L}_1 \, \bar{N}_1 \right] \frac{d\phi^m(\tau)}{d\tau} \, \eta_j(Q) \bigg) d\Gamma \, d\tau,$$

$$\bar{G}^{nm}_{iljk} = \int_0^{t_n} \int_\Gamma \left((\delta_{ik}\, \bar{L}_2 + F_{ik}\, \bar{L}_2^{-1} + J_{ik}\, \bar{L}_2\, \bar{N}_2) \right.$$

$$\left. - \frac{c_s}{c_d}\, (F_{ik}\, \bar{L}_1^{-1} + J_{ik}\, \bar{L}_1\, \bar{N}_1)\, \theta^m(\tau)\, v_j(Q) \right) d\Gamma\, d\tau \qquad (5.96)$$

and

$$\begin{aligned}
\bar{L}_1^\alpha &= L_1^\alpha(Q, t_n; S_l, \tau)\, H\,[c_d\, t' - r]\\
\bar{M}_1^\alpha &= M_1^\alpha(Q, t_n; S_l, \tau)\, H\,[c_d\, t' - r]\\
\bar{N}_1^\alpha &= N_1^\alpha(Q, t_n; S_l, \tau)\, H\,[c_d\, t' - r]\\
\bar{O}_1^\alpha &= O_1^\alpha(Q, t_n; S_l, \tau)\, H\,[c_d\, t' - r].
\end{aligned} \qquad (5.97)$$

$\bar{L}_2^\alpha$, $\bar{M}_2^\alpha$, $\bar{N}_2^\alpha$ and $\bar{O}_2^\alpha$ can be obtained from $\bar{L}_1^\alpha$, $\bar{M}_1^\alpha$, $\bar{N}_1^\alpha$ and $\bar{O}_1^\alpha$ respectively, replacing c_d by c_s in expression (5.97). It should be realised that α in expression (5.97) is an exponent, not an index.

Only constant time steps t_m will be considered in the two-dimensional transient elastodynamic numerical analysis.

The integrations over the Γ boundary are carried out numerically, using Gauss quadrature formulae for all time steps, but the first. When $n = m = 1$ and when it is necessary to integrate over the element in which the source point is ($j = l$), the integrand $\bar{G}^{ll}_{iljk}$ has a singularity of the type $\ln r$ when $r \to 0$. In this case it is advisable to carry out analytical integrations via the procedure outlined in appendix F of [9]. When $j = l$ the integrand of $\bar{H}^{ll}_{iljk}$ behaves like $1/r$ when $r \to 0$. This singularity is of the same type as the one which occurs when studying elastostatics. As constant elements were used the principal value of integrals that appear when computing $\bar{H}^{ll}_{iljk}$ ($j = l$) are equal to zero. However this is not the case when higher order elements are used to approximate displacements. In this situation, principal values that are not zero can be calculated analytically.

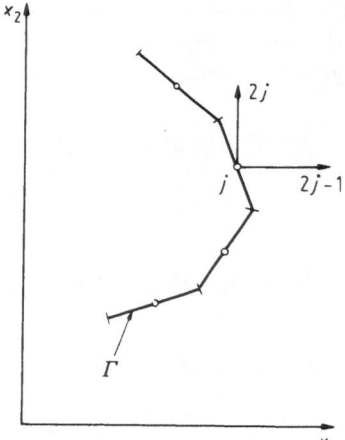

Fig. 5.7. Global numeration

It is now convenient to initial each node j, with numbers $2j - 1$ and $2j$ referring, respectively to directions 1 and 2 of that node, as shown in Fig. 5.7.

Consequently, the following relationships can be written

$$\bar{u}_{kj}^m = u_{2j+k-2}^{nm}$$

$$\bar{p}_{kj}^m = p_{2j+k-2}^{nm}$$

$$\bar{H}_{iljk}^{nm} = H_{(2l+i-2)(2j+k-2)}^{nm}$$ (5.98)

$$\bar{G}_{iljk}^{nm} = G_{(2l+i-2)(2j+k-2)}^{nm} \ .$$

Therefore, when constant elements are used,

$$c_{ik}(S_l)\, \bar{u}_{kl}^n = \tfrac{1}{2}\, \delta_{ik}\, \bar{u}_{kl}^n = \tfrac{1}{2}\, \bar{u}_{il}^n = \tfrac{1}{2}\, u_{2l+i-2}^n \ .$$ (5.99)

Taking full account of expressions (5.98) and (5.99), Eq. (5.94) can be written as

$$\tfrac{1}{2}\, u_i^n + \frac{1}{2\pi\varrho c_s}\, \sum_{j=1}^{2J} H_{ij}^{nn}\, u_j^n$$

$$= \frac{1}{2\pi\varrho c_s}\left(\sum_{j=1}^{2J} G_{ij}^{nn}\, p_j^n - \sum_{m=1}^{n-1} \sum_{j=1}^{2J} H_{ij}^{nm}\, u_j^m + \sum_{m=1}^{n-1} \sum_{j=1}^{2J} G_{ij}^{nm}\, p_j^m \right).$$ (5.100)

Equation (5.100) can be cast into

$$\mathbf{H}\,\mathbf{u} = \mathbf{G}\,\mathbf{p} + \mathbf{B}$$ (5.101)

where $\mathbf{H}$ and $\mathbf{G}$ are square matrices of order $(2J \times 2J)$ and $\mathbf{u}$, $\mathbf{p}$ and $\mathbf{B}$ are vectors.

When boundary conditions at a time t_n are considered and Eq. (5.101) is conveniently reordered Eq. (5.101) becomes

$$\mathbf{A}\,\mathbf{y} = \mathbf{c}$$ (5.102)

where, the vector $\mathbf{y}$ is formed by unknowns u_j and p_j at boundary nodes.

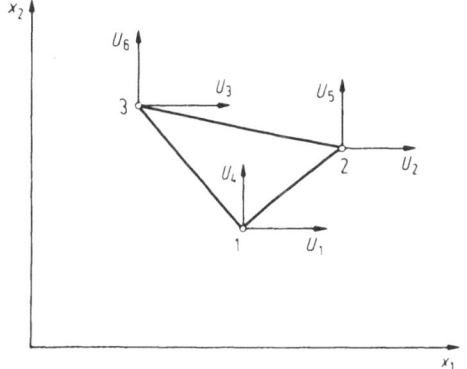

Fig. 5.8.   Triangular cell used to calculate stresses

After Eq. (5.102) has been solved displacements at internal points can be computed using the boundary equation for such points.

In order to use expression (5.41) to calculate internal stresses it is first necessary to calculate the derivatives of the displacement components with regard to the rectangular coordinates x_j. In this chapter this is accomplished numerically using triangular cells. Linear interpolation functions are used to approximate components of displacements u_k ($k = 1, 2$) inside each cell, i.e.,

$$u_1 = \mu_1 \, U_1 + \mu_2 \, U_2 + \mu_3 \, U_3$$
$$u_2 = \mu_1 \, U_4 + \mu_2 \, U_5 + \mu_3 \, U_6$$

(5.103)

where μ_α are triangular coordinates [21] and U_j ($j = 1, 6$) are the components of the displacements at the cell nodes as shown in Fig. 5.8.

When expressions (5.41) and (5.103) are used the following equation is obtained

$$\boldsymbol{\sigma} = \mathbf{D} \, \boldsymbol{\mu}' \, \mathbf{U}$$

(5.104)

where

$$\boldsymbol{\sigma} = \begin{bmatrix} \sigma_{11} \\ \sigma_{12} \\ \sigma_{22} \end{bmatrix}$$

$$\mathbf{D} = \begin{bmatrix} \lambda + 2G & 0 & \lambda \\ 0 & 2G & 0 \\ \lambda & 0 & \lambda + 2G \end{bmatrix}$$

(5.105)

$$\boldsymbol{\mu}' = \begin{bmatrix} \mu_{1,1} & \mu_{2,1} & \mu_{3,1} & 0 & 0 & 0 \\ \mu_{1,2}/2 & \mu_{2,2}/2 & \mu_{3,2}/2 & \mu_{1,1}/2 & \mu_{2,1}/2 & \mu_{3,1}/2 \\ 0 & 0 & 0 & \mu_{1,2} & \mu_{2,2} & \mu_{3,2} \end{bmatrix}$$

and

$$\mu_{i,j} = \frac{\partial \mu_i}{\partial x_j} \, .$$

(5.106)

5.10 Examples — Two-Dimensional Elastodynamics

In this section the numerical procedure previously discussed in Sect. 5.9 is illustrated by a series of examples comparing boundary elements with other numerical methods.

In all of the problems examined, the boundary integrations shown in Eqs. (5.95) and (5.96) were performed using a maximum of twenty Gauss points.

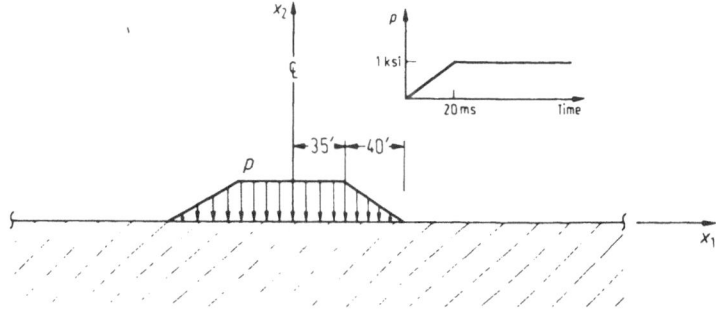

Fig. 5.9. Load for the half-plane under continuous prescribed stress distribution

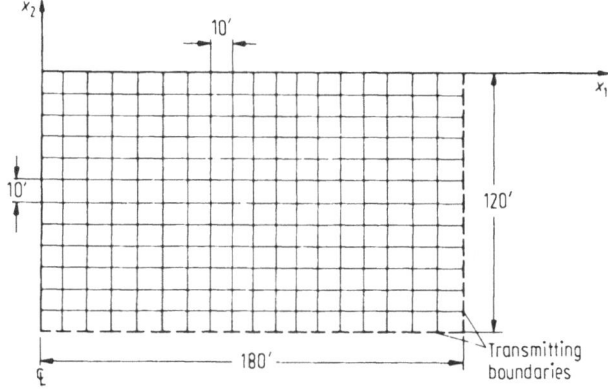

Fig. 5.10. Finite-difference mesh for the half-plane under continuous prescribed stress distribution

Further on in this section reference will be made to the parameter β given by

$$\beta = \frac{c_d \Delta t}{l_j}. \tag{5.107}$$

Example 1. *Half-plane under continuous prescribed stress distribution.* In this application the time-stepping technique discussed in this work is compared with the finite-difference model implemented by Tseng et al. [22]. In that report a transmitting boundary was developed and used together with the generalized lumped parameter model presented in [23–25].

The problem to be analysed is depicted in Fig. 5.9. The half-plane is initially at rest and its surface is disturbed by a vertical traction which is continuous in both time and space.

The following numerical values were adopted for the constants of the problem

$$E = 200 \text{ ksi}, \qquad v = 0.15,$$
$$c_d = 3.288 \times 10^4 \text{ ips}, \quad c_s = 2.112 \times 10^4 \text{ ips}.$$

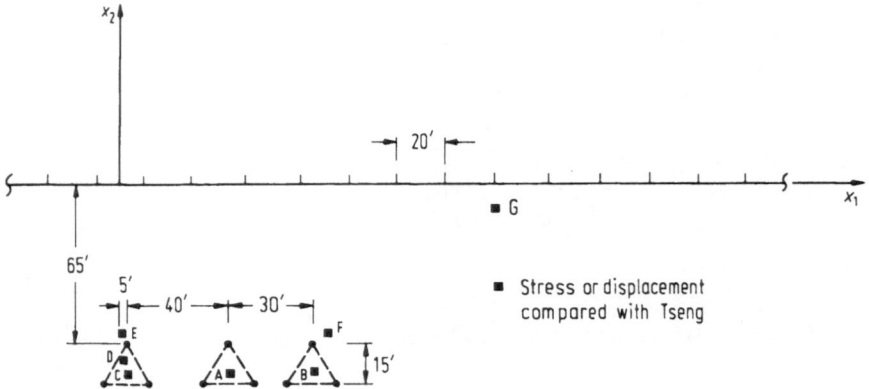

Fig. 5.11. Boundary element discretization for the half-plane under continuous prescribed stress distribution

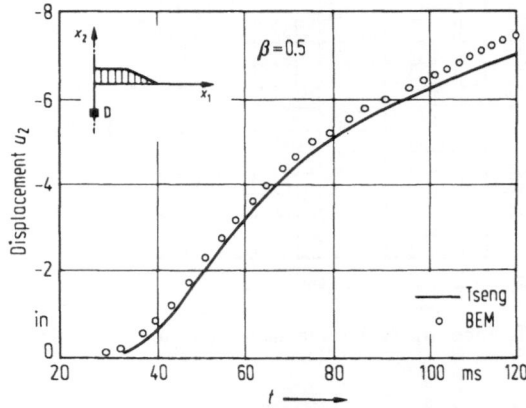

Fig. 5.12. Half-plane under continuous prescribed stress distribution. Displacement u_2 at the internal point $D(0', 70')$

The criterion given by Tseng [22] to choose the finite difference mesh requires that

$$t_r > 2 \frac{\Delta x}{c_d} \qquad (5.108)$$

where t_r is rise or decay time of the applied pressure and Δx gives the mesh refinement. When $t_r = 20$ msec, $\Delta x \leq 27.4$ ft is obtained. Tseng chose $\Delta x = 10$ ft and the discretization as depicted in Fig. 5.10 where the position selected for the cylindrical wave transmitting boundaries can also be seen.

The boundary element discretization and cells used in the analysis are shown in Fig. 5.11.

According to [22] the time increment Δt, used in this finite-difference analysis, must obey the following Eq. (5.109) and $\Delta t = 1$ msec was adopted.

$$\Delta t \leq 0.433 \frac{\Delta x}{c_d}. \qquad (5.109)$$

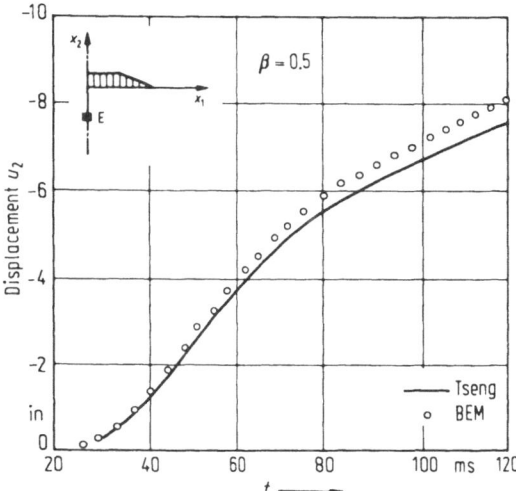

Fig. 5.13. Half-plane under continuous prescribed stress distribution. Displacement u_2 at the internal point $E(0', 60')$

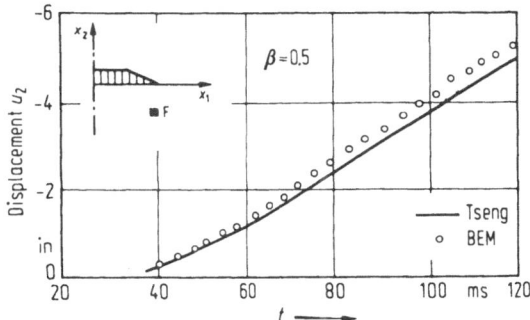

Fig. 5.14. Half-plane under continuous prescribed stress distribution. Displacement u_2 at the internal point $F(80', 60')$

For the boundary element analysis, β was taken to be equal to $\frac{1}{2}$, which gives

$$\Delta t = 3.65 \text{ msec}.$$

The time history of the vertical displacements plotted in Figs. 5.12–5.14 shows an acceptable agreement for the time interval considered.

In his research Tseng carried out another analysis using a pair of transmitting boundaries which enclosed a smaller rectangular region whose side lengths were equal to 90 ft and 150 ft. The two finite-difference analyses showed that the larger the region enclosed by the transmitting boundaries, the closer finite difference and boundary elements results were. Therefore it is justified to suppose that the major proportion of the difference between the displacements obtained with the two numerical methods under consideration is caused by errors generated at the transmitting boundaries.

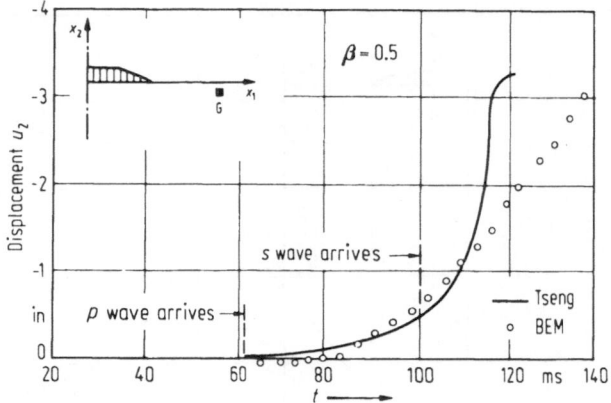

Fig. 5.15. Half-plane under continuous prescribed stress distribution. Displacement u_2 at the internal point $G(150', 10')$

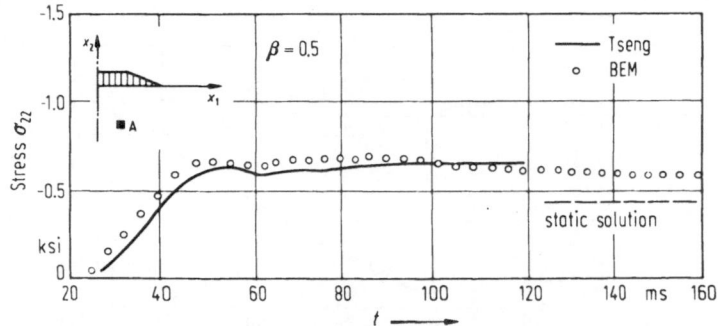

Fig. 5.16. Half-plane under continuous prescribed stress distribution. Stress σ_{22} at the internal point $A(45', 75')$

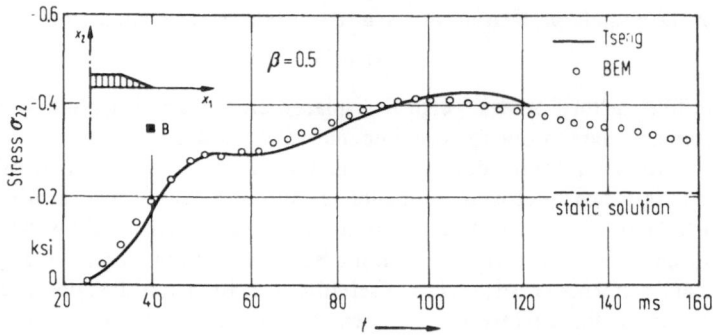

Fig. 5.17. Half-plane under continuous prescribed stress distribution. Stress σ_{22} at the internal point $B(75', 75')$

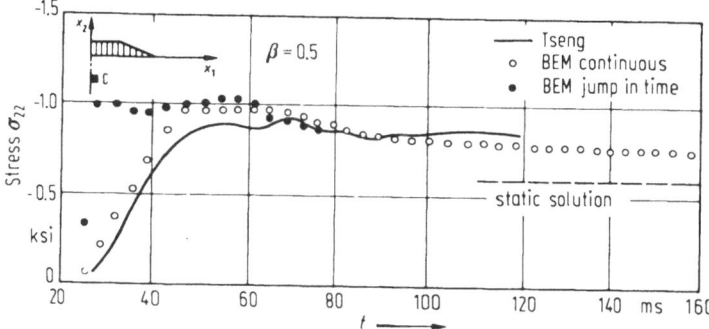

Fig. 5.18. Half-plane under continuous prescribed stress distribution. Stress σ_{22} at the internal point $C(5', 75')$

Fig. 5.19. Load applied as a Heaviside function in time for the half-plane under continuous prescribed stress distribution

Tseng also presented the time history of the vertical displacements for the point $G(150', 10')$ obtained with the $90' \times 150'$ rectangular region. As G is located exactly on the transmitting boundary it can be expected that finite-difference displacements at this point will have a low accuracy. The point G is also a critical one in the boundary element analysis because it is too close to the boundary of the half-plane. Results obtained with the two methods are shown in Fig. 5.15. As it was expected the agreement is not as close as recorded previously.

Figures 5.16–5.18 describe the time history of stresses at points $A(45', 75')$, $B(75', 75')$ and $C(5', 75')$.

When the load is applied as a step function in time, finite-differences can not be used because of the restrictions imposed by Eq. (5.108). A possbile way of overcoming this difficulty is by replacing the jump by a slope. In order to check the errors introduced by such a procedure the problem displaced in Fig. 5.9 was re-investigated using boundary elements, but this time the load was abruptly applied at $t = 0$ (see Fig. 5.19). The time history of stress plotted in Fig. 5.18 shows that a complete agreement occurs with the previous analysis during late times, but during early times the results are different.

Finally, for this example it can be concluded that

a) The solutions using both the finite difference and boundary element methods are in good agreement.
b) The time increment required by boundary elements was bigger than that necessary for finite differences.

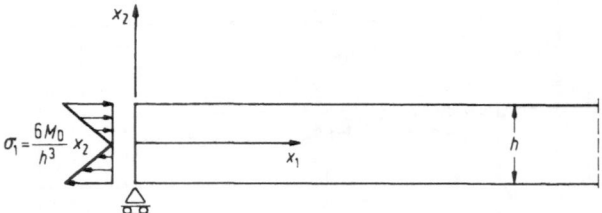

Fig. 5.20.   Geometry and loading of the semi-infinite beam

Fig. 5.21.   Boundary element mesh for the semi-infinite beam

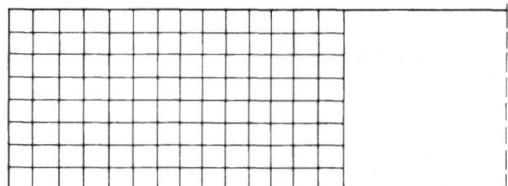

Fig. 5.22.   Finite element mesh for the semi-infinite beam

c) When the time variation of the load includes jumps, boundary elements are more suitable than finite differences.

Example 2. *Semi-infinite beam.* This application consists of a semi-infinite beam simply supported along its edge (see Fig. 5.20) and subjected to a suddenly applied bending moment

$$M_0 = M H (t - 0). \tag{5.110}$$

The Poisson ratio for this plane stress problem was taken to be $\frac{1}{3}$.

The boundary element mesh consisted of thirty six equal elements as depicted in Fig. 5.21 and β was taken as equal to $\frac{1}{2}$.

A finite element analysis of this problem was carried out by Fu [26] who used the mesh depicted in Fig. 5.22 in his numerical solution. Transverse displacements along the axes of the beam obtained with both numerical techniques are shown in Fig. 5.23. Within this same figure results obtained from the beam theory by Boley [27] are also plotted. The displacements depicted in Fig. 5.23 refer to

$$t = 5r/c_0 \tag{5.111}$$

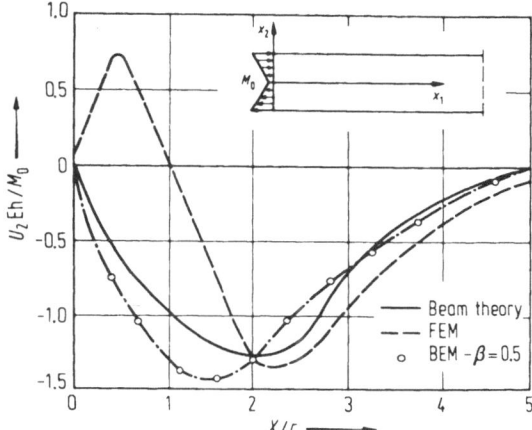

Fig. 5.23.   Transverse displacement along the semi-infinite beam at the time $t = 5r/c_0$

and Chao where r is the radius of gyration of the beam cross section and c_0 is the one-dimensional wave propagation speed [27].

None of the two-dimensional numerical analyses agreed completely with the analytical solution obtained from the beam theory. However the boundary element results show that the two-dimensional solution appears to be closer to the beam theory than initially indicated by Fu's finite element analysis.

5.11 Conclusions

In the present chapter the boundary element method has been applied to solve transient elastodynamics, with particular reference to two-dimensional solutions. The technique is based on a time dependent fundamental solution and integrating on time always with reference to the original initial state. In this manner it is possible to avoid carrying out domain integrations and all the boundary element expressions are refered to the boundary. The technique is specially useful for problems with infinite or semi-infinite domains.

Several numerical examples are presented to demonstrate the remarkable accuracy of the results obtained using interpolation functions in time and space. Care should be taken in the analysis on the choice of time intervals and boundary discretization in order to avoid contradicting the causality property too far, i.e. in each time step waves should not be allowed to travel between nodes far from each other.

Acknowledgement. This work has been partially supported by NATO grant 282/84, Double Jump Program.

References

1 Cruse, T.A., The Transient Problem in Classical Ealastodynamics Solved by Integral Equations. Ph.D. Thesis, University of Washington, 1967
2 Cruse, T.A. and Rizzo, F.J., A Direct Formulation and Numerical Solution of the General Transient Elastodynamic Problem, I. J. Math. Anal. Appl. **22**, 244–259, 1968
3 Cruse, T.A., A Direct Formulation and Numerical Solution of the General Transient Elastodynamic Problem, II. J. Math. Anal. Appl. **22**, 341–355, 1968
4 Papoulis, A., A New Method of Inversion of the Laplace Transform. Q. Appl. Math. **14**(4), 405–414, 1957
5 Manolis, G.D., Dynamic Response of Underground Structures. Ph.D. Thesis, University of Minnesota, 1980
6 Manolis, G.D. and Beskos, D.E., Dynamic Stress Concentration Studies by Boundary Integrals and Laplace Transform. Int. J. Num. Meth. Engng. **17**, 573–599, 1981
7 Niwa, Y., Fukui, T., Kato, S., and Fujiki, K., An application of the Integral Equation Method to Two-Dimensional Elastodynamics. Theor. Appl. Mech., University of Tokyo Press **28**, 281–290, 1980
8 Mansur, W.J. and Brebbia, C.A., Transient Elastodynamics using a Time-stepping Technique. Proceedings of the Fifth International Seminar on Boundary Element Methods in Engineering (C.A. Brebbia, ed.). Springer-Verlag, Berlin, Heidelberg, New York, 1983
9 Mansur, W.J., A Time-stepping Technique to Solve Wave Propagation Problems using the Boundary Element Method. Ph.D. Thesis, Southampton University, 1983
10 Brebbia, C.A. and Nardini, D., Dynamic Analysis in Solid Mechanics by an Alternative Boundary Element Procedure. Soil Dynamics and Earthquake Engineering Journal **2**(4), 1983
11 Fung, Y.C., *Foundations of Solid Mechanics*. Prentice-Hall, Inc., New Jersey, 1965
12 Eringen, A.C. and Suhubi, E.S., *Elastodynamics*. Vols. I and II, Academic Press, New York, San Francisco and London, 1975
13 Pao, Y.H. and Mow, C.C., *Diffraction of Elastic Waves and Dynamic Stress Concentrations*. Crane Russak, New York, 1973
14 Achenbach, J.D., *Wave Propagation in Elastic Solids*. North-Holland Publishing Company, Amsterdam and London, 1973
15 Miklowitz, J., *Elastic Waves and Waveguides*. North-Holland Publishing Company, Amsterdam, New York and Oxford, 1980
16 Venturini, W.S., Application of the Boundary Element formulation to Solve Geomechanical Problems. Ph.D. Thesis, University of Southampton, 1982
17 Mansur, W.J. and Brebbia, C.A., Formulation of the Boundary Element Method for Transient Problems Governed by the Scalar Wave Equation. Appl. Math. Modelling **6**, 307–311, 1982
18 Cole, D.M., Kosloff, D.D., and Minster, J.B., A Numerical Boundary Integral Equation Method for Elastodynamics, I. Bull. Seis. Soc. Amer. **68**(5), 1331–1357, 1978
19 Chapter 4 of this book
20 Manolis, G.D., A Comparative Study on Three Boundary Element Method Approaches to Problems in Elastodynamics. Int. J. Num. Meth. Engng. **19**, 73–91, 1983
21 Brebbia, C.A. and Connor, J.J., *Fundamentals of Finite Elements Techniques for Structural Engineers*. Butterworths, London, 1973
22 Tseng, M.N. and Robinson, A.R., A Transmitting Boundary for Finite-Difference Analysis of Wave Propagation in Solids. Project No. NR 064-183, University of Illinois, Urbana, Illinois, 1975
23 Ang, A.H.-S. and Newmark, N.M., Development of a Transmitting Boundary for Numerical Wave Motion Calculations. Report to Defense Atomic Support Agency, DASA 2631, 1971
24 Galloway, J.C. and Ang, A.H.-S., A Generalized Lumped-Parameter Model for Plane Problem of Solid Media. Civil Engineering Studies, Structural Research Series, No. 341, University of Illinois, Urbana, Illinois, 1968

25 Sameh, A.H.M. and Ang, A.H.-S., Numerical Analysis of Axisymmetric Wave Propaga-
    tion in Elastic-Plastic Layered Media. Civil Engineering Studies, Structural Research
    Series No. 335, University of Illionois, Urbana, Illinois, 1968
26 Fu, C.C., A Method for the Numerical Integration of the Equations of Motion Arising
    from a Finite-Element Analysis. Trans. ASME, Appl. Mech., series E **37**, 599−605, 1970
27 Boley, B.A. and Chao, C.C., An Approximate Analysis of Timoshenko Beams Under Dy-
    namic Loads. Trans. ASME, J. Appl. Mech., series E **25**, 31−36, 1958

Chapter 6

Propagation of Surface Waves

by L. C. Wrobel, S. H. Sphaier, and P. T. T. Esperança

6.1 Introduction

The study of propagation of surface waves and their interaction with fixed or floating bodies is of great interest to structural and ocean engineers. The problem is generally defined over a fluid region of infinite extent. This, together with the fact that the boundary conditions on the free surface of the fluid are nonlinear and the location of the free surface is time-dependent makes its solution rather involved.

Generally, linear wave theory is employed, with assumption of harmonic waves. This means that the boundary conditions are linearized and applied at the mean free surface of the fluid, since the wave elevations are small. Results of interest in this kind of analysis are usually forces, moments and hydrodynamic coefficients (added mass and damping) on the body surface. Thus, a numerical solution using the Boundary Element Method becomes very well suited for the problem. In fact, this is one of the main fields of application of the BEM, and one where its computational efficiency has long been recognized [1–3].

More recently, extensions for inclusion of second-order terms [4] or complete nonlinear analysis [5] have been carried out. In the latter case, the free surface becomes a moving boundary, the position of which changes in time. Again, the facility of the BEM for dealing with moving boundaries without reordering the whole arrangement of nodal points is appealing, and some efficient formulations have appeared in the literature [6–8].

This chapter starts by introducing the basic aspects of the general linear three-dimensional problem and its solution by the BEM. Since this formulation is generally costly to run in the computer due to its complexity, simpler particular approaches applicable to more restricted geometric configurations are discussed in Sects. 6.4–6.6. Section 6.7 presents a BEM solution for the propagation of waves which are no longer harmonic. Finally, the complete nonlinear problem is presented and the proposed schemes of solution discussed in detail.

6.2 Three-Dimensional Formulation

Consider a train of regular surface waves propagating in water of constant depth past one or more obstacles of arbitrary geometry (Fig. 6.1). Neglecting the viscous

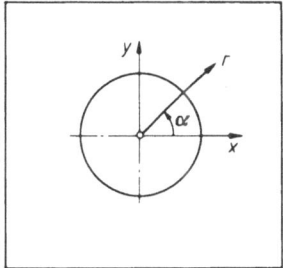

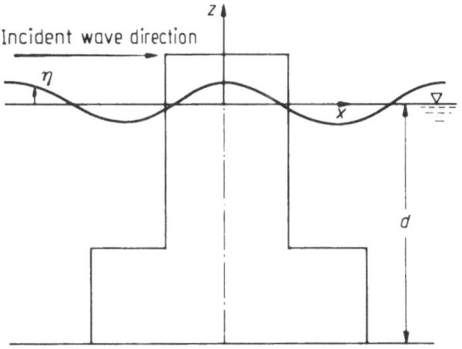

Fig. 6.1. Definition sketch of the problem

forces and assuming the fluid to be incompressible and the flow irrotational, the motion may be described by a velocity potential Φ which satisfies Laplace's equation

$$\frac{\partial^2 \Phi}{\partial x^2} + \frac{\partial^2 \Phi}{\partial y^2} + \frac{\partial^2 \Phi}{\partial z^2} = 0 \tag{6.1}$$

within the fluid region. In what follows, we shall consider the scalar function Φ to be harmonic, i.e.

$$\Phi\,(x, y, z, t) = \Phi\,(x, y, z)\,e^{-i\omega t} \tag{6.2}$$

where ω is the wave angular frequency.

The boundary conditions of the problem are as follows [9]:

$$\frac{\partial \Phi}{\partial z} = 0 \quad \text{at } z = -d \tag{6.3}$$

$$\frac{\partial \Phi}{\partial n} = 0 \quad \text{at the body surface} \tag{6.4}$$

$$\frac{\partial^2 \Phi}{\partial t^2} + g\,\frac{\partial \Phi}{\partial z} = 0 \quad \text{at } z = 0 \tag{6.5}$$

where n is the outward normal to the body surface, d is the water depth and g is the gravitational constant. Expressions (6.3) and (6.4) correspond to the kinematic boundary conditions at the ocean floor and at the immersed surfaces of the body, respectively. Expression (6.5) derive from the linearized kinematic and dynamic free-surface conditions and is applied at the mean free surface, assuming that the wave amplitudes are small with respect to the wave length. Once the potential Φ is determined, we can compute the free surface elevation above the still water level by the expression

$$\eta = -\frac{1}{g}\left(\frac{\partial \Phi}{\partial t}\right)_{z=0}. \tag{6.6}$$

Since the problem is linear, the potential Φ can be represented as the sum of an incident and a diffracted wave potentials:

$$\Phi = \Phi_I + \Phi_D. \tag{6.7}$$

Furthermore, to ensure that Φ_D has an outgoing behaviour at infinity, guaranteeing the unicity of the solution, we impose the Sommerfeld condition

$$\lim_{r \to \infty} r^{1/2}\left(\frac{\partial \Phi_D}{\partial r} - i\,k\,\Phi_D\right) = 0 \tag{6.8}$$

in which r is the radial ordinate and k is the wave number.

The expression of the velocity potential in the form of Eq. (6.7), involving a separation into indisturbed incident wave and scattered wave components, constitutes the basis of diffraction theory. The incident wave potential satisfies Eqs. (6.1) – (6.3), (6.5) and a spatial periodicity; it is specified in complex form as

$$\Phi_I(x, y, z) = -\frac{i\,g\,H}{2\,\omega}\frac{\cosh\left[k\,(z + d)\right]}{\cosh(k\,d)}\,e^{ik(x\cos\alpha + y\sin\alpha)} \tag{6.9}$$

where H is the wave height and α its angle of incidence. The wave number satisfies the dispersion relation

$$k \tanh(k\,d) = v = \frac{\omega^2}{g}. \tag{6.10}$$

Because all the equations of the problem are linear, the potential Φ_D also satisfies Eqs. (6.1)–(6.3) and (6.5), as well as the radiation condition (6.8). The body surface boundary condition (6.4) provides a link between Φ_D and Φ_I in the form

$$\frac{\partial \Phi_D}{\partial n} = -\frac{\partial \Phi_I}{\partial n}. \tag{6.11}$$

Equations (6.1), (6.3) and (6.5) applied to Φ_D, together with (6.8) and (6.11), define the problem in terms of Φ_D.

Solution of the above-defined problem through the Boundary Element Method can be obtained by using both the direct and indirect formulations. From the application of Green's third identity, the following boundary integral equation is obtained for the direct formulation [1]:

$$c\,(\xi, \eta, \zeta)\,\Phi_D(\xi, \eta, \zeta) \tag{6.12}$$

$$= -\int_{\Gamma}\left[G(x, y, z; \xi, \eta, \zeta)\,\frac{\partial \Phi_I}{\partial n}(x, y, z) + \Phi_D(x, y, z)\,\frac{\partial G(x, y, z; \xi, \eta, \zeta)}{\partial n(x, y, z)}\right]d\Gamma(x, y, z)$$

where the kinematic boundary condition (6.11) has been taken into account. The source and field points are defined by their coordinates (ξ, η, ζ) and (x, y, z), respectively. The value of the coefficient c is a function of the solid angle of the boundary at the source point. For a point on a smooth boundary, we have that $c = 2\pi$.

Alternatively, we can express the value of the velocity potential at a point in terms of a single-layer potential with unknown density σ, in the form [1]

$$\Phi_D(x, y, z) = \frac{1}{4\pi} \int_\Gamma \sigma(\xi, \eta, \zeta)\, G(x, y, z; \xi, \eta, \zeta)\, d\Gamma(\xi, \eta, \zeta). \qquad (6.13)$$

This expression is the starting one for the indirect BEM formulation.

In order to determine the unknown source density distribution, we take the derivative of Eq. (6.13) in the direction of the outward normal to Γ as the field point is taken to the boundary and apply the kinematic condition on the body surface, yielding the following boundary integral equation:

$$(6.14)$$

$$-\frac{1}{2}\sigma(x, y, z) + \frac{1}{4\pi}\int_\Gamma \sigma(\xi, \eta, \zeta)\, \frac{\partial G(x, y, z; \xi, \eta, \zeta)}{\partial n(x, y, z)}\, d\Gamma(\xi, \eta, \zeta) = -\frac{\partial \Phi_I}{\partial n}(x, y, z).$$

Once the above equation is solved for the source density distribution, values of Φ_D at a boundary point or any point in the fluid region can be calculated by a direct quadrature via Eq. (6.13).

There are two different schemes of solution of the previous integral equations within the BEM. The first one employs the so-called free space Green's function, $G = 1/R$, where R is the distance between the source and field points. Since no boundary condition is directly taken into account, this requires discretization of all boundaries of the fluid region (seabed, body surface, free surface and radiation boundaries). The main advantages of the above scheme are the simplicity of the fundamental solution and the possibility of incorporating a variable depth in the vicinity of the body. On the other hand, a large number of boundary elements is generally needed, what makes the data input cumbersome and leads to a large system of equations to be solved. This difficulty can, however, be partly alleviated by truncating the lateral boundaries at a small distance and taking the influence of the far field into account through a coupling with analytical series expansions, as discussed in Sect. 6.6.

The second and more common scheme of solution uses a particular fundamental solution, $G = 1/R + \bar{G}$, which directly satisfies the seabed, free surface and radiation boundary conditions. Thus, only the body surface needs be discretized. The function $\bar{G}$ is a regular one satisfying $\nabla^2\bar{G} = 0$ throughout the fluid region. Advantages of this scheme include a much simpler data input and a much smaller number of unknowns. The main disadvantage is the fact that the fundamental solution has a very complicated form and is valid only for constant depth.

An interesting discussion about the computational efficiency of several methods of solution of wave propagation problems can be found in [10].

6.2.1 Particular Fundamental Solutions

(i) Intermediate Waters (k d < 5)

The particular fundamental solution for the present case is obtained through the solution of the following boundary-value problem:

$$\nabla^2 G = - \delta (x, y, z; \xi, \eta, \zeta) \tag{6.15}$$

$$\frac{\partial G}{\partial z} = 0 \quad \text{at } z = -d \tag{6.16}$$

$$- \omega^2 G + g \frac{\partial G}{\partial z} = 0 \quad \text{at } z = 0 \tag{6.17}$$

$$\lim_{r \to \infty} r^{1/2} \left(\frac{\partial G}{\partial r} - i k G \right) = 0 \tag{6.18}$$

in which δ is the Dirac delta function.

As can be seen, function $G (x, y, z; \xi, \eta, \zeta)$ satisfies all boundary conditions apart from the kinematic one on the body surface.

Making use of Fourier transforms [3, 11], the following expression is obtained for G:

$$G = \frac{1}{R} + \frac{1}{R_1} + 2P \cdot V \cdot \int_0^\infty \frac{(\mu + v) \, e^{-\mu d} \cosh [\mu (z + d)] \cosh [\mu (\zeta + d)]}{\mu \sinh (\mu d) - v \cosh (\mu d)} \, J_0 (\mu r) \, d\mu \cdot$$
$$- i \, C_0 \cosh [k (z + d)] \cosh [k (\zeta + d)] \, J_0 (k r) \tag{6.19}$$

where J_0 is the Bessel function of the first kind of zero order and

$$r = [(x - \xi)^2 + (y - \eta)^2]^{1/2}$$
$$R = [r^2 + (z - \zeta)^2]^{1/2}$$
$$R_1 = [r^2 + (z + 2d + \zeta)^2]^{1/2}$$
$$C_0 = \frac{2 \pi (v^2 - k^2)}{(k^2 - v^2) \, d + v} .$$

The above expression may be modified to facilitate its computer implementation. Upon working with the principal value integral to reduce its interval of integration [12, 13] we end up with the following expression which is ready for numerical evaluation:

$$G = \frac{1}{R} + \frac{1}{R_1} + \frac{1}{R_0} + 2 \int_0^{\mu_1} \frac{g (\mu) - g (\mu_0)}{\mu - \mu_0} \, d\mu + 2 g (\mu_0) \ln \left(\frac{\mu_1 - \mu_0}{\mu_0} \right)$$

$$- \int_0^{\mu_1} e^{\mu (z + \zeta)} J_0 (\mu r) \, d\mu - 2 v \, e^{v(z + \zeta)} \gamma J_0 (v r)$$

$$- \frac{2 v}{\pi} e^{v(z + \zeta)} \left\{ \int_0^\pi \ln (\bar{\zeta}) \cos (vr \cos \theta) \, d\theta - \int_0^\pi \bar{\zeta} \sin (vr \cos \theta) \, d\theta \right.$$

$$+ \int\limits_0^\pi \sum_{n=1}^\infty \frac{(-1)^n}{n \, n!} \, s^n \cos{(n \, \bar\zeta)} \cos{(v \, r \cos{\theta})} \, d\theta$$

$$- \int\limits_0^\pi \sum_{n=1}^\infty \frac{(-1)^n}{n \, n!} \, s^n \sin{(n \, \bar\zeta)} \sin{(v \, r \cos{\theta})} \, d\theta \Bigg\}$$

$$- i \, C_0 \cosh{[k \, (z+d)]} \cosh{[k \, (\zeta + d)]} \, J_0 \, (k \, r) \qquad (6.20)$$

in which γ is the Euler constant, $\gamma = 0.57721 \ldots$, [14], and:

$$g \, (\mu) = \frac{\mu - \mu_0}{\mu \tanh{(\mu \, d)} - v} \, f \, (\mu)$$

$$f \, (\mu) = (\mu + v) \, e^{-\mu d} \, \frac{\cosh{[\mu \, (z+d)]} \cosh{[\mu \, (\zeta + d)]}}{\cosh{(\mu \, d)}} \, J_0 \, (\mu \, r)$$

$$g \, (\mu_0) = \lim_{\mu \to \mu_0} g \, (\mu) = \frac{\mu_0}{v + d \, (\mu_0^2 - v^2)} \, f \, (\mu_0)$$

$$R_0 = [r^2 + (z + \zeta)^2]^{1/2}$$

$$s = |\, \mu_1 - v \, | \, [(z + \zeta)^2 + (r \cos{\theta})^2]^{1/2}$$

$$\bar\zeta = \arctan{\left(\frac{r \cos{\theta}}{z + \zeta} \right)} \quad \text{such that} \ |\, \bar\zeta \, | < \pi$$

$$\mu_1 \, d \geqq 4.5$$

$$\mu_1 > \mu_0 = k \, .$$

An alternative series form for G is given by [15]

$$G = i \, C_0 \cosh{[k \, (z + d)]} \cosh{[k \, (\zeta + d)]} \, H_0^{(1)} \, (k \, r)$$

$$+ 4 \sum_{m=1}^\infty C_m \cos{[\mu_m \, (z + d)]} \cos{[\mu_m \, (\zeta + d)]} \, K_0 \, (\mu_m r) \qquad (6.21)$$

where K_0 denotes the modified Bessel function of the second kind of zero order, $H_0^{(1)}$ the Hankel function of the first kind of zero order, and

$$C_m = \frac{\mu_m^2 + v^2}{(\mu_m^2 + v^2) \, d - v}$$

in which μ_m are the real positive roots of the equation

$$\mu_m \tan{(\mu_m d)} + v = 0 \qquad (6.22)$$

taken in ascending order.

(ii) Deep Waters $(k \, d > 5)$

For the case of deep waters, the seabed boundary condition is modified; the fundamental solution can then be written in the following integral form [11]:

$$G = \frac{1}{R} + P \cdot V \cdot \int\limits_0^\infty \frac{\mu + v}{\mu - v} \, e^{\mu (z + \zeta)} \, J_0 \, (\mu \, r) \, d\mu + i \, 2 \, \pi \, v \, e^{v(z + \zeta)} \, J_0 \, (v \, r) \, . \qquad (6.23)$$

Alternative expressions are given by Hogben and Standing [16] and Hearn [17].

(iii) Asymptotic Form for High Frequencies

The free surface boundary condition may be simplified in the case of high frequencies, leading to the following fundamental solution [50]

$$G = \frac{1}{R} + \frac{1}{R_0} + \sum_{n=1}^{\infty} (-1)^n \left(\frac{1}{R_{2n}} - \frac{1}{R_{3n}} + \frac{1}{R_{4n}} - \frac{1}{R_{5n}} \right) \qquad (6.24)$$

in which:

$$R_{2n} = [r^2 + (z - 2nd - \zeta)^2]^{1/2}$$
$$R_{3n} = [r^2 + (z + 2nd + \zeta)^2]^{1/2}$$
$$R_{4n} = [r^2 + (z + 2nd - \zeta)^2]^{1/2}$$
$$R_{5n} = [r^2 + (z - 2nd + \zeta)^2]^{1/2} .$$

(iv) Asymptotic Form for Low Frequencies

In the case of low frequencies, the free surface condition becomes the boundary condition for a rigid horizontal wall

$$\frac{\partial G}{\partial z} = 0 \qquad (6.25)$$

and the asymptotic form of the fundamental solution can be written as [50]

$$G = \frac{1}{R} + \frac{1}{R_0} + \sum_{n=1}^{\infty} \left(\frac{1}{R_{2n}} + \frac{1}{R_{3n}} + \frac{1}{R_{4n}} + \frac{1}{R_{5n}} \right). \qquad (6.26)$$

Expressions (6.24) and (6.26) have advantages over the complete form of the fundamental solution because they are simpler and have no convergence problems.

6.2.2 Numerical and Computational Aspects

As a guideline on the choice of the form the fundamental solution to be employed, integral or series, Garrison and Chow [18] recommend that when $kr > 0.1$ the series form (6.21) converges sufficiently rapid for it to be used. For $kr \leq 0.1$, there appears some convergence problems due to the singular behaviour of the Bessel function K_0 at the origin, and the integral form (6.20) is more convenient to use. Both expressions, however, are quite complex and some numerical aspects have to be considered in detail.

With relation to the series form, it is important to determine the number of terms that have to be taken to keep the truncation error small. The truncation error can be estimated by the asymptotic behaviour of the function K_0 [14]:

$$K_0 (\mu_m r) \sim \left(\frac{\pi}{2 \mu_m r} \right)^{1/2} e^{-\mu_m r} \quad \text{when } \mu_m r \to \infty . \qquad (6.27)$$

Thus, by truncating the series with a number of terms $N > 10\, d/(\pi r)$, we keep the error less than 10^{-5}.

To calculate the roots μ_m of the transcendental Eq. (6.22), Hogben and Standing [16] suggest using the Newton-Raphson scheme. As a starting point, the following

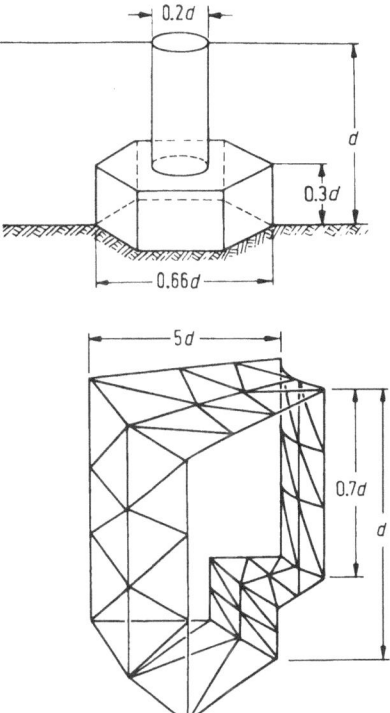

Fig. 6.2. Submerged storage tank and column: geometry and discretization

values can be assumed:

$$\mu_m\, d = (m - \tfrac{1}{2})\,\pi + \frac{1}{v\,d} \quad \text{if } v\,d > 4.8\,m$$

$$\mu_m\, d = m\,\pi - \frac{v\,d}{m\,\pi} \qquad \text{if } v\,d \leq 4.8\,m\,.$$

All integrals in expression (6.20) can be evaluated numerically by using standard schemes, for instance Gauss quadrature. For the first one, however, care should be taken in order not to place an integration point at $\mu = \mu_0$ because the integrand, although non-singular, is not defined at this point. The series expansions that appear in the expression converge quickly; the use of recurrence relations for calculating their terms is recommended for computational efficiency.

The imaginary term of expression (6.20) presents some round-off problems for large values of the water depth d. In these cases, we have that $k \rightarrow v$ [see Eq. (6.10)] and the value of the coefficient C_0, which envolves $(v^2 - k^2)$ in its numerator, becomes very small. Thus, large errors may be introduced particularly in the calculation of the hydrodynamic coefficients which are strongly dependent on the imaginary term, as is the case of the damping coefficients. In order to avoid this problem, Wybro [19] suggests substituting the factor $(v^2 - k^2)$ by $- k^2/\cosh^2(k\,d)$, obtained through the dispersion relation (6.10).

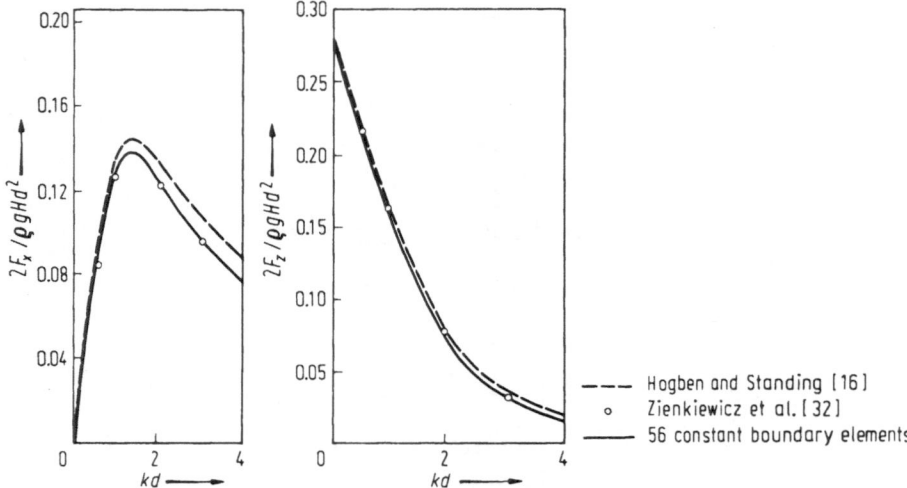

Fig. 6.3. Horizontal and vertical forces on submerged tank with column

Numerical results for several different diffraction problems have been presented in the literature [12, 18, 20, 21], using the particular fundamental solution. A comparison between the results obtained for both BEM schemes and a finite element solution is presented by Au and Brebbia [22] for the case of a submerged tank surmounted by a vertical column (Fig. 6.2). The wave forces on the column, obtained by all three schemes, are displayed and compared in Fig. 6.3.

6.3 Floating Bodies

For the case of floating bodies there appear new wave components due to the oscillating motions of the body, which radiate from the body to infinity. Thus, the linear solution for the velocity potential presents an additional component and is given by

$$\Phi = \Phi_I + \Phi_D + \Phi_R \tag{6.28}$$

in which Φ_I is the incident wave potential, Φ_D is the diffracted wave potential due to the interaction of the incident waves with the fixed body, and Φ_R is the radiated wave potential, due to the body oscillating in still water.

A rigid body possesses six degrees of freedom and its oscillating motions are defined in the form (Fig. 6.4):

$$\eta_1 = \text{surge} ; \quad \eta_2 = \text{sway} ; \quad \eta_3 = \text{heave} ;$$
$$\eta_4 = \text{roll} ; \quad \eta_5 = \text{pitch} ; \quad \eta_6 = \text{yaw} .$$

The classical linear form for the radiated potentials is [23]

$$\Phi_R = \sum_{j=1}^{6} \Phi_j \, \dot{\eta}_j (t) . \tag{6.29}$$

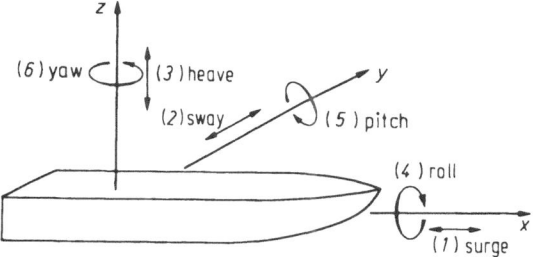

Fig. 6.4. Definition of the six components of motion of a floating body

The potential Φ_R should satisfy the same conditions as the diffracted potential Φ_D apart from the kinematic boundary condition on the body surface which assumes the form

$$\frac{\partial \Phi_j}{\partial n} = n_j, \quad j = 1, 2, \ldots, 6 \tag{6.30}$$

where

$$n_1 = n_x; \qquad\qquad n_2 = n_y; \qquad\qquad n_3 = n_z;$$
$$n_4 = z\, n_y - y\, n_z; \quad n_5 = x\, n_z - z\, n_x; \quad n_6 = y\, n_x - x\, n_y$$

in which n_x, n_y, n_z are direction cosines of the normal at a point (x, y, z).

Therefore we can use the same integral equation and fundamental solutions for dealing with floating bodies. It is interesting to note that since the function $\partial \Phi / \partial n$ may assume seven different forms, corresponding to the diffraction and the six radiation problems, we have to solve seven different integral equations. However, all of them have the same kernels and this greatly simplifies the numerical solution from the computational viewpoint.

After the solution of the boundary integral equation for the velocity potential, we can compute the wave elevation using Eq. (6.6), the pressure through Bernoulli's equation, forces, moments, added mass and damping coefficients, and the motions.

Before closing this section, it should be pointed out that one difficulty that the integral equation approach sometimes encounters is that it breaks down at certain "irregular" wave frequencies for which the solution of the integral equation is not unique. This phenomenon was first demonstrated by John [15] and was later

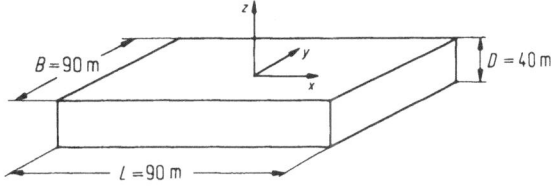

Fig. 6.5. Geometry of floating box

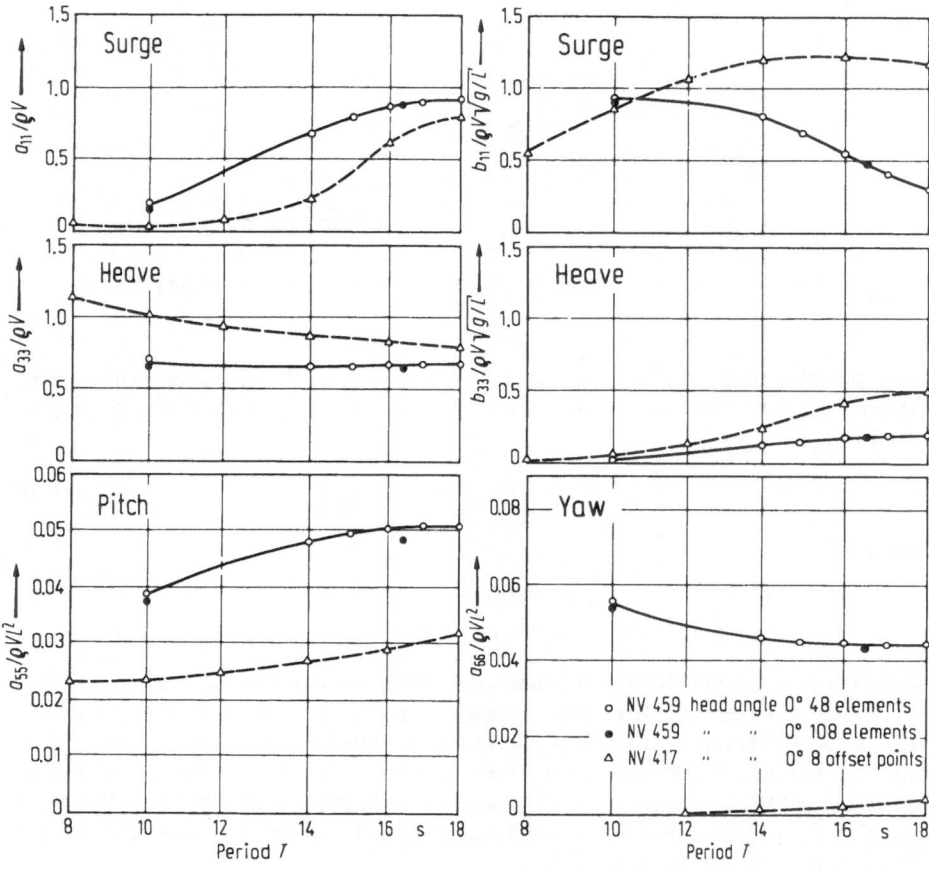

Fig. 6.6. Added mass and damping coefficients for floating box in infinite water depth; NV 459: 3-D elements; NV 417: 2-D elements; V: volume of the box

studied by Frank [24], Murphy [25] and others. These irregular frequencies may be identified as the eigenvalues of the equivalent interior problem. In practical terms, the relatively short wavelengths corresponding to the irregular frequencies are not usually critical in design [2].

There exist numerical results for the oscillating motions and hydrodynamic coefficients of many different types of structures [18–22]. A typical case is the floating tank with a box-like form analysed by Faltinsen [12], shown in Figs. 6.5 and 6.6.

6.4 Vertical Axisymmetric Bodies

The approach described in the previous sections is generally costly to run in the computer because of the complexity of the fundamental solution. Thus, whenever possible, more economical alternative methods restricted to particular configurations should be employed.

One such method can be developed for the case of vertical axisymmetric bodies by deriving an axisymmetric fundamental solution. This may be done by writing the three-dimensional one in cylindrical polar coordinates and integrating it analytically with respect to the circumferential direction. However, the boundary conditions of the problem are arbitrary, nonaxisymmetric. Thus, the series form (6.21) of the three-dimensional fundamental solution (more convenient to work with because the independent variables are separated) is initially expressed as a Fourier series in terms of the azimuthal angle about the vertical axis of symmetry as [26]

$$G = \sum_{j=0}^{\infty} \left(\sum_{m=0}^{\infty} G_{jm} \right) (2 - \delta_{j0}) \cos [j (\theta - \Theta)] \qquad (6.31)$$

with

$$G_{j0} = i\, C_0 \cosh [k\, (z + d)] \cosh [k\, (Z + d)]\, J_j\, (k_r^R)\, H_j^{(1)}\, (k_R^r)$$
$$G_{jm} = 4\, C_m \cos [\mu_m\, (z + d)] \cos [\mu_m\, (Z + d)]\, K_j\, (\mu_m k_R)\, I_j\, (\mu_m r_R^R) \quad \text{for } m \geq 1.$$

In the above equation, δ_{j0} is the Kronecker delta, I_j denotes the modified Bessel function of the first kind of jth order, and the field and source points are now defined by their cylindrical coordinates (r, z, θ) and (R, Z, Θ), respectively. The upper value of the alternative argument is used if $r \geq R$ and the lower otherwise.

We now use the boundary integral Eq. (6.14) together with the Fourier expansion of the fundamental solution (6.31) to obtain a Fourier series, the coefficients of which are the integral equations valid on the arc AA' of Fig. 6.7.

The source density σ is a function of position on the body and may be written as $\sigma(s, \theta)$, where the coordinate $s(r, z)$ specifies a point on the curve AA'. Since the flow is symmetric about the x-axis, we may expand σ as an even Fourier series in θ:

$$\sigma(s, \theta) = \sum_{\ell=0}^{\infty} \sigma_\ell(s) \cos (\ell \theta) . \qquad (6.32)$$

The term involving the incident potential Φ_I in Eq. (6.14) may likewise be expressed as an even Fourier series:

$$\frac{\partial \Phi_I}{\partial n} = \sum_{\ell=0}^{\infty} \Phi_\ell(s) \cos (\ell \theta) \qquad (6.33)$$

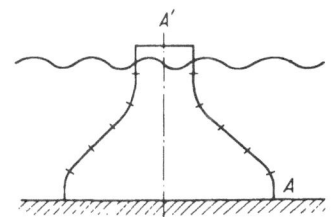

Fig. 6.7. Vertical axisymmetric body

in which

$$\Phi_l = \frac{i\,g\,H\,k}{2\,\omega\,\cosh(k\,d)}\,(2 - \delta_{l0})\,i^l\{r'\sinh[k\,(z+d)]\,J_l(k\,r)$$
$$- z'\cosh[k\,(z+d)]\,J_l'(k\,r)\}$$
$$r' = dr/ds\,;\quad z' = dz/ds\,.$$

Substituting Eqs. (6.31)−(6.33) into (6.14) and noting that $d\Gamma = R\,d\Theta\,dS$, where dS is an element of AA′, the surface integral in Eq. (6.14) can now be integrated with respect to Θ to give an even Fourier series in θ with each coefficient involving only an integral over the boundary contour S. All of the Fourier coefficients for each value of l can now be equated to obtain an infinite number of one-dimensional integral equations, each independent of θ, in the form

$$- \sigma_l(s) + \int_S \sigma_l(S)\left(\sum_{m=0}^{\infty} \frac{\partial G_{lm}}{\partial n}\right)R\,dS = -\,2\,\Phi_l(s),\quad l = 0, 1, 2, \dots\,. \qquad (6.34)$$

Each such equation can be solved in the usual boundary element way. Expressions for the coefficients of the matrices obtained by discretizing the boundary contour and integrating over each element are given by Black [27], Fenton [26], Eatock-Taylor and Dolla [28] and Isaacson [29]. These expressions are very lengthy and consist of the sum of several series, but all series converge very quickly and all terms are finite.

Fenton [26] also developed expressions for the forces and moments on the body in terms of the velocity potential. Of these components, the vertical force involves only the case $l = 0$, the horizontal force and overturning moment (in the $x - z$ plane) involve only the case $l = 1$, and the three remaining orthogonal components of force and moment are zero. Thus, only the two one-dimensional integral equations corresponding to $l = 0$ and $l = 1$ need be solved to determine the wave loads on the body (but not the detailed pressure distribution around the body).

Eatock-Taylor and Dolla [28] and Isaacson [29] extended the previous analysis to the case of floating bodies. Figure 6.8 presents the results obtained by Isaacson [29] for the added-mass and damping coefficients of a floating sphere with its center at the still water level for the case $a/d = 0.1$, and $k\,a$ up to 3, being a the sphere radius. The results are compared with those of Kim [30] which were obtained for deep water wave conditions.

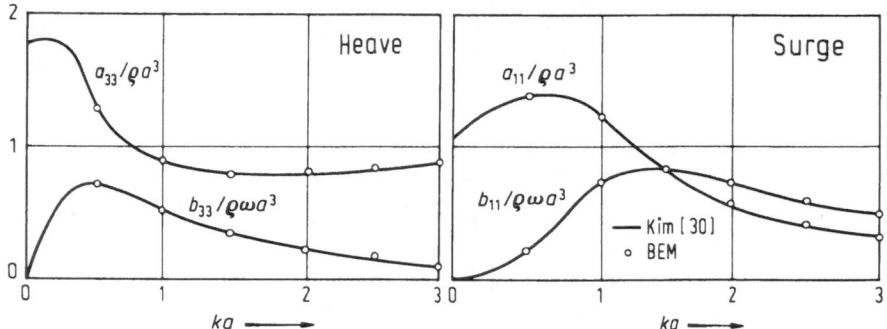

Fig. 6.8. Added mass and damping coefficients for floating sphere

6.5 Vertical Cylinders of Arbitrary Section

If we consider that a train of surface waves propagating in water of constant depth encounter one or more vertical cylinders of arbitrary but constant cross-section extending from the seabed and piercing the free surface (Fig. 6.9), the diffracted wave potential $\Phi_D (x, y, z, t)$ may be decomposed as follows:

$$\Phi_D (x, y, z, t) = -\frac{i g H}{2 \omega} \frac{\cosh [k (z + d)]}{\cosh (k d)} \phi_D (x, y) e^{-i\omega t} . \tag{6.35}$$

This directly satisfies the seabed and linearized free surface boundary conditions (6.3) and (6.5), since we consider valid the dispersion relation (6.10). Substituting expression (6.35) into Laplace's equation (6.1) for Φ_D gives

$$\frac{\partial^2 \phi_D}{\partial x^2} + \frac{\partial^2 \phi_D}{\partial y^2} + k^2 \phi_D = 0 . \tag{6.36}$$

Thus the actual problem is to compute the reduced wave potential ϕ_D which satisfies the Helmholtz equation (6.36) together with the boundary condition

$$\frac{\partial \phi_D}{\partial n} = -\frac{\partial \phi_I}{\partial n} \quad \text{at the body surface } \Gamma \tag{6.37}$$

and the radiation condition (6.8) which may be written in simplified form as [31], [32]

$$\frac{\partial \phi_D}{\partial n} - i k \phi_D = 0 \quad \text{at } \Gamma_\infty \tag{6.38}$$

being Γ_∞ the surface of a circle of infinitely large radius, enclosing all the obstacles (Fig. 6.10). Note that the reduced incident wave potential $\phi_I (x, y)$ is given as

$$\phi_I (x, y) = e^{ik(x\cos\alpha + y\sin\alpha)} \tag{6.39}$$

according to expression (6.9).

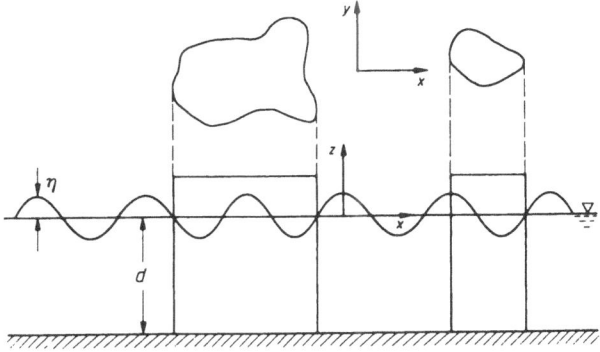

Fig. 6.9.   Vertical cylinders of arbitrary section

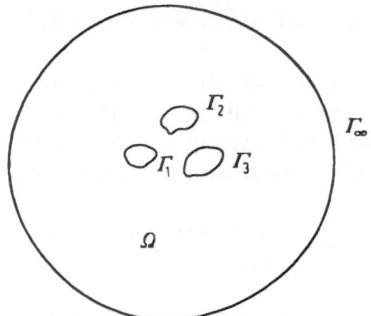

Fig. 6.10.   Fluid region Ω bounded by surfaces Γ and Γ_∞

Once ϕ_D is known, we can compute the wave elevation around the cylinders using Eq. (6.6) and the pressure through the linearized Bernoulli equation

$$p = -\varrho g z - \varrho \frac{\partial \Phi}{\partial t} \tag{6.40}$$

where ϱ is mass density; since the velocity potential has a hyperbolic cosine variation with depth, these variables plus the forces and moments that act on the cylinders are calculated from the formulas:

$$\eta = \frac{H}{2} \phi e^{-i\omega t} \tag{6.41}$$

$$p = -\varrho g z + \frac{1}{2} \varrho g H \frac{\cosh [k (z + d)]}{\cosh (k d)} \phi e^{-i\omega t} \tag{6.42}$$

$$F_z = \frac{\varrho g H}{k} \frac{\cosh [k (z + d)]}{\cosh (k d)} \chi e^{-i\omega t} \tag{6.43}$$

$$F = \frac{\varrho g H d}{k} \frac{\tanh (k d)}{k d} \chi e^{-i\omega t} \tag{6.44}$$

$$M = \frac{\varrho g H d^2}{k} \left[\frac{k d \sinh (k d) + 1 - \cosh (k d)}{(k d)^2 \cosh (k d)} \right] \chi e^{-i\omega t} \tag{6.45}$$

in which $\phi = \phi_I + \phi_D$, and χ is a vector in the horizontal plane given as

$$\chi = -\frac{k}{2} \int_\Gamma \phi \mathbf{n} \, d\Gamma . \tag{6.46}$$

The boundary integral equation equivalent to the problem defined by Eqs. (6.36), (6.37) and (6.38) is similar in form to Eq. (6.12). The fundamental solution to Helmholtz equation is of the form

$$G (x, y; \xi, \eta) = \frac{i}{4} H_0^{(1)} (k r) \tag{6.47}$$

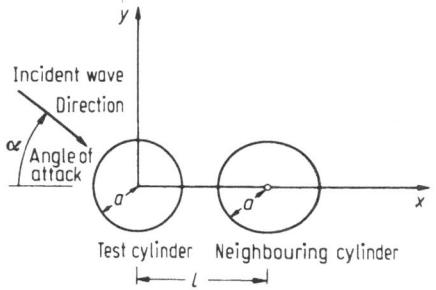

Fig. 6.11. Interaction between a pair of circular cylinders

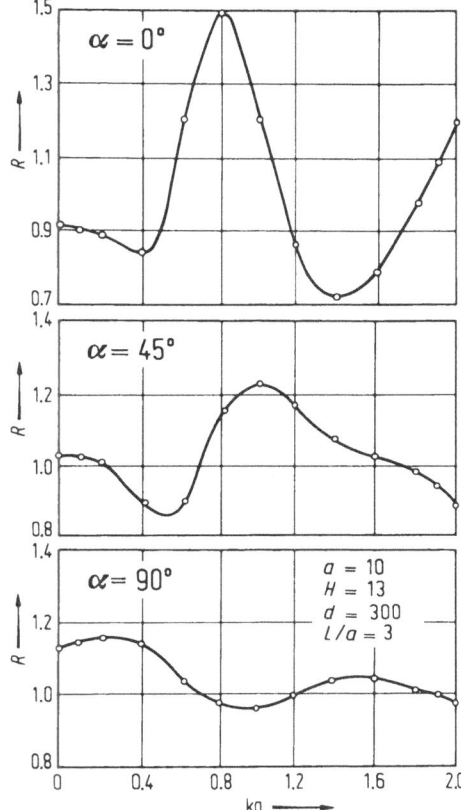

Fig. 6.12. Variation of force ratio for a pair of circular cylinders; R = ratio of horizontal force on test cylinder to force on isolated one

and identically satisfies the radiation condition. To show this, we consider asymptotic expansions of the Hankel functions as $r \to \infty$ [14] as follows:

$$G \sim \frac{i}{4} \left(\frac{2}{\pi k r} \right)^{1/2} \exp \left[i \left(k r - \pi/4 \right) \right] \tag{6.48}$$

$$\frac{\partial G}{\partial n} = \frac{\partial G}{\partial r} = -\frac{i k}{4} H_1^{(1)} (k r) \sim -\frac{k}{4} \left(\frac{2}{\pi k r} \right)^{1/2} \exp \left[i \left(k r - \pi/4 \right) \right] \tag{6.49}$$

thus

$$\frac{\partial G}{\partial n} - i k G = 0 \quad \text{at } \Gamma_\infty . \tag{6.50}$$

Some particularities of the numerical evaluation of the integrals over each boundary element, after the discretization process has been carried out, are discussed in [33–35]. For computer efficiency, since

$$G = \tfrac{1}{4} \left[- Y_0 (k r) + i J_0 (k r) \right] \tag{6.51}$$

$$\frac{\partial G}{\partial n} = \frac{k}{4} \left[Y_1 (k r) - i J_1 (k r) \right] \frac{\partial r}{\partial n} \tag{6.52}$$

t is usual to employ polynomial expansions of the Bessel functions, as given by Abramowitz and Stegun [14].

Masetti and Wrobel [35] have applied the present formulation for the study of the interaction effects between close-placed circular cylinders of equal radii (Fig. 6.11). Results obtained using 24 constant elements for each cylinder are plotted in Fig. 6.12 for three different angles of attack, $\alpha = 0°$, 45° and 90°. As can be seen from the figure, the increase in force is most significant when the neighbouring cylinder is directly behind the test one, i.e., $\alpha = 0°$; in this case, the force on the test cylinder may be increased of up to 50%.

6.6 Horizontal Cylinders of Arbitrary Section

Another case of practical interest is the two-dimensional wave motion in the $y - z$ plane past an infinite horizontal cylinder whose axis is parallel to the x-axis (Fig. 6.13). One main aspect of the problem is the treatment of freely floating bodies oscillating harmonically on or near the free surface. The first solution of this problem in the field of naval hydrodynamics was developed by Frank [24] for an infinitely deep fluid region and is known as the Frank close-fit method.

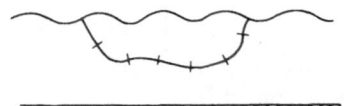

Fig. 6.13. Horizontal cylinders of arbitrary section

The governing equations for this problem are essentially the same as for the three-dimensional one, and can be written for the reduced radiated potential $\phi(y, z, t)$ as follows:

$$\frac{\partial^2 \phi}{\partial y^2} + \frac{\partial^2 \phi}{\partial z^2} = 0 \quad \text{in the fluid region} \tag{6.53}$$

$$\frac{\partial^2 \phi}{\partial t^2} + g\,\frac{\partial \phi}{\partial z} = 0 \quad \text{at } z = 0 \tag{6.54}$$

$$\lim_{r \to \infty} \left(\frac{\partial \phi}{\partial r} - i\,k\,\phi \right) = 0 \tag{6.55}$$

$$\frac{\partial \phi_j}{\partial n} = V_j \quad \text{at the body surface} \tag{6.56}$$

in which the body can oscillate in sway, heave or roll about an axis through the point $(0, z_0)$, in the form

$$V_2 = -\,\omega A_2 \sin \alpha \sin \omega t$$

$$V_3 = \omega A_3 \cos \alpha \sin \omega t$$

$$V_4 = \omega A_4 \left[(z - z_0) \sin \alpha + y \cos \alpha \right] \sin \omega t$$

for the swaying, heaving and rolling modes, respectively, where the direction cosines for the three modes of motion are

$$n_2 = - \sin \alpha \; ; \quad n_3 = \cos \alpha \; ; \quad n_4 = (z - z_0) \sin \alpha + y \cos \alpha$$

and A_j are the oscillation amplitudes in the jth mode.

`Frank's method employs an indirect BEM formulation, such that the boundary-value problem is converted into an integral equation similar to Eq. (6.13), in the form

$$\phi (y, z, t) = \int_\Gamma \sigma (\eta, \zeta) \, G^* (y, z; \eta, \zeta; t) \, d\Gamma (\eta, \zeta) \,. \tag{6.57}$$

According to Wehausen and Laitone [11], the complex two-dimensional fundamental solution which satisfies all boundary conditions apart from that on the body surface is given by

$$G^* (y, z; \eta, \zeta; t) = \frac{1}{2 \pi} \left[\ln (x - \xi) - \ln (x - \bar{\xi}) + 2 P \cdot V \cdot \int_0^\infty \frac{e^{-i\mu(x-\bar{\xi})}}{v - \mu} \, d\mu \right] \cos \omega t$$
$$- e^{-iv(x-\bar{\xi})} \sin \omega t \tag{6.58}$$

where

$$x = y + i z$$
$$\xi = \eta + i \zeta$$
$$\bar{\xi} = \eta - i \zeta$$

so that the real point-source potential is

$$H (y, z; \eta, \zeta; t) = \mathrm{Re} \, [G^* (y, z; \eta, \zeta; t)] \,. \tag{6.59}$$

If we let

$$G (y, z; \eta, \zeta) = \mathrm{Re} \, \frac{1}{2 \pi} \left[\ln (x - \xi) - \ln (x - \bar{\xi}) + 2 P \cdot V \cdot \int_0^\infty \frac{e^{-i\mu(x-\bar{\xi})}}{v - \mu} \, d\mu \right]$$
$$- i \, \mathrm{Re} \, [e^{-iv(x-\bar{\xi})}] \tag{6.60}$$

then

$$H (y, z; \eta, \zeta; t) = \mathrm{Re} \, [G (y, z; \eta, \zeta) \, e^{-i\omega t}] \tag{6.61}$$

so that

$$\phi (y, z, t) = \mathrm{Re} \left[\int_\Gamma \sigma (\eta, \zeta) \, G (y, z; \eta, \zeta) \, d\Gamma (\eta, \zeta) \, e^{-i\omega t} \right]. \tag{6.62}$$

Application of the boundary conditions (6.56) on the cylinder surface then yields

$$\mathrm{Re} \int_\Gamma \sigma (\eta, \zeta) \frac{\partial G (y, z; \eta, \zeta)}{\partial n (y, z)} \, d\Gamma (\eta, \zeta) = 0 \tag{6.63}$$

$$\mathrm{Im} \int_\Gamma \sigma (\eta, \zeta) \frac{\partial G (y, z; \eta, \zeta)}{\partial n (y, z)} \, d\Gamma (\eta, \zeta) = \omega A_j \, n_j \,. \tag{6.64}$$

A more convenient expression to evaluate the principal value integral is

$$P \cdot V \cdot \int_0^\infty \frac{e^{-i\mu(x-\bar{\xi})}}{v - \mu} \, d\mu = e^{v(z+\zeta)} \, [\cos v \, (y - \eta) - i \sin v \, (y - \eta)]$$
$$\times \left\{ \gamma + \ln r + \sum_{n=1}^\infty \frac{r^n \cos \theta}{n \, n!} + i \left[\theta + \sum_{n+1}^\infty \frac{r^n \sin n \theta}{n \, n!} \right] \right\} \tag{6.65}$$

where

$$r = |-iv(x - \bar{\xi})|$$

$$\theta = \tan^{-1}\{\text{Im}[-iv(x - \bar{\xi})]/\text{Re}[-iv(x - \bar{\xi})]\} + \pi.$$

Using the linearized Bernoulli equation we obtain the hydrodynamic pressure at a point on the cylinder surface

$$p(y, z, t) = -\varrho\frac{\partial\phi(y, z, t)}{\partial t} = p_a(y, z)\cos\omega t + p_v(y, z)\sin\omega t \qquad (6.66)$$

in which p_a and p_v are the hydrodynamic pressure on phase with the displacement and velocity of the cylinder, respectively. The hydrodynamic forces acting on the cylinder are obtained by integrating $p \cdot \mathbf{n}$ over the submerged portion of the cylinder surface, so that the added mass and damping coefficients are

$$M_j = \int_\Gamma p_a \cdot n_j \, d\Gamma \qquad (6.67)$$

$$D_j = \int_\Gamma p_v \cdot n_j \, d\Gamma \qquad (6.68)$$

Frank observed that this procedure also presents the problem of "irregular" frequencies discussed in Sect. 6.3, and that they become more critical in the two-dimensional approach. To avoid this difficulty, two procedures have been presented. The first was developed by Ohmatsu [36], following a suggestion of Pauling [37] of modifying the interior problem by extending the source distribution along the free surface where a rigid-wall boundary condition is applied. Ohmatsu [36] showed that the problem may, in fact, be removed but the boundary condition to be applied can be an arbitrary Neumann condition. Another way of removing the irregular frequencies was presented by Ursell [38] by adding a source at the origin. Ogilvie and Shin [39] have generalized this approach by modifying the fundamental solution in a simple way.

The method presented here can also be applied to determine the diffracted potential due to incident waves, and the related exciting force. In this case, we substitute the body boundary condition (6.56) by expression (6.11).

In the case of finite depth, the same procedures can be used if we provide a fundamental solution that satisfies all the previous conditions plus a seabed condition of zero normal velocity, expression (6.16). According to Wehausen and Laitone [11], this fundamental solution is of the form:

$$G^* = \ln\frac{rr_2}{d^2} - 2P \cdot V \cdot \int_0^\infty\left(\frac{(\mu + v)e^{-\mu d}\cosh[\mu(z + d)]\cosh[\mu(\zeta + d)]\cos[\mu(y - \eta)]}{\mu[\mu\sinh(\mu d) - v\cosh(\mu d)]} - \frac{e^{-\mu d}}{\mu}\right)$$

$$\times d\mu\cos\omega t$$

$$-2\frac{v + k}{k}e^{-kd}\frac{\sinh(kd)\cosh[k(z + d)]\cosh[k(\zeta + d)]\cos[k(y - \eta)]}{vd + \sinh^2(kd)}\sin\omega t \qquad (6.69)$$

where:

$$r = [(y - \eta)^2 + (z - \zeta)^2]^{1/2}$$

$$r_2 = [(y - \eta)^2 + (z + 2d + \zeta)^2]^{1/2}.$$

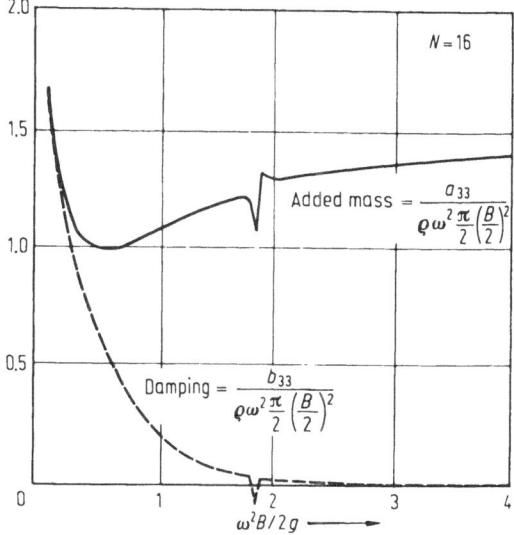

Fig. 6.14. Heave added mass and damping coefficients for rectangular cylinder; $B/D = 2.5$

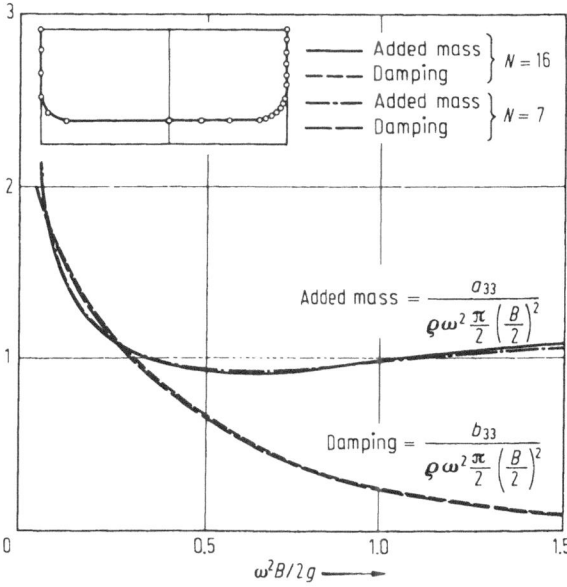

Fig. 6.15. Heave added mass and damping coefficient for ship-like section

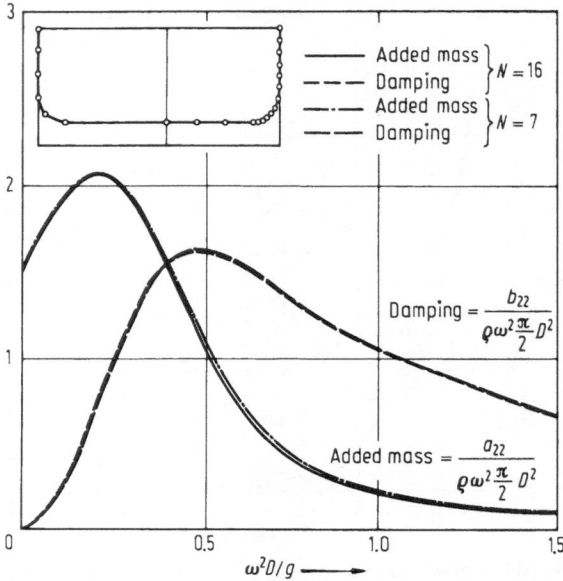

Fig. 6.16. Sway added mass and damping coefficients for ship-like section

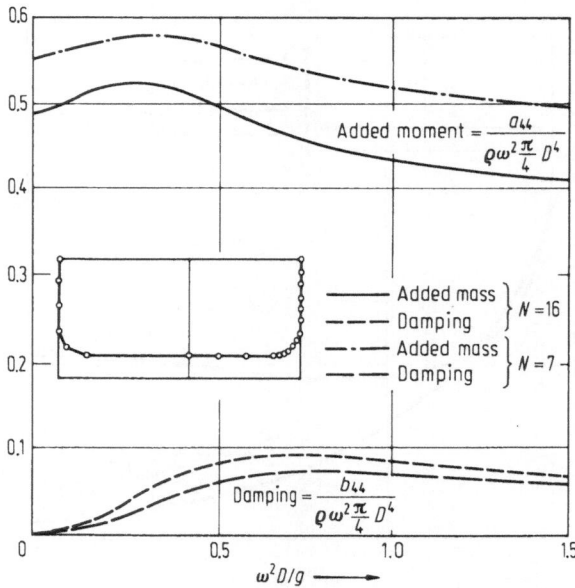

Fig. 6.17. Roll added moment and damping coefficients for ship like section

An alternative series form for G^* is

$$G^* = -\frac{C_0}{k} \cosh[k(z+d)]\cosh[k(\zeta+d)]\sin[k(y-\eta)-\omega t]$$

$$-2\pi \sum_{m=1}^{\infty} \frac{C_m}{\mu_m} \cos[\mu_m(z+d)]\cos[\mu_m(\zeta+d)]e^{-\mu_m(y-\eta)}\sin\omega t. \qquad (6.70)$$

Figure 6.14 shows results for heave added mass and damping coefficients for a rectangular cylinder for which the relation breadth/draft is equal to 2.5 [24]. From this figure, we can see the behaviour of the solution near the irregular frequencies. Figures 6.15–6.17 present the added mass and damping coefficients for heave, sway and roll for a ship-like section [24]. From them, we note the influence of the boundary discretization and the convergence of the solution.

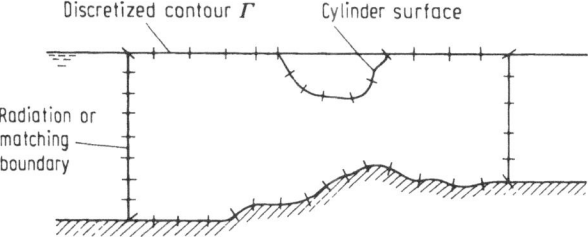

Fig. 6.18.   Fluid boundaries for a vertical plane problem

As discussed in Sect. 6.2 for the three-dimensional case, a different scheme of solution can be attempted for the problem by employing the free space Green's function $G = \ln 1/r$ and discretizing the whole boundary of the truncated domain depicted in Fig. 6.18. Very good results obtained by using this scheme have been reported in [40–42].

6.6.1 Obliquely Incident Waves

In the case of obliquely incident waves making an angle α with the y-axis, the governing equation for the problem becomes the Helmholtz equation [43]

$$\frac{\partial^2 \phi}{\partial y^2} + \frac{\partial^2 \phi}{\partial z^2} - \beta^2 \phi = 0 \qquad (6.71)$$

for the reduced potential $\phi(y, z)$, assuming that

$$\Phi(x, y, z, t) = F(x)\phi(y, z)e^{-i\omega t}. \qquad (6.72)$$

In Eq. (6.71), we have that $\beta = k\sin\alpha$.

The fundamental solution for this problem is

$$G(y, z; \eta, \zeta) = \frac{1}{4}H_0^{(1)}(i\beta r) = \frac{1}{2\pi}K_0(\beta r) \qquad (6.73)$$

in which r is the distance between the source and field points.

The boundary conditions of the diffraction problem are given by expressions (6.3), (6.5), (6.37) and (6.55), where the reduced incident wave potential has the form

$$\phi_I(y, z) = \frac{\cosh[k(z+d)]}{\cosh(kd)} e^{iky\cos\alpha}.$$ (6.74)

Since the fundamental solution (6.73) is a free space Green's function, there is a necessity of discretizing all the boundary contour. Georgiadis and Hartz [44] solved this problem by extending the lateral boundaries at a large distance and directly applying the Sommerfeld condition (6.55). In their scheme, they employed the direct formulation with constant boundary elements. By defining

$$h_{ij} = \int_{\Gamma_j} \frac{\partial G(y, z; \eta_i, \zeta_i)}{\partial n(y, z)} d\Gamma(y, z)$$ (6.75)

$$d_{ij} = \int_{\Gamma_j} G(y, z; \eta_i, \zeta_i) d\Gamma(y, z)$$ (6.76)

in which (η_i, ζ_i) are the coordinates of the source point i, we can write the boundary integral equation in discretized form as follows

$$\frac{1}{2}\phi_i + \sum_{j=1}^{N} h_{ij}\phi_j = \sum_{j=1}^{N} d_{ij} q_j$$ (6.77)

where $q_j = (\partial\phi/\partial n)_j$ and N is the number of boundary elements.

Modifying the coefficients h_{ii} by adding the free term $\frac{1}{2}$, Eq. (6.77) becomes

$$\sum_{j=1}^{N} h_{ij}\phi_j = \sum_{j=1}^{N} d_{ij} q_j .$$ (6.78)

Applying the above equation at each source point, we can write the resulting system of equations in matricial form as

$$[H_1\ H_2\ H_3\ H_4]\begin{Bmatrix}\phi_1\\\phi_2\\\phi_3\\\phi_4\end{Bmatrix} = [D_1\ D_2\ D_3\ D_4]\begin{Bmatrix}q_1\\q_2\\q_3\\q_4\end{Bmatrix}.$$ (6.79)

The vectors ϕ_k and q_k contain the values of ϕ_j and q_j along the contour Γ_k ($k = 1, 2, 3, 4$), respectively (see Fig. 6.19).

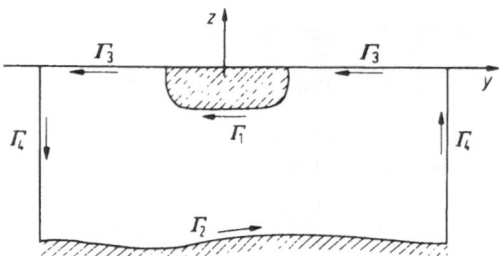

Fig. 6.19.   Truncated fluid region

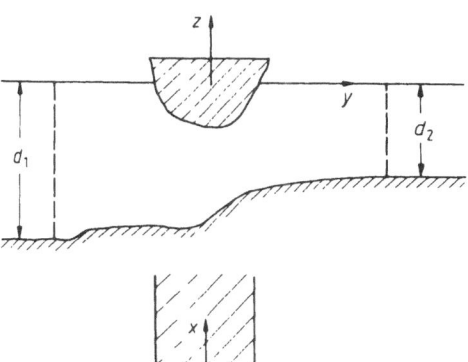

Fig. 6.20. Fluid region divided into three parts

By applying the boundary conditions of the problem, the system (6.79) can be reordered as follows

$$[\mathbf{H}_1 \ \mathbf{H}_2 \ (\mathbf{H}_3 - v \ \mathbf{D}_3) \ (\mathbf{H}_4 - i \ k \ \mathbf{D}_4)] \begin{Bmatrix} \boldsymbol{\phi}_1 \\ \boldsymbol{\phi}_2 \\ \boldsymbol{\phi}_3 \\ \boldsymbol{\phi}_4 \end{Bmatrix} = [\mathbf{D}_1] \{\mathbf{q}_1\} . \tag{6.80}$$

Using for $\mathbf{q}_1$ the specified values on the body surface, the above system can be solved for the unknown ϕ values.

A different approach for the problem was developed by Liu and Abbaspour [45] using the so-called hybrid integral equation method. Instead of directly applying the radiation condition on contour Γ_4, they divided the fluid region into three parts (Fig. 6.20); the two external ones are of constant (but not equal) depth and extend to infinity, such that general series solutions can be applied.

Assuming that the incident wave propagates from $y = -\infty$ over a constant water depth d_1, we can write, similarly to Eq. (6.74)

$$\phi_I = \frac{\cosh [k_1 (z + d_1)]}{\cosh (k_1 d_1)} e^{i k_1 y \cos \alpha} . \tag{6.81}$$

The presence of the body and the irregular bottom creates disturbances on the flow. The Sommerfeld radiation condition then requires that

$$\lim_{k_1 y \cos \alpha \to -\infty} \left(\frac{\partial \phi}{\partial y} + i k_1 \cos \alpha \ \phi \right) = 0 \tag{6.82}$$

$$\lim_{k_2 y \cos \alpha \to \infty} \left(\frac{\partial (\phi - \phi_I)}{\partial y} - i k_2 \cos \alpha \ (\phi - \phi_I) \right) = 0 . \tag{6.83}$$

In the present case, the general series solutions for the regions of constant depth are as follows

$$\phi_j = A_j^0 \, e^{\pm i k_j y \cos \alpha} \frac{\cosh [k_j (z + d_j)]}{\cosh (k_j d_j)}$$

$$+ \sum_{n=1}^{\infty} A_j^n \, e^{\pm \gamma_j^n y} \frac{\cos [K_j^n (z + d_j)]}{\cos (K_j^n d_j)} + \begin{pmatrix} 0 \\ \phi_I \end{pmatrix}$$ (6.84)

where $j = 1, 2$, the upper values of the alternative choices are used if $j = 1$ and the lower otherwise. The coefficients k_j and K_j^n are the real and imaginary roots of the dispersion relation, i.e.

$$\omega^2 = g \, k_j \tanh (k_j d_j) ; \quad \omega^2 = g \, K_j^n \tan (K_j^n d_j)$$

and γ_j^n are defined as

$$\gamma_j^n = [(K_j^n)^2 + \beta^2]^{1/2} \, .$$

The coefficients A_j^0 and A_j^n in Eq. (6.84) are to be determined.

Now, the internal boundary-value problem must match the external one over the contour Γ_4. This means that the values of ϕ and q over Γ_4 can be written as

$$\phi = R A ; \quad q = T A \, .$$ (6.85)

Introducing these values into the system (6.79) and reordering, we finally obtain

$$[H_1 \; H_2 \; (H_3 - v \, D_3) \; (H_4 \, R - D_4 \, T)] \begin{Bmatrix} \phi_1 \\ \phi_2 \\ \phi_3 \\ A \end{Bmatrix} = [D_1] \{q_1\} \, .$$ (6.86)

It should be noted that, for numerical computations, the infinite series in expression (6.84) are truncated into finite series with M unknown coefficients, and that the number of nodes along Γ_4 must be the same as that of the unknown coefficients.

6.7 Transient Problems

In this section, we shall apply the BEM to the case of linear transient wave problems, i.e. assume that the waves are no longer harmonic. Considering, for simplicity, only two-dimensional propagation in tanks (Fig. 6.21), the problem can be mathematically described by the following equations

$$\frac{\partial^2 \phi}{\partial x^2} + \frac{\partial^2 \phi}{\partial y^2} = 0 \quad \text{in the fluid region}$$ (6.87)

$$\frac{\partial \phi}{\partial n} = v \, (x, y, t) \quad \text{at solid boundaries}$$ (6.88)

$$\frac{\partial \phi}{\partial y} = \frac{\partial \phi}{\partial n} = \frac{\partial \eta}{\partial t} \quad \text{at } y = 0$$ (6.89)

$$\frac{\partial \phi}{\partial t} + g \, \eta = 0 \quad \text{at } y = 0$$ (6.90)

where v is a prescribed normal velocity ($v = 0$ for fixed boundaries).

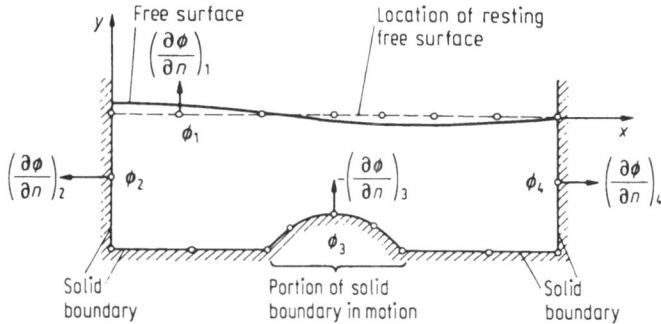

Fig. 6.21. Schematic drawing showing two-dimensional wave propagation in tank and discretization

A boundary integral equation equivalent to Laplace's equation was already derived in Sect. 6.2, Eq. (6.12). Discretizing the whole boundary as shown in Fig. 6.21, applying the discretized equation at each boundary node and integrating, we end up with a system of equations similar to (6.79) which is valid at any instant in time:

$$[\mathbf{H_1\ H_2\ H_3\ H_4}] \begin{Bmatrix} \phi_1^{k+1} \\ \phi_2^{k+1} \\ \phi_3^{k+1} \\ \phi_4^{k+1} \end{Bmatrix} = [\mathbf{D_1\ D_2\ D_3\ D_4}] \begin{Bmatrix} q_1^{k+1} \\ q_2^{k+1} \\ q_3^{k+1} \\ q_4^{k+1} \end{Bmatrix} \tag{6.91}$$

where $k+1$ represents the time $t = (k+1)\,\Delta t$, with Δt a prescribed time step.

The free surface boundary condition may be approximated in finite difference form, maintaining second-order accuracy, as [46, 47]:

$$\eta^{k+1} = \eta^k + \frac{\Delta t}{2} [q^{k+1} + q^k] \tag{6.92}$$

$$\phi^{k+1} = \phi^k - g\,\Delta t\,\eta^k - \tfrac{1}{2}g\,\Delta t^2\,[\theta\,q^{k+1} + (1-\theta)\,q^k] \tag{6.93}$$

in which θ is a weighting factor that positions the terms between an explicit scheme $(\theta = 0)$ and a fully implicit one $(\theta = 1)$. A stability analysis carried out in [46] shows that value $\theta = 0.17$ produces a small frequency distortion over a wide range of frequencies.

Using Eq. (6.93), the system (6.91) can be rearranged in the form

$$[(-\tfrac{1}{2}g\,\Delta t^2\,\theta\,\mathbf{H_1} - \mathbf{D_1})\ \mathbf{H_2\ H_3\ H_4}] \begin{Bmatrix} q_1^{k+1} \\ \phi_2^{k+1} \\ \phi_3^{k+1} \\ \phi_4^{k+1} \end{Bmatrix}$$

$$= [-\mathbf{H_1\ D_2\ D_3\ D_4}] \begin{Bmatrix} \phi_1^k - g\,\Delta t\,\eta^k - \tfrac{1}{2}g\,\Delta t^2\,(1-\theta)\,q_1^k \\ q_2^{k+1} \\ q_3^{k+1} \\ q_4^{k+1} \end{Bmatrix} \tag{6.94}$$

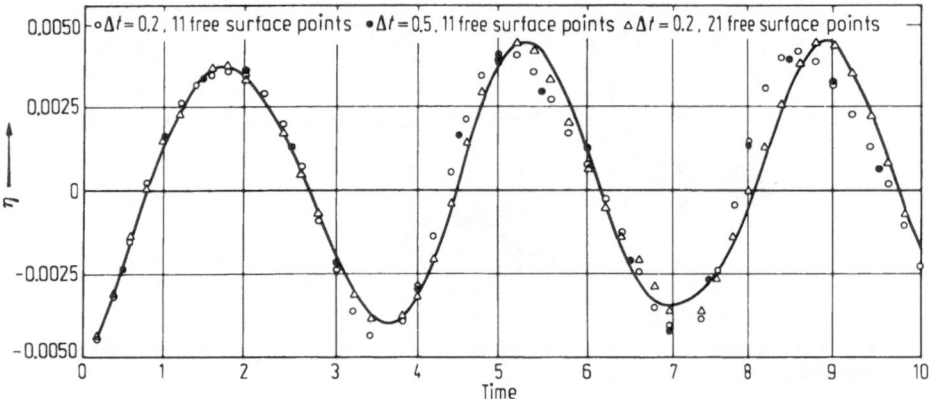

Fig. 6.22.   Time variation of surface displacement of wall points

in which all quantities on the right side are prescribed or have been calculated at previous time steps. The solution of the transient problem then proceeds from known initial conditions by forward time stepping with time increment Δt.

Note that if there is no part of the solid boundary in motion, the matrices **H** and **D** in (6.94) do not change in time so they can be computed a single time and stored. Furthermore, for a constant Δt, the coefficient matrix can be triangularized only once. At each time step, then, only a matrix-vector multiplication need be carried out, the resulting right-hand side decomposed and back-substitution performed; after obtaining the solution for q^{k+1} on the free surface from the system (6.94), then ϕ^{k+1} for free surface nodes is obtained from Eq. (6.93) and η^{k+1} from Eq. (6.92), and the solution can proceed to the next time level.

Results obtained by Salmon et al. [46] through the previously-described scheme for a triangular wave in a box are presented in Figs. 6.22 and 6.23, compared to an analytical solution. Figure 6.22 shows the time history of the wall points while Fig. 6.23 depicts the free surface profiles at four different times.

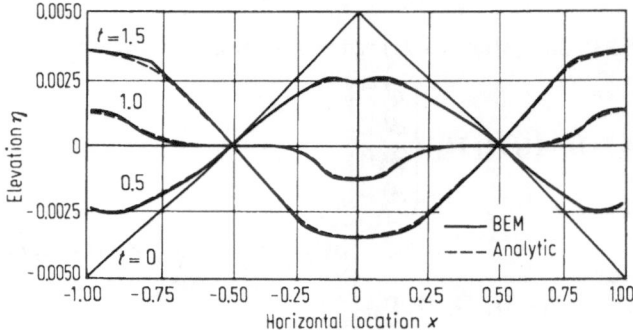

Fig. 6.23.   Free surface profiles at four different time instants

6.8 Nonlinear Problems

The nonlinearities of the problem arise on account of the nonlinear terms in the two free surface boundary conditions and the fact that the position of the free surface is unknown and its determination becomes part of the solution of the problem. Considering, as in the previous section, only two-dimensional propagation in tanks, the general problem is defined by the same governing equation and boundary conditions as before, except that the complete nonlinear free surface conditions now apply instead of Eqs. (6.89) and (6.90):

$$\frac{\partial \phi}{\partial n} = n_y \frac{\partial \eta}{\partial t} \qquad\qquad \text{at } y = \eta \qquad\qquad (6.95)$$

$$\frac{\partial \phi}{\partial t} + \frac{1}{2}\left[\left(\frac{\partial \phi}{\partial x}\right)^2 + \left(\frac{\partial \phi}{\partial y}\right)^2\right] + g\,\eta = 0 \quad \text{at } y = \eta \qquad (6.96)$$

in which n_y is the direction cosine of the normal with respect to the y-axis.

Liu [6] proposed a scheme of solution in which the two free surface boundary conditions are approximated in the following finite difference forms:

$$\eta^{k+1} = \eta^k + \frac{\Delta t\,[\theta_1 q^{k+1} + (1 - \theta_1)\,q^k]}{n_y^k} \qquad\qquad (6.97)$$

$$\phi^{k+1} = \phi^k - \Delta t \left\{\frac{1}{2}\left[\left(\frac{\partial \phi}{\partial x}\right)^2 + \left(\frac{\partial \phi}{\partial y}\right)^2\right]^k + \theta_2\,g\,\eta^{k+1} + (1 - \theta_2)\,g\,\eta^k\right\}. \quad (6.98)$$

The nonlinear terms in Eqs. (6.97) and (6.98) are calculated at time level k even though the equations are written for time level $k + 1$. Liu [6] states that although this problem can be avoided by iteration, a small time step value may provide sufficient accuracy.

A stability analysis to determine the best choice for the weighting factors θ_1 and θ_2 is generally difficult to carry out for nonlinear problems, and usually the value $\theta_1 = \theta_2 = \frac{1}{2}$ is assumed.

Another important point refers to the determination of the nonlinear terms in Eq. (6.98), which involve derivatives of the potential function along the free surface. The simplest way of computing them is by writing

$$\left(\frac{\partial \phi}{\partial x}\right)^2 + \left(\frac{\partial \phi}{\partial y}\right)^2 = \left(\frac{\partial \phi}{\partial n}\right)^2 + \left(\frac{\partial \phi}{\partial s}\right)^2 \qquad\qquad (6.99)$$

where s is the tangential direction along the free surface. The tangential derivative can then be computed by using a finite difference approximation; this does not affect the overall accuracy, since the $\partial \phi/\partial s$ terms are small and thus can be computed more crudely than the other terms.

Consider the same problem represented in Fig. 6.21, for which the system of Eqs. (6.91) is obtained. Substituting (6.97) into (6.98) produces a relation between ϕ^{k+1} and q^{k+1}:

$$\qquad\qquad\qquad\qquad\qquad\qquad\qquad\qquad\qquad\qquad\qquad (6.100)$$

$$\phi^{k+1} = \phi^k - \Delta t \left\{\frac{1}{2}\left[\left(\frac{\partial \phi}{\partial x}\right)^2 + \left(\frac{\partial \phi}{\partial y}\right)^2\right]^k + g\,\eta^k + g\,\Delta t\,\theta_2\,[\theta_1\,q^{k+1} + (1 - \theta_1)\,q^k]/n_y^k\right\}.$$

which, upon substitution in (6.91), gives as a result a system of equations which right-hand side involves only known quantities. Its solution produces the values of q^{k+1} on the free surface, from which values of η^{k+1} and ϕ^{k+1} may be obtained through Eqs. (6.97) and (6.98), respectively. Note that the same remarks of the previous section regarding computer efficiency also apply to this case.

Recently, Kim et al. [8] proposed a more refined scheme in which the nonlinear free surface conditions are approximated as follows:

$$\eta^{k+1} = \eta^k + \Delta t \left[\theta_1 \, q^{k+1}/n_y^{k+1} + (1 - \theta_1) \, q^k/n_y^k\right] \tag{6.101}$$

$$\phi^{k+1} = \phi^k - \Delta t \left\{\frac{1}{2}(1 - \theta_3)\left[\left(\frac{\partial \phi}{\partial x}\right)^2 + \left(\frac{\partial \phi}{\partial y}\right)^2\right]^k\right.$$

$$\left. + \frac{1}{2}\theta_3\left[\left(\frac{\partial \phi}{\partial x}\right)^2 + \left(\frac{\partial \phi}{\partial y}\right)^2\right]^{k+1} + \theta_2 \, g \, \eta^{k+1} + (1 - \theta_2) \, g \, \eta^k\right\}. \tag{6.102}$$

The nonlinear terms are linearized by estimating the unknowns at time level $k + 1$ and then improving the estimate by iteration. Kim et al. [8] report that, for the examples they studied, usually three or four iterations were required for the solution to converge within each time step, and also that the value of the weighting factors were taken as $\theta_1 = \theta_2 = \theta_3 = \frac{1}{2}$.

A rather different numerical scheme for the solution of the problem was derived by Nakayama [7], using Galerkin's method. In this scheme, the errors introduced in the satisfaction of the dynamic free surface boundary condition (6.96) are required to vanish over the free surface Γ_1 in the sense of a weighted mean:

$$\int_{\Gamma_1} W\left\{\frac{\partial \phi}{\partial t} + \frac{1}{2}\left[n_y^2\left(\frac{\partial \eta}{\partial t}\right)^2 + \left(\frac{\partial \phi}{\partial s}\right)^2\right] + g \, \eta - \frac{\partial D}{\partial t}\right\} d\Gamma = 0 \tag{6.103}$$

where $W(s)$ is a weighting factor and $D(s, t)$ is the so-called error correcting term [48], which is introduced in order to avoid the accumulation of errors generally associated with time-stepping schemes. The term D is forced to be zero at the actual time level, as will be shown later.

The solution procedure is described as follows, referring to Fig. 6.21:

Step 1. The problem boundaries $\Gamma = \Gamma_1 + \Gamma_2 + \Gamma_3 + \Gamma_4$ are discretized into a number of elements;

Step 2. Within each element, ϕ, η, W and D are assumed to be linear functions with respect to s, which is a local coordinate measured along the element; this may be expressed as follows:

$$\phi = \mathbf{N}^T \boldsymbol{\phi}$$

$$\eta = \mathbf{N}^T \boldsymbol{\eta}$$

$$W = \mathbf{N}^T \mathbf{W}$$

$$D = \mathbf{N}^T \mathbf{D}$$

$$\mathbf{N}^T = \frac{1}{\ell}[\ell - s \; s], \quad 0 \leq s \leq \ell$$

where ℓ denotes the length of the line element;

Step 3. The boundary integral Eq. (6.12), which can be written in this case as

$$c_i \phi_i + \int_\Gamma \phi \frac{\partial}{\partial n} \left(\ln \frac{1}{r} \right) d\Gamma - \int_{\Gamma_3} v_n \ln \frac{1}{r} d\Gamma - \int_{\Gamma_1} n_y \frac{\partial \eta}{\partial t} \ln \frac{1}{r} d\Gamma = 0 \quad (6.104)$$

for the collocation point i, is discretized as follows:

$$c_i \phi_i + \sum_\Gamma \mathbf{A}^T \boldsymbol{\phi} - \sum_{\Gamma_3} \mathbf{B}^T \mathbf{v}_n - \sum_{\Gamma_1} n_y \mathbf{C}^T \dot{\boldsymbol{\eta}} = 0 \quad \text{for } i = 1, 2, ..., N \quad (6.105)$$

in which the dot stands for time derivative; $\mathbf{A}$, $\mathbf{B}$ and $\mathbf{C}$ are vectors composed of influence coefficients;

Step 4. The weighted residual Eq. (6.103) is discretized by using the procedure described in Step 2. Then, arbitrariness of the weighting function vector $\mathbf{W}$ yields the following nonlinear algebraic equations with respect to ϕ and η:

$$\sum_{\Gamma_1} \left[\ell \, \mathbf{F} \dot{\boldsymbol{\phi}} + \tfrac{1}{2} n_y^2 \, \mathbf{H}(\dot{\eta}) \, \dot{\boldsymbol{\eta}} + \frac{1}{2\ell} \mathbf{P} \boldsymbol{\phi}^T \mathbf{E} \boldsymbol{\phi} + \ell \, \mathbf{F}(g \, \boldsymbol{\eta} - \dot{D}) \right] = 0 \quad (6.106)$$

where:

$$\mathbf{E} = \begin{bmatrix} 1 & -1 \\ -1 & 1 \end{bmatrix}; \qquad \mathbf{F} = \tfrac{1}{6} \begin{bmatrix} 2 & 1 \\ 1 & 2 \end{bmatrix}$$

$$\mathbf{H}(\dot{\eta}) = \tfrac{1}{12} \begin{bmatrix} 3 \dot{\eta}_j + \dot{\eta}_{j+1} & \dot{\eta}_j + \dot{\eta}_{j+1} \\ \dot{\eta}_j + \dot{\eta}_{j+1} & \dot{\eta}_j + 3 \dot{\eta}_{j+1} \end{bmatrix}; \qquad \mathbf{P}^T = \tfrac{1}{2}[1 \quad 1]$$

and $\mathbf{0}$ denotes a zero vector. The quantities in matrix H represent values at the nodes of a general element j.

The nonlinear Eqs. (6.105) and (6.106) are solved by using an incremental method of solution. Denoting values of ϕ and η at two successive time levels by ϕ^k, η^k, ϕ^{k+1} and η^{k+1}, increments $\Delta\phi$ and $\Delta\eta$ are defined as follows:

$$\phi^{k+1} = \phi^k + \Delta\phi \quad (6.107)$$

$$\eta^{k+1} = \eta^k + \Delta\eta . \quad (6.108)$$

Assuming that higher order product terms of $\Delta\phi$ and $\Delta\eta$ can be neglected, linearized equations with respect to $\Delta\phi$ and $\Delta\eta$ are obtained. By using relations (6.107) and (6.108), all variables at the time level $k + 1$ can be expressed in terms of the increments $\Delta\phi$, $\Delta\eta$ and values at time level k.

Nakayama [7] pointed out that when node i or element j is located on the free surface, the vectors $\mathbf{A}$, $\mathbf{B}$ and $\mathbf{C}$ at the time level $k + 1$ are related to the unknown quantity η. He noted that these vectors can be calculated approximately by the use of the free surface profile at the previous time level.

The time derivatives $\dot{\phi}$ and $\dot{\eta}$ were assumed to change as linear functions with respect to time during the interval Δt and expressed as follows:

$$\dot{\phi}^{k+1} = \frac{2}{\Delta t} \Delta\phi - \dot{\phi}^k \quad (6.109)$$

$$\dot{\eta}^{k+1} = \frac{2}{\Delta t} \Delta\eta - \dot{\eta}^k . \quad (6.110)$$

The derivative $\dot{D}$ is approximated with the backward finite difference scheme

$$D = \frac{D^{k+1} - D^k}{\Delta t} \tag{6.111}$$

but, as mentioned earlier, it is requested that D^{k+1} vanishes, thus

$$D = -\frac{D^k}{\Delta t}. \tag{6.112}$$

Substituting variables at time level $k+1$ expressed in terms of $\Delta\phi$, $\Delta\eta$ and variables at time level k into Eqs. (6.105) and (6.106), and linearizing these equations with respect to $\Delta\phi$ and $\Delta\eta$, the following linear system of algebraic equations is obtained:

$$\begin{bmatrix} G_{11} & G_{12} \\ G_{21} & G_{22} \end{bmatrix} \begin{Bmatrix} \Delta\phi \\ \Delta\eta \end{Bmatrix} = \begin{Bmatrix} R_1 \\ R_2 \end{Bmatrix}. \tag{6.113}$$

The values of ϕ, η, $\dot{\phi}$ and $\dot{\eta}$ at time level $k+1$ can then be calculated through Eqs. (6.107) – (6.110). The value of D at the time level $k+1$ is calculated at each nodal point on the free surface by the use of the following relation:

$$D^{k+1} = D^k + \Delta t \left\{ \frac{\partial\phi}{\partial t} + \frac{1}{2}\left[n_y^2 \left(\frac{\partial\eta}{\partial t}\right)^2 + \left(\frac{\partial\phi}{\partial s}\right)^2 \right] + g\,\eta \right\} \tag{6.114}$$

where all quantities between brackets refer to time level $k+1$.

Another numerical scheme was developed by Faltinsen [5] for the nonlinear waves radiating from a body oscillating on the free surface of a fluid of infinite extent and infinite depth. He discretized only the body surface and free surface, since for this case the integrals along the boundaries at infinity vanish. Furthermore, the free surface discretization is cut at a point $x = b(t)$, where b is a large number dependent on time, and the contribution of the integral from $b(t)$ to ∞ calculated analytically by using a far field solution (the same applies to the contribution of the integral from $-\infty$ to $-b(t)$).

The expressions for the velocity potential ϕ on the free surface and the free surface elevation are obtained by the following time-stepping procedure. By using the material derivative of ϕ and the Bernoulli equation, we can write

$$\frac{D\phi}{Dt} = \frac{\partial\phi}{\partial t} + \frac{\partial\phi}{\partial x}\frac{\partial\phi}{\partial x} + \frac{\partial\phi}{\partial y}\frac{\partial\phi}{\partial y} = \frac{1}{2}\left[\left(\frac{\partial\phi}{\partial x}\right)^2 + \left(\frac{\partial\phi}{\partial y}\right)^2 \right] - g\,\eta. \tag{6.115}$$

Further, from the kinematic boundary condition at the free surface,

$$\frac{D\eta}{Dt} = \frac{\partial\phi}{\partial y} \tag{6.116}$$

and from the material derivative of a particle

$$\frac{D x_F}{Dt} = \frac{\partial\phi}{\partial x} \tag{6.117}$$

where x_F is the x-coordinate of a fluid particle.

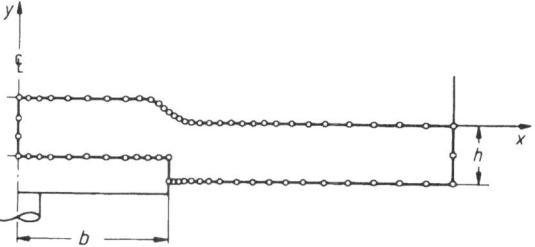

Fig. 6.24.   Discretization of two-dimensional tank

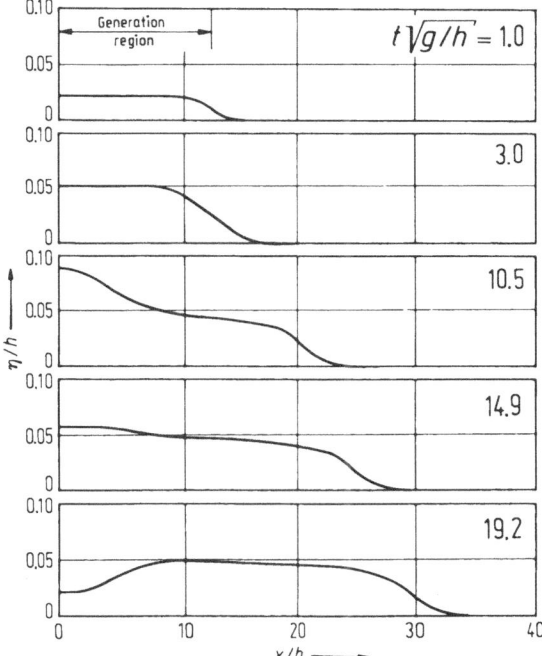

Fig. 6.25.   Free surface profiles at several time instants

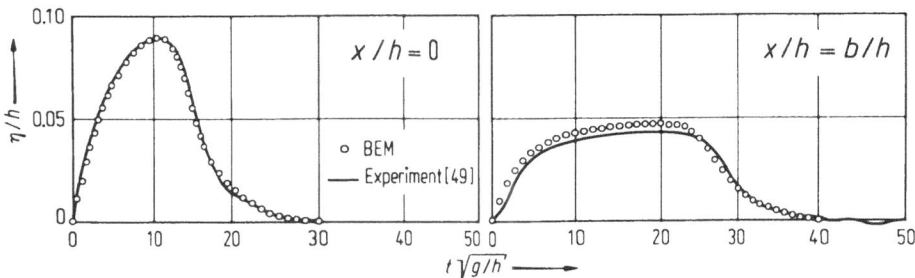

Fig. 6.26.   Time variation of free surface displacement

Thus, knowing the velocity potential at the free surface and the coordinates of particles on the free surface at some instant of time, Eqs. (6.115)–(6.117) provide means to find the change with time of these variables.

Figure 6.24 shows the discretization employed by Nakayama [7] for the problem of waves generated by vertical ground movements. Figure 6.25 presents the profiles of the free surface at different time instants, while Fig. 6.26 shows the time histories of the nondimensional elevation η/d at $x = 0$ and $x = \ell$. The numerical results are compared with the experimental data obtained by Hammack [49], with very good agreement. This problem was also analyzed by Liu [6]. In the time history of η/d at $x = \ell$ which he obtained, a strange zig-zag behaviour of the computed values is observed.

References

1 Brebbia, C.A., Telles, J.C.F., and Wrobel, L.C., *Boundary Element Techniques: Theory and Applications in Engineering*. Springer-Verlag, Berlin, 1984

2 Sarpkaya, T. and Isaacson, M., *Mechanics of Wave Forces on Offshore Structures*. Van Nostrand Reinhold, New York, 1981

3 Susbielles, G. and Bratu, C., *Vagues et Ouvrages Pétroliers en Mer*. Editions Techip, Paris, 1981

4 Masuda, K. and Kato, W., Hybrid B.E.M. for Calculating Nonlinear Wave Forces on Three-Dimensional Bodies, *Boundary Elements* (Ed. C.A. Brebbia et al.). Springer-Verlag, Berlin, 1983

5 Faltinsen, O.M., Numerical Solutions of Transient Nonlinear Free-Surface Motion Outside or Inside Moving Bodies. Second Int. Conf. on Numerical Ship Hydrodynamics, Berkeley, Calif., 1977

6 Liu, P.L.F., Integral Equation Solutions to Nonlinear Free-Surface Flows. Second Int. Conf. on Finite Elements in Water Resources, Pentech Press, London, 1978

7 Nakayama, T., Boundary Element Analysis of Nonlinear Water Wave Problems. Int. J. Num. Meth. Engng. **19**, 953 – 970, 1983

8 Kim, S.K., Liu, P.L.F., and Liggett, J.A., Boundary Integral Equation Solutions for Solitary Wave Generation, Propagation and Run-up. Coastal Engineering **7**, 299 – 317, 1983

9 Lamb, H., *Hydrodynamics*. Dover Publications, 1945

10 Bai, K.J. and Yeung, R., Numerical Solutions to Free Surface Flow Problems. Proc. Tenth Symp. Naval Hydrodynamics, Cambridge, Mass., 1974

11 Wehausen, J.V. and Laitone, E.V., Surface Waves, in *Handbuch der Physik IX*, Springer-Verlag, Berlin, 1960

12 Faltinsen, O.M. and Michelsen, F.C., Motions of Large Structures in Waves at Zero Froude Number. Proc. Int. Symp. Dynamics of Marine Vehicles and Structures in Waves, London, 1974

13 Monacella, V.J., The Disturbance due to a Slender Ship Oscillating in Waves in a Fluid of Finite Depth. Journal of Ship Research **10**, 242 – 252, 1966

14 Abramowitz, M. and Stegun, I.A., *Handbook of Mathematical Functions*. Dover, New York, 1965

15 John, F., On the Motion of Floating Bodies II. Comm. Pure and Applied Math. **3**, 45 – 101, 1950

16 Hogben, N. and Standing, R.G., Wave Loads on Large Bodies. Proc. Int. Symp. Dynamics of Marine Vehicles and Structures in Waves, London, 1974

17 Hearn, G.E., Alternative Methods of Evaluating Green's Functions in Three-Dimensional Ship-Wave Problems. Journal of Ship Research **21**, 89 – 93, 1977

18 Garrison, C.J. and Chow, P.Y., Wave Forces on Submerged Bodies. J. Waterways, Harbours and Coastal Eng. Div., ASCE **98**, 375 – 392, 1972

19 Wybro, P.G., On the Dynamics of Column – Stabilized Platforms Including Three-Dimensional Interaction Effects. Marine Technology **17**, 174–198, 1980

20 Lebreton, J.C. and Cormault, P., Wave Action on Slightly Immersed Structures: Some Theoretical and Experimental Considerations. Proc. Symp. Research on Wave Action, Delft, 1969

21 Sphaier, S.H. and Esperança, P.T.T., Motions and Forces on Large Bodies in the Presence of a Free Surface. A. M.C. Journal **2**, 1982 (in Portuguese)

22 Au, M.C. and Brebbia, C.A., Computation of Wave Forces on Three-Dimensional Offshore Structures, *Boundary Element Methods in Engineering* (Ed. C.A. Brebbia). Springer-Verlag, Berlin, 1982

23 Haskind, M.D., The Hydrodynamical Theory of the Oscillation of a Ship in Waves. Technical and Research Bulletin No. 1–12, SNAME, 1953

24 Frank, W., On the Oscillation of Cylinders in or below the Free Surface of Deep Fluids. Report 2375, NSRDC, Washington, D.C., 1967

25 Murphy, J.E., Integral Equation Failure in Wave Calculations. J. Waterways, Port, Coastal and Ocean Eng. Div., ASCE **104**, 330–334, 1978

26 Fenton, J.D., Wave Forces on Vertical Bodies of Revolution. J. Fluid Mech. **85**, 241–255, 1978

27 Black, J.L., Wave Forces on Vertical Axisymmetric Bodies. J. Fluid Mech. **67**, 369–376, 1975

28 Eatock-Taylor, R. and Dolla, J.P., Hydrodynamic Loads on Vertical Bodies of Revolution. Report No. OEG/78/6, Dept. of Mech. Engng., University College, London, 1978

29 Isaacson, M. de St.Q., Fixed and Floating Axisymmetric Structures in Waves. J. Waterways, Port, Coastal and Ocean Eng. Div., ASCE **108**, 180–199, 1982

30 Kim, W.D., On a Free-Floating Ship in Waves. Journal of Ship Research **10**, 182–191, 1966

31 Brebbia, C.A. and Walker, S., *Dynamic Analysis of Offshore Structures.* Newnes-Butterworths, London, 1979

32 Zienkiewicz, O.C., Bettess, P., and Kelly, D.W., The Finite Element Method for Determining Fluid Loadings on Rigid Structures: Two- and Three-Dimensional Formulations, in *Numerical Methods in Offshore Engineering* (Ed. O.C. Zienkiewicz et al.). Wiley, Chichester, 1978

33 Isaacson, M. de St.Q., Vertical Cylinders of Arbitrary Section in Waves. J. Waterways, Port, Coastal and Ocean Eng. Div., ASCE **104**, 309–324, 1978

34 Au, M.C. and Brebbia, C.A., Diffraction of Water Waves for Vertical Cylinders Using Boundary Elements. Applied Math. Modelling **7**, 106–114, 1983

35 Masetti, I.Q. and Wrobel, L.C., A Study of the Interaction Between Cylinders in Waves Using Boundary Elements. Offshore Engineering **4** (Ed. F.L.L.B. Carneiro et al.), Pentech Press, London, 1984

36 Ohmatsu, S., On the Irregular Frequencies in the Theory of Oscillating Bodies in a Free Surface. Papers Ship Res. Inst. No. 48, Tokyo, 1975

37 Paulling, J.R., Stability and Ship Motion in a Seaway. Summary Report, U.S. Coast Guard, 1970

38 Ursell, F., On the Heaving Motion of a Circular Cylinder on the Surface of a Fluid. Quart. J. Mech. Appl. Math. **2**, 218–231, 1949

39 Ogilvie, T.F. and Shin, Y.S., Integral Equation Solutions for Time-Dependent Free Surface Problems. J.S.N.A. Japan **143**, 86–96, 1978

40 Ijima, T., Chou, C. R., and Yoshida, A., Method of Analysis for Two-Dimensional Water Wave Problems. Proc. Fifteenth Coastal Eng. Conf., Honolulu, 1976

41 Yamamoto, T. and Yoshida, A., Elastic Mooring of Floating Breakwaters. Proc. Seventh Int. Harbour Congress, Antwerp, 1978

42 Au, M.C. and Brebbia, C.A., Numerical Prediction of Wave Forces Using the Boundary Element Method. Applied Math. Modelling **6**, 218–228, 1982

43 Mei, C.C., Numerical Methods in Water Wave Diffraction and Radiation. Ann. Rev. Fluid Mech. **10**, 393–416, 1978

44 Georgiadis, G. and Hartz, B.J., A Boundary Element Program for the Computation of Three-Dimensional Hydrodynamic Coefficients. Proc. Int. Conf. Finite Element Methods, Shangai, 1982

45 Liu, P.L.F. and Abbaspour, M., An Integral Equation Method for the Diffraction of Oblique Waves by an Infinite Cylinder. Int. J. Num. Meth. Engng. **18,** 1497–1504, 1982
46 Salmon, J.R., Liu, P.L.F., and Liggett, J.A., Integral Equation Method for Linear Water Waves. J. Hydr. Div., ASCE **106,** 1995–2010, 1980
47 Lennon, G.P., Liu, P.L.F., and Liggett, J.A., Boundary Integral Solutions of Water Wave Problems. J. Hydr. Div., ASCE **108,** 921–931, 1982
48 Nakayama, T. and Washizu, K., The Boundary Element Method Applied to the Analysis of Two-Dimensional Nonlinear Sloshing Problems. Int. J. Num. Meth. Engng. **17,** 1631–1646, 1981
49 Hammack, J.L., A Note on Tsunamis: Their Generation and Propagation in an Ocean of Uniform Depth. J. Fluid Mech. **60,** 769–799, 1973
50 Garrison, C.J., Hydrodynamic Loading of Large Offshore Structures. Three-Dimensional Source Distribution Methods, in *Numerical Methods in Offshore Engineering* (Ed. O.C. Zienkiewicz et al.). Wiley, Chichester, 1978

Chapter 7

Boundary Integral Formulation of Mass Matrices for Dynamic Analysis

by D. Nardini and C. A. Brebbia

7.1 Introduction

A novel alternative method for dynamic analysis in solid mechanics using boundary elements is presented in this chapter. The basic principles of vibrations of solids are briefly reviewed in order to provide the necessary background for the derivation of the method and interpretation of the numerical results. For a more detailed study of elastodynamics, [1–3] are recommended.

The first boundary integral formulation for the dynamic analysis of solids was presented by Rizzo and Cruse [7] in 1968. This formulation, which has since then been used by many other authors, requires the solution of the problem in the frequency domain rather than the time domain. The dynamic response of the problem is obtained by sweeping the frequency range of interest, and time dependent solutions can only be found by applying back-transformation procedures using sometimes Laplace rather than Fourier transform.

The above formulation requires the use of a frequency dependent fundamental solution and the eigenvalues of the system are found by applying different forcing frequencies until resonance occurs. The technique has been applied by a series of authors [5, 6] but can be uneconomic and inaccurate. Manolis and Beskos [7] extended Cruse's work for steady state dynamic stress concentration, together with a modification of the Laplace transform inversion scheme. One of the main problems of this approach is the numerical inversion of the Laplace transform although some useful guidelines can be found in [8].

This chapter describes a new method which permits the formulation of a mass matrix in function of the boundary nodes only. The approach is developed for two and three dimensional bodies and permits to find their transient response when subjected to time dependent external tractions and support excitations. The new procedure allows for elastodynamic problems to be treated in a similar way as in finite differences or finite elements, i.e. the problem is reduced to a set of time dependent differential equations expressed in matrix form. The problem of free vibrations is then derived as a special case of the transient formulation [9–11]. The main advantage of the new approach is that the boundary integrals need to be computed only once as they are frequency independent. Hence the procedure is extremely economic for free vibrations when compared to previous techniques.

The implementation of the new approach for two dimensional elastodynamics is described in detail. The method is applied for transient and eigenvalue analysis and a series of important numerical and computational aspects are discussed.

The procedure is well suited for Computer Aided Design applications as only the surface of the solid needs to be described. The elegance of the approach and the accuracy of the numerical results is demonstrated by some of the examples presented at the end of the Chapter.

7.2 Formulation of the Dynamical Problem

Let us consider an elastic body Ω enclosed by a boundary Γ (Fig. 7.1), subjected to the given body forces b and external boundary tractions.

The conditions of dynamical equilibrium of an infinitesimal volume are expressed in cartesian tensor notation as

$$\sigma_{ij,j} + b_i - \varrho\,\ddot{u}_i = 0 \tag{7.1}$$

where

σ_{ij}: cartesian stress tensor u_i: displacement field
b_i: volume body force ϱ: mass density.

Standard tensor notation is employed, where a repeated subscript implies summation (1, 2, 3) and the comma denotes a co-ordinate differentiation. The temporal differentiation is denoted by an overhead dot.

The relationship between the stress tensor σ_{ij} and the displacement field u_i depends on the constitutive law of the material. In the case of the homogeneous elastic material, considered in this chapter, the following relations hold

$$\sigma_{ij} = \lambda\,\delta_{ij}\,u_{k,k} + \mu\,(u_{i,j} + u_{j,i}) \tag{7.2}$$

where λ and μ are Lamés constants. They are expressed in terms of the more familiar Young's modulus and Poisson's ratio as

$$\lambda = \frac{vE}{(1+v)(1-2v)} \tag{7.3}$$

$$\mu = \frac{E}{2(1+v)} \tag{7.4}$$

while δ_{ij} represents the Kronecker delta symbol.

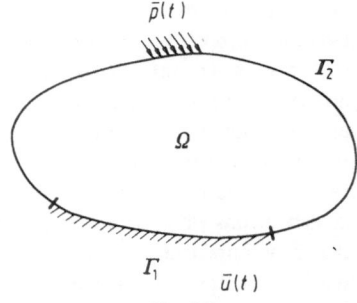

$\bar{p}(t)$

Γ_2

Ω

Γ_1 $\bar{u}(t)$

Fig. 7.1

Differentiating the stress tensor of Eq. (7.2) with respect to the jth co-ordinate and summing yields

$$\sigma_{ij,j} = \mu u_{i,jj} + (\lambda + \mu) u_{j,ji} . \tag{7.5}$$

Using this result, Eq. (7.1) is now expressed in terms of the displacement field as

$$\mu u_{i,jj} + (\lambda + \mu) u_{j,ji} + b_i - \varrho \ddot{u}_i = 0 \tag{7.6}$$

which represents the dynamic equilibrium equation of elasticity.

In order to formulate uniquely the dynamic problem of an elastic body Ω enclosed by a boundary Γ, one has to define the boundary and the initial conditions. We shall assume to have the boundary divided into two distinct parts: Γ_1 with the imposed displacements, and Γ_2 with the imposed tractions (see Fig. 7.1). The boundary conditions are thus expressed as

$$u_i = \bar{u}_i(t) \quad \text{on } \Gamma_1 \tag{7.7}$$

$$p_i = \bar{p}_i(t) \quad \text{on } \Gamma_2 \tag{7.8}$$

where p_i denotes components of the tractions, and values with a bar are externally imposed fields. Tractions at the boundary are related to stresses by a co-ordinate transformation using components of the outward normal n_j:

$$p_i = \sigma_{ij} n_j . \tag{7.9}$$

Finally, the initial conditions specify the state of displacements and velocities at time zero, i.e.

$$u_i = u_i^0, \quad \dot{u}_i = \dot{u}_i^0 \quad \text{in } \Omega \text{ at time } t = 0. \tag{7.10}$$

The object of dynamic analysis is to solve Eq. (7.6) for displacements, subjected to the boundary conditions (7.7) and (7.8), and the initial conditions (7.10).

7.3 Various Boundary Integral Formulations

In order to reduce the problem of elastodynamics to the boundary, one can take a number of different approaches, the two main categories being indirect and direct methods (see, for example, [12, 13]). They are both based on employing a singular solution that satisfies the equations of motion, then using the Betti-Rayleigh reciprocal identity, and finally expressing the resulting volume integrals by the equivalent boundary integrals. Two such methods, presented in [12], will be reviewed here only briefly.

(i) Impulse Response

The dynamical solution in the unbounded medium due to a point impulse at an arbitrary point is established. Using this solution in the Betti-Rayleigh dynamic reciprocal relationship results in a boundary integral expression (Wheeler and Sternberg formula). The final expression, from which a numerical solution procedure can be established, has been derived by Cole et al. [14], and presents a dynamic counterpart of the Somigliana identity in statics. This approach has also been employed by Mansur and Brebbia to solve two-dimensional elastodynamic problems [15, 16].

(ii) Harmonic Response

Here, the solution is sought in the frequency domain by use of the Laplace or Fourier transform and the fundamental solution resulting from the harmonic excitation in the unbounded domain is employed. This approach has been first derived by Cruse and Rizzo [4]. Using the Betti-Rayleigh reciprocal relationship, one arrives at the boundary integral expression, which describes the response of the medium for the given frequency component of the forcing function.

7.4 Boundary Integral Formulation Using the Statical Fundamental Solution

An alternative approach to the solution of equations of motion will be described, as developed by Nardini and Brebbia [9–11].

We shall confine the external influences to the boundary only, in which case the volume forces are only those due to inertia, Eq. (7.1) is thus reduced to

$$\sigma_{ij,j} - \varrho \ddot{u}_i = 0. \tag{7.11}$$

The generalised Maxwell-Betti reciprocal theorem, relating two independent displacement modes can be written as follows,

$$\int_\Omega (u_i^* \, \sigma_{ij,j} - u_i \, \sigma_{ij,j}^*) \, d\Omega = \int_\Gamma (u_i^* \, p_i - u_i \, p_i^*) \, d\Gamma \tag{7.12}$$

with u_i being the first displacement field, σ_{ij} and p_i its stress tensor and traction vector, respectively. The second displacement field u_i^* has its stresses and tractions denoted by σ_{ij}^* and p_i^*. This field can be assumed to be virtual and we shall associate the solution to the static equation under point load, i.e. the solution of

$$\sigma_{kij,j}^* + \delta_{ki} \, \Delta \, (x - \xi) = 0 \tag{7.13}$$

where

$$\Delta \, (x - \xi) = 0 \quad \text{for } x \neq \xi$$
$$\int_\Omega \Delta \, (x - \xi) \, d\Omega = 1. \tag{7.14}$$

The solution to Eq. (7.13) is denoted by u_{ki}^*, and the corresponding tractions at the boundary by p_{ki}^*. Taking into account Eqs. (7.13) and (7.14), Eq. (7.12) yields

$$\int_\Omega \sigma_{ij,j} \, u_{ki}^* \, d\Omega = - c_{ki} u_i + \int_\Gamma u_{ki}^* \, p_i \, d\Gamma - \int_\Gamma p_{ki}^* \, u_i \, d\Gamma \tag{7.15}$$

which is a well known boundary integral expression in elastostatics. The constants c_{ki} are uniquely defined by the position of the point x with respect to the boundary.

Substituting $\sigma_{ij,j}$ from Eq. (7.11), the density ϱ being a constant, Eq. (7.15) becomes

$$\varrho \int_\Omega \ddot{u}_i \, u_{ki}^* \, d\Omega = - c_{ki} u_i + \int_\Gamma u_{ki}^* \, p_i \, d\Gamma - \int_\Gamma p_{ki}^* \, u_i \, d\Gamma. \tag{7.16}$$

At this point, in order to transform the inertial domain integral, an approximation to the accelerations $\ddot{u}_i$ within the domain will be employed. The time dependent displacements $u_i(\xi, t)$ are expressed by the sum of m co-ordinate functions $f_j^j(\xi)$ multiplied by the unknown time dependent functions $\alpha_i^j(t)$

$$u_i(\xi, t) = \alpha_i^j(t) \, f^j(\xi) \tag{7.17}$$

with summation on $j = 1$ to m implied.

The accelerations are derived from this expression as

$$\ddot{u}_i = \ddot{\alpha}_i^j(t) \, f^j(\xi). \tag{7.18}$$

With this substitution, the domain integral of Eq. (7.16) yields

$$\int_\Omega \ddot{u}_i \, u_{ki}^* \, d\Omega = \ddot{\alpha}_i^j \int_\Omega f^j \, u_{ki}^* \, d\Omega = \ddot{\alpha}_i^j \int_\Omega \delta_{\ell i} \, f^j \, u_{ki}^* \, d\Omega. \tag{7.19}$$

The resulting domain integral can now be transformed into equivalent boundary integrals as will be described in what follows. The first step is to look for the solution to the static problem in the unbounded domain, i.e. the solution of the following equation:

$$\bar{\sigma}_{\ell i m,m} + \delta_{\ell i} f^j = 0. \tag{7.20}$$

The displacement solution to this equation will be denoted by $\psi_{\ell i}^j$, and the corresponding tractions by $\eta_{\ell i}^j$. Using the transformation given by Eq. (7.15), one can reduce the domain integral (7.19) to equivalent boundary integrals as follows

$$\int_\Omega \delta_{\ell i} \, f^j \, u_{ki}^* \, d\Omega = - \int_\Omega \bar{\sigma}_{\ell i m,m}^j \, u_{ki}^* \, d\Omega = c_{ki} \, \psi_{\ell i}^j - \int_\Gamma u_{ki}^* \, \eta_{\ell i}^j \, d\Gamma + \int_\Gamma p_{ki}^* \, \psi_{\ell i}^j \, d\Gamma. \tag{7.21}$$

Substituting this result into formula (7.19) and rearranging Eq. (7.16) gives

$$c_{ki} \, u_i + \int_\Gamma p_{ki}^* \, u_i \, d\Gamma - \int_\Gamma u_{ki}^* \, p_i \, d\Gamma + \varrho \left(c_{ki} \, \psi_{\ell i}^j + \int_\Gamma p_{ki}^* \, \psi_{\ell i}^j \, d\Gamma - \int_\Gamma u_{ki}^* \, \eta_{\ell i}^j \, d\Gamma \right) \ddot{\alpha}_\ell^j = 0. \tag{7.22}$$

This represents a boundary integral expression from which a numerical solution procedure can be derived.

7.5 The Numerical Solution Procedure

To solve Eq. (7.22) numerically the boundary element method can now be employed. The boundary is divided into elements, including a finite number of nodal points in the usual manner [17].

When the unit force producing the displacement field u_{ki}^* is applied at each node in turn, the expression (7.22) yields a system of equations of the form

$$c_{ki}(A_n) \, u_i(A_n) + \int_\Gamma p_{ki}^*(A_n, B) \, u_i(B) \, d\Gamma - \int_\Gamma u_{ki}^*(A_n, B) \, p_i(B) \, d\Gamma \tag{7.23}$$

$$+ \varrho \left(c_{ki}(A_n) \, \psi_{\ell i}^j(A_n) + \int_\Gamma p_{ki}^*(A_n, B) \, \psi_{\ell i}^j(B) \, d\Gamma - \int_\Gamma u_{ki}^*(A_n, B) \, \eta_{\ell i}^j(B) \, d\Gamma \right) \ddot{\alpha}_\ell^j = 0$$

where A_n coincide with the n-th nodal point, while B represents a point of integration on the boundary. This expression can be written in a more compact form using matrix notation as follows:

$$c_n\,u_n + \int_\Gamma p_n^*\,u\,d\Gamma - \int_\Gamma u_n^*\,p\,d\Gamma + \varrho\left(c_n\,\psi_n^j + \int_\Gamma p_n^*\,\psi^j\,d\Gamma - \int_\Gamma u_n^*\,\eta^j\,d\Gamma\right)\ddot{\alpha}^j = 0 \quad (7.24)$$

(with sum on $j = 1, \ldots, m$ and no sum on $n = 1, \ldots, m$).

To obtain a numerical solution to this system of integral equations, an approximation to displacements and tractions within the boundary elements is established. Denoting the displacement vector of the nodes contained within the element e by u_e, and the corresponding tractions by p_e, the values of displacements and tractions are approximated using a matrix of interpolation functions Φ, i.e.

$$u = \Phi\,u_e \quad (7.25)$$

$$p = \Phi\,p_e. \quad (7.26)$$

Notice that the interpolation functions used for u and p could be different, but being the same simplifies somewhat the computations. For a chosen class of co-ordinate functions f^j, the functions ψ and η are known at the boundary, so that those integrals containing them can be computed as given in Eq. (7.24). However, approximating these functions along the boundary using the same set of Φ inter-polation functions as (7.25) or (7.26) will enable us to use the same coefficients as previously computed and simplify the amount of integration required to run a problem.

Hence we will propose:

$$\psi^j = \Phi\,\psi_e^j \quad (7.27)$$

$$\eta^j = \Phi\,\eta_e^j. \quad (7.28)$$

Substituting Eqs. (7.25) through (7.28) into (7.24), results in the following set of equations:

$$c_n\,u_n + h_{ne}\,u_e - g_{ne}\,p_e + \varrho(c_n\,\psi_n^j + h_{ne}\,\psi_e^j - g_{ne}\,\eta_e^j)\ddot{\alpha}^j = 0 \quad (7.29)$$

(with sum on $e = 1, \ldots, n_e$ (number of elements), $j = 1, \ldots, m$ and no sum on $n = 1, \ldots, m$).

Submatrices h and g are well known from elastostatics and are given by

$$h_{ne} = \int_{\Gamma_e} p_n^*\,\Phi\,d\Gamma \quad (7.30)$$

$$g_{ne} = \int_{\Gamma_e} u_n^*\,\Phi\,d\Gamma \quad (7.31)$$

where Γ_e denotes the part of the boundary belonging to element e.

The system of Eqs. (7.29) is now written in assembled matrix form as

$$H\,U - G\,P + \varrho(H\,\psi - G\,\eta)\,\ddot{\alpha} = 0. \quad (7.32)$$

It is evident that the global matrix H is assembled from individual element matrices h_{ne} and the diagonal submatrices c_n. The global matrix G is obtained from individual matrices g_{ne}, while the matrices ψ and η simply contain values of functions ψ and η at nodal points.

The remaining step is to select one of the two bases U and α, present in Eq. (7.32). A natural choice is U, since the boundary conditions are expressed through it. The transformation from U to α is given by Eq. (7.24). Applying this equation to every nodal point, a linear relationship results in the form

$$U = F\alpha \qquad (7.33)$$

with the coefficients of the matrix F being the values of functions f^j at nodes. A choice of linearly independent set of functions, their number equal to the number of nodal points, ensures that the matrix F possesses an inverse, i.e. $E = F^{-1}$, resulting in the inverse transform

$$\alpha = EU \qquad (7.34)$$

and consequently,

$$\ddot{\alpha} = E\ddot{U}. \qquad (7.35)$$

Using this substitution for $\ddot{\alpha}$, Eq. (7.32) yields

$$HU + M\ddot{U} = GP \qquad (7.36)$$

where the generalised mass matrix M is given by

$$M = \varrho(H\psi - G\eta)\,E. \qquad (7.37)$$

Equation (7.36) represents a system of linear differential equations with constant coefficients. In order to be able to obtain the solution for displacements and tractions, a distinction between the two types of boundaries have to be made. We shall denote the variables at Γ_1 by the subscript 1 and at Γ_2 by 2, i.e.,

$$U = \begin{bmatrix} U_1 \\ U_2 \end{bmatrix} \qquad (7.38)$$

$$P = \begin{bmatrix} P_1 \\ P_2 \end{bmatrix}. \qquad (7.39)$$

Partitioning the global matrices into the corresponding submatrices, Eq. (7.36) is rewritten as

$$H_{11}\,U_1 + H_{12}\,U_2 + M_{11}\,\ddot{U}_1 + M_{12}\,\ddot{U}_2 = G_{11}\,P_1 + G_{12}\,P_2 \qquad (7.40)$$

$$H_{21}\,U_1 + H_{22}\,U_2 + M_{21}\,\ddot{U}_1 + M_{22}\,\ddot{U}_2 = G_{21}\,P_1 + G_{22}\,P_2. \qquad (7.41)$$

Inverting G_{11} from the first set of equations, the tractions P_1 at Γ_1 can be expressed in terms of other variables, finally obtaining

$$\hat{H}\,U_1 + \hat{M}\,\ddot{U}_2 = \hat{G}\,P_2 + \tilde{H}\,U_1 + \tilde{M}\,\ddot{U}_1 \qquad (7.42)$$

with the modified matrices defined as follows,

$$\hat{H} = H_{22} - G_{21}\,G_{11}^{-1}\,H_{12} \qquad (7.43)$$

$$\hat{M} = M_{22} - G_{21}\,G_{11}^{-1}\,M_{12} \qquad (7.44)$$

$$\hat{G} = G_{22} - G_{21}\,G_{11}^{-1}\,G_{12} \qquad (7.45)$$

$$\tilde{H} = G_{21}\,G_{11}^{-1}\,H_{11} - H_{21} \qquad (7.46)$$

$$\tilde{M} = G_{21}\,G_{11}^{-1}\,M_{11} - M_{21}. \qquad (7.47)$$

The right hand side of the system of Eqs. (7.42) contains the externally applied displacements U_1 and tractions P_2, so the system can be solved for U_2 using a direct time integration procedure.

Internal Degrees of Freedom

Confining Eq. (7.42) to the boundary produces a limited number of dynamical degrees of freedom, which may result in a less accurate approximation of higher modes of vibration. In order to be able to overcome this limitation, internal points may be assigned, resulting in additional equations of motion.

7.6 Derivation of Different Types of Dynamical Problems

Free Vibrations

The analysis of natural modes and frequencies can be deduced from the general transient system of Eq. (7.42) by setting the external influences (right hand side) to zero, thus

$$\hat{H} U_2 + \hat{M} \ddot{U}_2 = 0. \tag{7.48}$$

Using the fact that in free vibrations the displacements are harmonic functions of time, i.e.

$$\ddot{U}_2 = - \omega^2 U_2 \tag{7.49}$$

with ω being the natural circular frequency, Eq. (7.48) is reduced to

$$\hat{H} U_2 = \omega^2 \hat{M} U_2 \tag{7.50}$$

which represents the generalized algebraic enigenvalue problem. It should be pointed out that neither $\hat{H}$ nor $\hat{M}$ are symmetric, positive definite, so that care should be taken in the choice of the appropriate eingenvalue solution algorithm.

Forced Vibrations

It is seldom the case in practical application to have simultaneously both, the external tractions and support movements, applied to the system. Therefore, we shall consider the two cases separately although the two effects could be taken into consideration simultaneously if required.

In the absence of support excitations, Eq. (7.42) is simplified into

$$\hat{H} U_2 + \hat{M} \ddot{U}_2 = \hat{G} P_2 \tag{7.51}$$

where P_2 is a vector of externally applied, time dependent tractions, at nodal points of Γ_2. The distribution of tractions along the boundary elements is governed by the interpolation function Φ, as given by Eq. (7.26).

It is often the case that at the free boundary, especially at corners, tractions between the neighbouring elements are not continuous (Fig. 7.2).

When boundary elements are of linear or higher order, values of tractions at common nodes are generally different. The introduction of double nodes at these points should be avoided due to the number of undesirable side-effects that they produce.

One way out of this difficulty is to compute the right hand side of Eq. (7.51) on element level, rather than on node level. Namely, the **G** matrix should never be assembled, but its element submatrices should be multiplied by the element vectors of P_2. In that way the external tractions on Γ_2 are expressed, and taken into the account, element wise, which permits the discontinuities of tractions at element boundaries without any constraints.

Such a procedure presents an elegant way out of an otherwise serious difficulty of the stress discontinuity problem, present in higher order boundary elements.

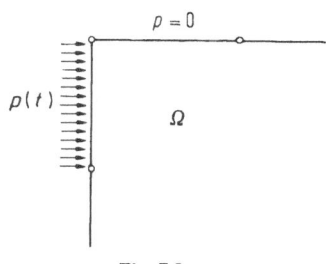

Fig. 7.2

Support Excitations

When support movements are given as general functions of time, Eq. (7.42) should be employed, which in the absence of external tractions gives

$$\hat{\mathbf{H}}\,\mathbf{U}_2 + \hat{\mathbf{M}}\,\ddot{\mathbf{U}}_2 = \overset{\scriptscriptstyle\approx}{\mathbf{H}}\,\mathbf{U}_1 + \overset{\scriptscriptstyle\approx}{\mathbf{M}}\,\ddot{\mathbf{U}}_1 . \tag{7.52}$$

This expression employs the enforced displacements $\mathbf{U}_1$ and accelerations $\ddot{\mathbf{U}}_1$ at Γ_1 part of the boundary, and is valid for any compatible variation of $\mathbf{U}_1$ with time.

However, a case of support movements, most frequently encountered in practice, is the kinematic excitation, in which all the points belonging to Γ_1 move as a rigid body. The most obvious example of such a movement is the effect of an earthquake on a structure.

The resulting body force in the relative coordinate system (i.e. with respect to the support) for a translational acceleration a_i in the i-th direction is

$$b_i = -\varrho a_i . \tag{7.53}$$

Therefore the governing differential Eq. (7.1) becomes

$$\sigma_{ij,j} - \varrho a_i - \varrho \ddot{u}_i = 0 . \tag{7.54}$$

Applying the same solution procedure described in this chapter, in place of Eq. (7.36) we obtain

$$\mathbf{H}\,\mathbf{U} + \mathbf{M}\,\ddot{\mathbf{U}} = \mathbf{G}\,\mathbf{P} + \mathbf{B} \tag{7.55}$$

with the coefficients of the vector **B** given by

$$b_k = \left[\varrho \int_{\Omega} u_{ki}^* \, d\Omega\right] a_i . \tag{7.56}$$

This volume integral can be expressed by the equivalent boundary integrals using the Galerkin tensor, the procedure being described by Danson [18]. Another way to take this integral to the boundary is by using the transformation (7.21), with the special case of f^j being a unit function (described in the next section).

Finally, taking into the account that $U_1 = 0$ and $P_2 = 0$, Eq. (7.39) is reduced to

$$\hat{H} U_2 + \hat{M} \ddot{U}_2 = \hat{B} \tag{7.57}$$

with

$$\hat{B} = B_2 - G_{21} G_{11}^{-1} B_1 . \tag{7.58}$$

7.7 Two-Dimensional Formulation

The Fundamental Solution

The static fundamental solution for the plane strain problem is given by (see for example Brebbia et al. [17]).

$$u_{ki}^* = \frac{1}{8 \pi \mu (1 - v)} [\delta_{ki}(4v - 3) \ln R + r_k r_i] \tag{7.59}$$

with

$R =$ distance from the source to a field point
$r_k = x_k/R$ x_k is the k-th coordinate from the source to a field point.
μ, v shear modulus and Poisson's ratio.

The corresponding tractions are derived as

$$p_{ki}^* = \frac{1}{4\pi(1 - v) R} [(1 - 2v)(r_k n_i - r_i n_k) - \{\delta_{ki}(1 - 2v) + 2r_k r_i\} r_m n_m] \tag{7.60}$$

with n_i being components of the outward normal to the boundary.

Inertia Interpolation Functions

The choice of a class of functions f^j, used for the approximation of the inertial terms, will strongly influence the accuracy of the results. Fortunately, the absence of derivatives in the mass integral imposes very few restrictions on the class of available functions.

It is to be expected that even simple interpolation functions will be sufficient for computation of the lower natural frequencies and modes. To obtain higher modes accurately, one would have to employ a more sophisticated approximation, coupled with the introduction of some internal degrees of freedom. Therefore, this choice of interpolation function will depend on the particular application of the method.

Besides these considerations, the class of functions f^j must produce a fairly simple closed solution to the problem given by Eq. (7.20) in the unbounded domain, for only then can the domain integral, due to inertia, be reduced to the boundary, as shown in Eq. (7.21).

A number of examples of choices of functions f^j follows:

a) A constant

$$f(\xi) = 1 \tag{7.61}$$

in which case the function ψ, being the solution to Eq. (7.20) is simply

$$\psi_{\ell i} = \frac{1 - 2v}{5 - 4v} r_\ell r_i R^2. \tag{7.62}$$

b) A linear function in terms of a coordinate ξ_k

$$f(\xi) = \xi_k \tag{7.63}$$

in which case

$$\psi_{\ell i} = a_1 \delta_{ik} \xi_\ell + a_2 \delta_{k\ell} \xi_i + a_3 \delta_{\ell i} \xi_k \tag{7.64}$$

with the constants:

$$a_1 = -\frac{1}{16(3 - 4v)}$$

$$a_2 = \frac{1 - 2v}{2(1 - v)} a_1 \tag{7.65}$$

$$a_3 = -(5 - 8v) a_1.$$

c) A conical function defined by a point A_j

$$f^j(\xi) = R(A_j, \xi) \tag{7.66}$$

where $R(A_j, \xi)$ is the distance between the point A_j which is defining the function, and a point ξ. The function ψ is again found to be quite simple, given by

$$\psi_{\ell i} = [(3 - 10v/3) \delta_{\ell i} - r_\ell r_i] \frac{R^3}{30(1 - v)} \tag{7.67}$$

where R, r_ℓ, r_i now refer to the distance between A_j and ξ.

d) A harmonic function of a coordinate ξ_k

$$f(\xi) = \sin\left[\frac{m\pi}{L} \xi_k\right] \tag{7.68}$$

where L is the basic half-period of the sin, and m is a number of waves. In this case the function ψ is found to be harmonic too, given by the expression

$$\psi_{\ell i} = \left[\left(\frac{L}{m\pi}\right)^2 \left(-\delta_{\ell i} + \frac{1}{2(1 - v)} \delta_{ik} \delta_{\ell k}\right)\right] \sin\left(\frac{m\pi}{L} \xi_k\right)$$

with no sum over k.

The possible classes of functions f^j and their corresponding functions ψ are by no means exhausted by the ones presented above.

7.8 Computer Implementation

a) Boundary Element Type

It had been shown that constant boundary elements do not yield a satisfactory approximation in solid mechanics, so that either linear or quadratic elements should be employed. This is due to the fact that constant elements do not impose sufficient continuity of the displacements along the boundary. The elements used in the applications described in this chapter are straight, but with quadratic interpolation functions, thus having three nodes.

The accuracy in computing the boundary integrals is critical for the numerical stability of the solution. Therefore, one should attempt to work out the analytical expressions for the integrals involved, and resort to the numerical integration only where this is not feasible. In the numerical integration, the Gauss quadrature is generally accepted as most efficient, but in order to satisfy both, the accuracy and the efficiency criteria, integration with varying number of Gauss points is essential. This is needed because in the vicinity of a singularity, the fundamental solution has very high gradients. Also, functions containing logarithmic or hyperbolic singularities should be integrated by modified Gauss quadratic procedures, which take the singularity into account [17].

b) The Algebraic Eigenvalue Problem Solution

As shown in Eq. (7.50), the case of free vibrations can be reduced to a generalised algebraic eigenvalue problem. Since both the matrices $\mathbf{H}$ and $\mathbf{M}$ are neither symmetric, nor positive definite, the choice of a method of solution is narrowed. Nevertheless, as the order of the matrices is much smaller than in the use of domain formulations (finite elements or finite differences), the efficiency of the eigenvalue solver is not as critical.

A method chosen here was to reduce the generalised eigenvalue problem to a standard one by the inversion of matrix $\mathbf{H}$, obtaining

$$\mathbf{A}\mathbf{U}_2 = \lambda \mathbf{U}_2 \tag{7.70}$$

with

$$\mathbf{A} = \hat{\mathbf{H}}^{-1}\hat{\mathbf{M}}, \quad \lambda = 1/\omega^2 .$$

The matrix $\mathbf{A}$ is transformed into the three-diagonal form by the Householder algorithm, and then the eigenvalues and eigenvectors of the transformed matrix are found by the $Q-R$ algorithm (see for example, Wilkinson [19]).

It should be pointed out that the resulting eigenvalues are generally complex. However, the examples worked out have shown that complex frequencies, if present, appear in higher modes, which are not usually required in the analysis. In this respect, one must also bear in mind that inaccurate results in higher modes of vibrations are inevitable when approximating a continuum problem by a finite number of degrees of freedom.

c) The Transient Response Solution

Induced transient vibrations may result either from inforced displacements on Γ_1 or from applied tractions on the Γ_2 part of the boundary, as shown in Eq. (7.42).

Solution to either of these cases may be obtained by a direct time integration procedure.

However, in the choice of the appropriate method, due to unreliable higher modes, one should bear in mind the following criteria:

(i) The method should be unconditionally stable, and therefore implicit.
(ii) The method should provide numerical damping for high frequencies.

One of the direct time integration procedures meeting these requirements is the Houbolt method [20], and this has been chosen for the transient analysis presented in this chapter.

d) Computer Storage Considerations

From Eqs. (7.36) and (7.37) it is evident that six global matrices are involved in the computations, these being:

H, G elastostatic matrices obtained by boundary integration
F   the $u-\alpha$ transformation matrix, obtained by evaluation the functions f^j at nodes
ψ, η   matrices obtained by evaluating functions $\psi^j_{t_i}$ and $\eta^j_{t_i}$ at nodes
M   the generalised mass matrix.

In the computer implementation, it is possible to sequence the operations in such a way as to accommodate all the required data in only two global matrices. Therefore, the working area of the program is much the same as for the case of many elastostatic boundary element programs.

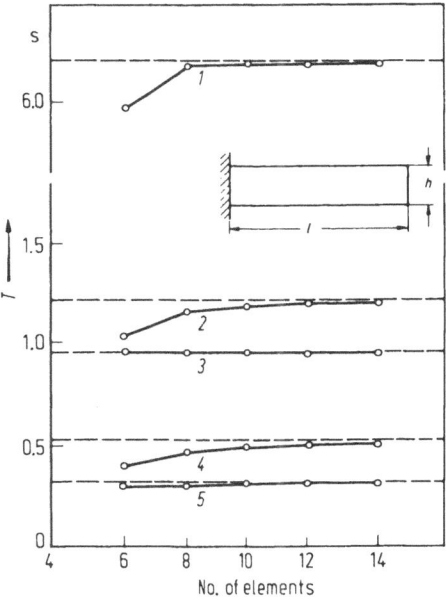

Fig. 7.3

7.9 Applications

A number of plane elastodynamic cases will be discussed here. They range from eigenvalue analysis to transient dynamics under external tractions or excitations at the supports. The boundary elements employed in all the examples are quadratic and the domain interpolation functions used are of class c. In all cases an internal node has been added and a constant type function – type a) of Sect. 7.7 – associated with it.

Example 1. *Eigenvalue Analysis of a Cantilever Beam.* The first five modes of vibrations of the deep cantilever beam shown in Fig. 7.3 were computed using different boundary element meshes. The properties of the beam were $h = 6$, $\ell = 24$, $E/\varrho = 10^4$, $v = 0.2$.

The results shown in the same figure are compared against a finite element solution using rectangular four-noded elements and containing 451 nodes. It can be

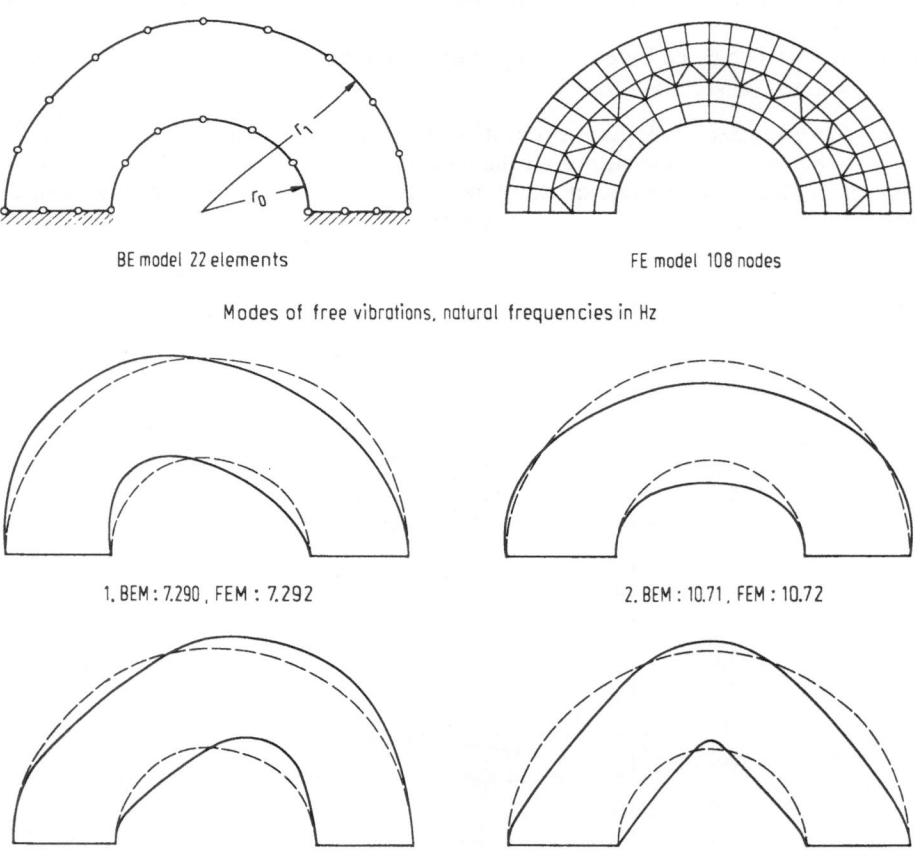

BE model 22 elements   FE model 108 nodes

Modes of free vibrations, natural frequencies in Hz

1. BEM : 7.290 , FEM : 7.292   2. BEM : 10.71 , FEM : 10.72

3. BEM : 16.79 , FEM : 16.49   4. BEM : 18.89, FEM : 18.35

Fig. 7.4

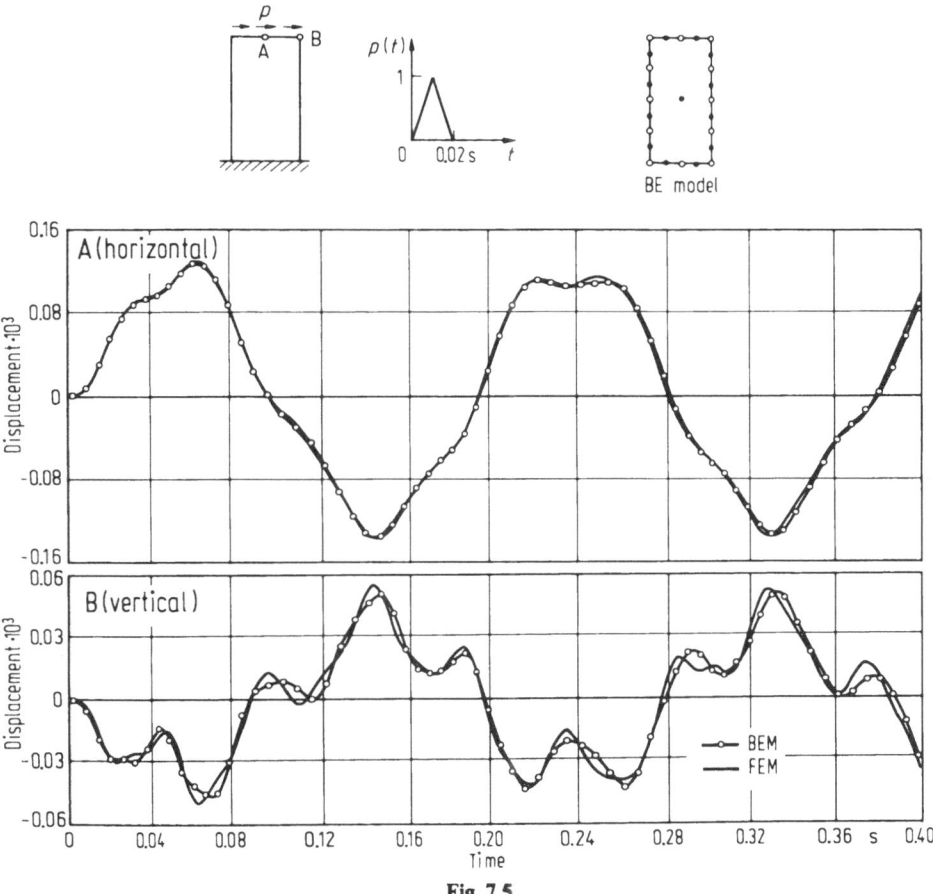

Fig. 7.5

seen that even for a small number of boundary elements the eigenvalues are
accurate when compared against the Finite Element Method fine mesh solution.
They also show a good rate of convergence.

Example 2. *Eigenvalue Analysis of an Arch.* The free vibrations of the arch shown in
Fig. 7.4 were analysed using boundary and finite elements. The dimensions and
constants for the arch were $r_0 = 2.5$; $r_1 = 5.0$; $E/\varrho = 10^4$ and $v = 0.25$. The two
numerical models used in the computations are also shown in the figure, together
with the first four modes of vibrations. Notice that the computed values for the
natural frequencies using BEM or FEM are in good agreement despite the small
number of boundary elements used in this analysis.

Example 3. *Rectangle Subjected to a Shear Impulse.* A simple rectangle cantilever
under a uniform shear impulse — Fig. 7.5 — was analysed. The material constants
were taken as $E = 10^5$, $\varrho = 1$, $v = 0.25$. The horizontal and vertical displacements
for points A and B in the figure are plotted against time for the FEM and the BEM

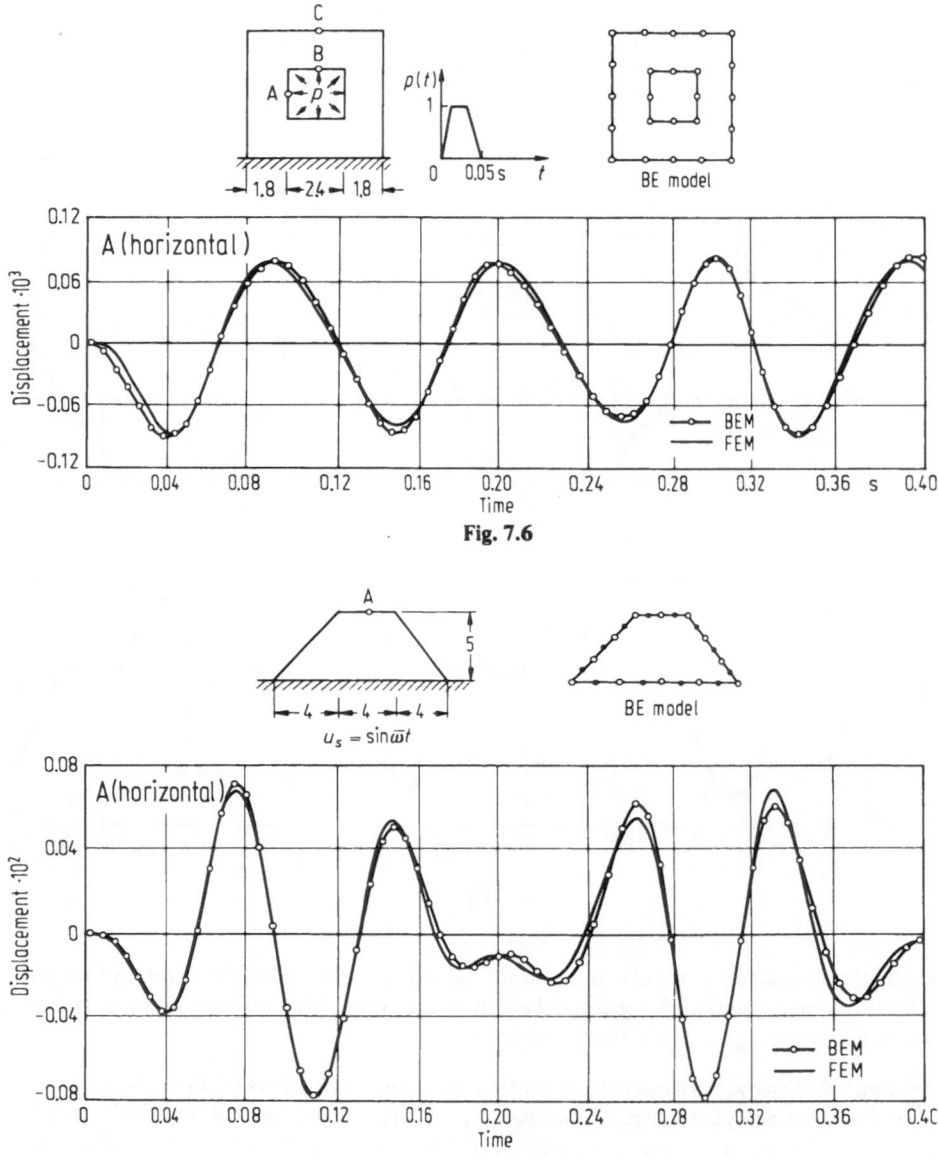

Fig. 7.6

Fig. 7.7

solutions. The agreement is excellent although only twelve boundary elements were used.

Example 4. *Hollow Square under Internal Pressure.* The structure shown in Fig. 7.6 was analysed when subjected to a sudden internal pressure of short duration. The material constants were taken to be the same as for example 3. Horizontal displacements of point A are compared by the two methods (FEM and BEM), and the results obtained show excellent agreement.

Example 5. *Dam under Harmonic Excitation.* A dam-like structure is analysed subject to a sinusoidal excitation at its base (Fig. 7.7). The forcing frequency was chosen to be 16 Hz, which is roughly a mean between the first ($f_1 = 11.04$ Hz) and the second ($f_2 = 20.72$ Hz) natural frequencies of the structure. Horizontal displacements at the crest are compared for a BEM and the FEM solution and they are shown to be in agreement.

Conclusions

This chapter describes a new technique to solve dynamic problems using the boundary element method. The technique allows the determination of natural frequencies and modes of vibrations of the system by applying the well known static fundamental solution without using internal cells. The approach effectively permits the mass integrals which originally were obtained by integrating over the domain, to be reduced to boundary integrals. The method can also be used to solve transient cases and a wide variety of time dependent problems.

Acknowledgement. This work has been partially supported by NATO grant 282/84, Double Jump Program.

References

1 Achenbach, J.D., *Wave Propagation in Elastic Solids.* North Holland, 1973
2 Love, A.E.H., *A Treatise on the Mathematical Theory of Elasticity.* Dover, 1944
3 Clough, R.W. and Penzien, J., *Dynamics of Structures.* McGraw-Hill, 1975
4 Cruse, T.A. and Rizzo, F.J., A Direct Formulation and Numerical Solution of the General Transient Elastodynamic Problem. J. Math. Anal. Appl. **22**, 1968
5 Wong, G.I.K. and Hutchinson, J.R., An Improved Boundary Element Method for Plate Vibrations, in *Boundary Element Methods.* Ed. C.A. Brebbia, Springer-Verlag, Berlin, 1981
6 Niwa, Y., Kobayashi, S., and Kitahara, M., Application of the Boundary Integral Equation Method to Eigenvalue Problems in Elastodynamics, in *Boundary Element Methods in Engineering.* Ed. C.A. Brebbia, Springer-Verlag, Berlin, 1982
7 Manolis, G.D. and Beskos, D.E., Dynamic Stress Concentration Studies by Boundary Integral and Laplace Transform. Int. J. Numerical Methods in Engineering **17**, 573−599, 1981
8 Bellman, R.E., Kalaba, R.E., and Lockett, J., *Numerical Inversion of the Laplace Transform.* Elsevier, New York, 1966
9 Nardini, D. and Brebbia, C.A., A New Approach to Free Vibration Analysis Using Boundary Elements, in *Boundary Element Methods in Engineering.* Ed. C.A. Brebbia, Springer-Verlag, 1982
10 Brebbia, C.A. and Nardini, D., Dynamic Analysis in Solid Mechanics by an Alternative Boundary Element Procedure. Int. Jour. Soil Dyn. Earthquake Eng. **2**, 1983
11 Nardini, D. and Brebbia, C.A., Transient Dynamic Analysis by the Boundary Element Method, in *Boundary Elements.* Ed. C.A. Brebbia, Springer-Verlag, Berlin, 1983
12 Dominguez, J. and Alarcon, E., Elastodynamics, in *Progress in Boundary Elements* **1**, Ed. C.A. Brebbia, Pentech Press and Wiley, 1981

13 Geers, T.L., Boundary Element Methods for Transient Response Analysis, in *Computational Methods for Transient Analysis*. Ed. T. Belytschko and T.J.R. Hughes, North Holland, 1983

14 Cole, D.M., Kosloff, D.D., and Minster, J.B., A Numerical Boundary Integral Equation Method for Elastodynamics. Bull. Seism. Soc. Amer. **68,** 1978

15 Mansur, W. and Brebbia, C.A., Transient Dynamic Analysis by the Boundary Element Method, in *Boundary Elements*. Ed. C.A. Brebbia, Springer-Verlag, Berlin and New York, 1983

16 Mansur, W. and Brebbia, C.A., Elastodynamics, Chapt. 5 in this volume

17 Brebbia, C.A., J. Telles, and L. Wrobel, *Boundary Element Techniques – Theory and Applications in Engineering*. Springer-Verlag, Berlin and New York, 1984

18 Danson, D.J., A Boundary Element Formulation of Problems in Linear Isotropic Elasticity with Body Forces, in *Boundary Element Methods*. Ed. C.A. Brebbia, Springer-Verlag, 1983

19 Wilkinson, J.H., *The Algebraic Eigenvalue Problem*. Oxford University Press, 1965

20 Bathe, K.J. and Wilson, E.L., Stability and Accuracy Analysis of Direct Integration Methods. Int. Journ. Earthq. Eng. Struc. Dyn. **1,** 1973

Chapter 8

Boundary Element Method for Laminar Viscous Flow and Convective Diffusion Problems

by K. Onishi, T. Kuroki, and M. Tanaka

8.1 Introduction

The viscous flow problems has been analyzed extensively by Wu and his co-workers (1976) using boundary integral equation method. The application of the direct boundary element method for viscous flow problems was discussed by Brebbia and Wrobel (1978) based on Laplace-Poisson equation formulation. Khader (1983) showed boundary element solutions of laminar developed duct flows of the viscous fluid.

For convection-diffusion problems using direct boundary elements, Ikeuchi (1983) presented boundary element solutions of a steady state. Recently, Matsunashi (1983) developed the boundary element method to the solution of two-dimensional convection-diffusion equations in a transient state with given potential fluid flows. Farooq and Kuwabara (1983) introduced the boundary integral equation method to heat convection using Green's function subject to boundary conditions.

In this paper, the potentialities of the direct formulation of the boundary element method for the approximate solution of viscous and thermal fluid flows are discussed. The governing equations describing the thermal fluid flow are based on the Boussinesq approximation. The convection-diffusion in the viscous fluid flow is also presented. New type of boundary condition on vorticity is presented for boundary elements. Unknown stream function, vorticity, and temperature are staggered in the application of the boundary element method. Nonlinear flow equations are solved by simple iterations. Boundary element upwind technique is presented to increase the stability of the computational scheme. Boundary element results for two-dimensional models with low Reynolds numbers were compared favorably with exact or finite element solutions.

8.2 Governing Equations

We consider unsteady flow and diffusion problems in two space dimensions with rectangular coordinates x_i $(i = 1, 2)$. The x_1 axis is directed toward the horizontal direction, and the x_2 is directed upward to the opposite direction of gravitational acceleration. Time variable is denoted by t. We shall summarize here a simplified set of viscous flow and convection-diffusion equations in nondimensional forms for a thermal fluid.

8.2.1 Field Equations

Let us denote by u_j $(j = 1, 2)$ the velocity component of the fluid. If the velocity components are related to the unknown stream function ψ as

$$u_1 = \frac{\partial \psi}{\partial x_2} \tag{8.1}$$

$$u_2 = -\frac{\partial \psi}{\partial x_1} \tag{8.2}$$

the continuity equation of the fluid is identically satisfied. The physical significance of the stream function is that the contour lines present streaklines, particularly in steady flow, they show the streamlines actually traced out by the particle of the fluid.

Scalar vorticity ω in two dimensions is defined by the equation:

$$\omega = \frac{\partial u_2}{\partial x_1} - \frac{\partial u_1}{\partial x_2}. \tag{8.3}$$

By substituting Eqs. (8.1) and (8.2) into this equation, we can obtain:

$$\nabla^2 \psi = -\omega. \tag{8.4}$$

This can be regarded as the Poisson equation for the stream function, provided that the vorticity on the right hand side is known.

For the fluid under consideration, we can write equation of motion by the vorticity transport equation:

$$\frac{\partial \omega}{\partial t} + u_j \frac{\partial \omega}{\partial x_j} = \frac{1}{Re} \nabla^2 \omega + (Gr/Re^2) \frac{\partial T}{\partial x_1} \tag{8.5}$$

in which Re denotes the Reynolds number, and Gr the Grashoff number.

If the viscous dissipation can be neglected, heat conduction equation is expressible in terms of the unknown temperature T as follows:

$$\frac{\partial T}{\partial t} + u_j \frac{\partial T}{\partial x_j} = \frac{1}{Re \cdot Pr} \nabla^2 T \tag{8.6}$$

in which Pr is the Prandtl number. Moreover, the mass transport can be described in terms of the unknown concentration C as:

$$\frac{\partial C}{\partial t} + u_j \frac{\partial C}{\partial x_j} = \frac{1}{Pe} \nabla^2 C \tag{8.7}$$

in which Pe is the Peclet number. This equation has the same form as in the thermal diffusion equation.

8.2.2 Boundary Conditions

We shall consider here two kinds of boundary fluid surfaces: One is a rigid surface on which no slip occures, the other is a free surface on which tangential stresses are prescribed. The boundary surface is also subject to thermal conditions as well as the conditions on the mass concentration. These boundary conditions depend on the particular problem of interest. Their specific form will be illustrated later, when numerical examples are presented. Only general aspects of the boundary conditions are summarized in this section.

Conditions on Stream Function

Note that the stream function $\psi(P, t)$ as a function of the spatial position P can also be defined by the line integral:

$$\psi(P, t) = \oint_{P_0}^{P} u_j n_j \, ds \tag{8.8}$$

in which C is the path of integration from the fixed point P_0 to an arbitrary point P, n_j the jth component of the unit normal drawn to the right along the path, and ds the infinitesimal line element. Therefore, if velocity components are known along a part of the boundary, we can specify the value of the stream function along the part. The artificial boundary of an inlet upstream is often of this kind.

The first type of the boundary condition is the Dirichlet boundary condition of the form:

$$\psi = \bar{\psi} \tag{8.9}$$

in which $\bar{\psi}$ is the specified value of the stream function. If $\bar{\psi}$ is constant, this condition implies that the part of the boundary on which Eq. (8.9) is specified constitutes a stream line.

The second type is the Neumann boundary condition of the form:

$$\frac{\partial \psi}{\partial n} = \bar{\psi}_n \tag{8.10}$$

in which n is the outward unit normal to the boundary, and $\bar{\psi}_n$ denotes the specified slope of the stream function in the normal direction. From the definition of the stream function, we notice that $\bar{\psi}_n$ represents the velocity component in the direction which has a right angle counterclockwise to the n direction. In particular, we often encounter the case $\bar{\psi}_n = 0$ on the artificial boundary downstream or on an outlet which is sufficiently remote from an obstacle.

Conditions on Vorticity

The first type of the boundary condition is the Dirichlet boundary condition of the form:

$$\omega = \bar{\omega} \tag{8.11}$$

in which $\bar{\omega}$ is the specified vorticity. When the velocity distribution of both u_1 and u_2 are known along a part of the boundary, this type of boundary condition applies to Eq. (8.3).

The second type of boundary condition on the vorticity becomes more subtle. For the sake of simplicity, let us suppose that the boundary coincides with the x_1 axis and the flow domain occupies the half plane $x_2 > 0$. The derivation of the condition proposed here is based on the Taylor expansion of the stream function around a boundary point Q:

$$\psi\,(P,t) = \psi\,(Q,t) + \frac{\partial \psi}{\partial x_2}\,PQ + \frac{1}{2}\,\frac{\partial^2 \psi}{\partial x_2^2}\,PQ^2 + \frac{1}{6}\,\frac{\partial^3 \psi}{\partial x_2^3}\,PQ^3 + O\,(PQ^4) \quad (8.12)$$

in which P is a skin-deep point inside the domain just on the line drawn with a rect-angle to the x_1 axis from the point Q, and PQ the distance between these two points.

From the condition $\partial u_2/\partial x_1 = 0$ along the fixed boundary wall, the vorticity on the boundary Γ can be evaluated as follows:

$$\omega\,|_\Gamma = -\frac{\partial u_1}{\partial x_2} = -\frac{\partial^2 \psi}{\partial x_2^2} \quad (8.13)$$

which gives the second order coefficient of the Taylor expansion. The first derivative of the vorticity in x_2 on the boundary is evaluated as

$$\frac{\partial \omega}{\partial x_2}\bigg|_\Gamma = -\frac{\partial^2 u_1}{\partial x_2^2} = -\frac{\partial^3 \psi}{\partial x_2^3}. \quad (8.14)$$

Since the positive x_2 direction is opposite to the direction of the outward unit normal n, this equation is recast into

$$\frac{\partial \omega}{\partial n}\bigg|_\Gamma = \frac{\partial^3 \psi}{\partial x_2^3} \quad (8.15)$$

which gives the third order coefficient of the Taylor expansion.

By neglecting higher order terms, we can obtain the relation:

$$\psi\,(P,t) = \psi\,|_\Gamma - \frac{\partial \psi}{\partial n}\bigg|_\Gamma PQ - \frac{1}{2}\,\omega\,|_\Gamma\,PQ^2 + \frac{1}{6}\,\frac{\partial \omega}{\partial n}\bigg|_\Gamma PQ^3. \quad (8.16)$$

The derivative $\partial \psi/\partial n\,|_\Gamma$ implies the tangential velocity along the boundary wall. If we solve this equation for $\partial \omega/\partial n$, we have

$$-(1/Re)\,\frac{\partial \omega}{\partial n}\bigg|_\Gamma = \gamma \left\{ \omega - \frac{2}{PQ}\left(\frac{\psi\,|_\Gamma - \psi\,(P,t)}{PQ} - \frac{\partial \psi}{\partial n}\bigg|_\Gamma \right) \right\} \quad (8.17)$$

in which $\gamma = -3\,(1/Re)/PQ$. This is the boundary condition of the second kind, which has the same form as in the linear radiation condition of the Newton cooling in thermal problems. The above condition is a linear combination of all the boundary unknowns. This condition requires the knowledge of internal values of the stream function.

Conditions on Temperature and Concentration

The boundary condition of the first kind for heat equation is the isothermal condition:

$$T = \bar{T} \quad (8.18)$$

and for mass diffusion equation:

$$C = \bar{C} \quad (8.19)$$

in which $\bar{T}$ and $\bar{C}$ are the specified temperature and concentration, respectively.

The boundary condition of the second kind on temperature is the radiation condition in the form:

$$- k \frac{\partial T}{\partial n} + u_j n_j T = h (T - \bar{T}) + \bar{T}_n \tag{8.20}$$

in which k is the coefficient of heat conduction, h the coefficient of heat transfer, n_j the component of the outward unit normal, and $\bar{T}_n$ the specified heat flux on the surface. If $h = 0$, then this condition reduces to the total thermal flux condition as the sum of conductive and convective fluxes. In addition, if $\bar{T}_n = 0$ and also $u_j = 0$, then it reduces to the adiabatic boundary condition.

The second boundary condition on the concentration can be written in the form:

$$- D \frac{\partial C}{\partial n} + u_j n_j C = \bar{C}_n \tag{8.21}$$

as the sum of diffusive and convective fluxes, in which D is the diffusion coefficient, and $\bar{C}_n$ the specified value of the total mass flux across the surface.

8.3 Boundary Integral Equations

8.3.1 BIE of the Stream Function

Suppose that the right hand side of Eq. (8.4) is known at some instant. To transform the Poisson equation into an integral equation, we consider the fundamental solution ψ^* satisfying

$$\nabla^2 \psi^* = - \delta (P) \tag{8.22}$$

in the whole of interest, where the Dirac delta function $\delta (P)$ denotes a unit source at the spatial position P. An explicit form of the solution in two dimensions is known as the logarithmic potential which is given by

$$\psi^* (Q, P) = \frac{1}{2\pi} \ln \frac{1}{PQ} . \tag{8.23}$$

Multiplying Eq. (8.4) by ψ^* and integrating by parts twice over the domain Ω, we can obtain the boundary integral equation for the stream function at the time level t_k $(k = 0, 1, 2, \ldots)$ as follows:

$$c (P) \psi (P, t_k) + \int_\Gamma \psi (Q, t_k) \frac{\partial \psi^*}{\partial n} (Q, P) d\Gamma$$

$$= \int_\Gamma \frac{\partial \psi}{\partial n} (Q, t_k) \psi^* (Q, P) d\Gamma + \int_\Omega \omega (Q, t_{k-1}) \psi^* (Q, P) d\Omega \tag{8.24}$$

in which Γ is the bounding surface of the domain Ω, and n is the outward unit normal to the boundary. The factor $c (P)$ is determined by the rule $c = \Theta / 2\pi$ in two dimensions, where Θ denotes the inner angle (in radians) centered at the point P with respect to the geometry of the domain. If the point P is an internal point of the domain, we have $c(P) = 1$ and this equation gives the integral representation of the value of the stream function at the time t_k in terms of the boundary values at t_k, the boundary fluxes (the tangential velocities at the t_k), and the distribution of the vorticity at the time t_{k-1} over the domain Ω.

In order to calculate the velocity components from the integral equation, we have to differentiate Eq. (8.24) with respect to the internal variable point P. For an arbitrary direction n', we have

$$\frac{\partial \psi}{\partial n'} (P, t_k) = - \int_\Gamma \psi(Q, t_k) \frac{\partial^2 \psi^*}{\partial n' \partial n} (Q, P) \, d\Gamma$$

$$+ \int_\Gamma \frac{\partial \psi}{\partial n} (Q, t_k) \frac{\partial \psi^*}{\partial n'} (Q, P) \, d\Gamma$$

$$+ \int_\Omega \omega(Q, t_{k-1}) \frac{\partial \psi^*}{\partial n'} (Q, P) \, d\Omega . \tag{8.25}$$

According to Eqs. (8.1) and (8.2), we can calculate two velocity components.

8.3.2 BIE of the Vorticity

We consider the time dependent fundamental solution which satisfies

$$\frac{\partial \omega^*}{\partial t} + v \nabla^2 \omega^* = - \delta (P, t_k) \tag{8.26}$$

in the infinite domain, in which the Dirac delta function $\delta(P, t_k)$ denotes the unit load at a spatial position P at the time t_k. An explicit form of the fundamental solution is given by

$$\omega^* (Q, t: P, t_k) = \begin{cases} 0 & (t_k < t) \\ \dfrac{1}{4 v \pi (t_k - t)} \exp \left[- \dfrac{P \, Q^2}{4 v (t_k - t)} \right] & (t < t_k) . \end{cases} \tag{8.27}$$

This is known as a free space Green's function for heat equation, or heat potential.

Let us consider the vorticity transport equation in the conventional form:

$$\frac{\partial \omega}{\partial t} - v \nabla^2 \omega = - u_j \frac{\partial \omega}{\partial x_j} + \alpha \frac{\partial T}{\partial x_1} \tag{8.28}$$

in which $v = 1/Re$, $\alpha = Gr/Re^2$. The convective term and the buoyancy term on the right hand side are dealt with as if they constitute pseudo driving forces. It is assumed that the distribution of the vorticity at some instant can be determined as the combined effect of convection and buoyancy at the time shortly before that instant. Multiplying this equation by the fundamental solution, and integrating by parts twice over the domain and also over the time interval $t_{k-1} < t < t_k$, we can obtain by the method of collocation that

$$c(P)\,\omega(P,t_k) = \int_\Gamma d\Gamma \int_{t_{k-1}}^{t_k} \omega(Q,t)\left\{-v\frac{\partial\omega^*}{\partial n}(Q,t\colon P,t_k)\right\} dt$$

$$-\int_\Gamma d\Gamma \int_{t_{k-1}}^{t_k} \left\{-v\frac{\partial\omega}{\partial n}(Q,t)\right\}\omega^*(Q,t\colon P,t_k)\,dt$$

$$+\int_\Omega \omega(Q,t_{k-1})\,\omega^*(Q,t_{k-1}\colon P,t_k)\,d\Omega$$

$$-\int_\Omega d\Omega \int_{t_{k-1}}^{t_k} u_j(Q,t)\frac{\partial\omega}{\partial x_j}(Q,t)\,\omega^*(Q,t\colon P,t_k)\,dt$$

$$+\int_\Omega d\Omega \int_{t_{k-1}}^{t_k} \alpha\frac{\partial T}{\partial x_1}(Q,t_{k-1})\,\omega^*(Q,t\colon P,t_k)\,dt. \tag{8.29}$$

The first order derivative of the vorticity at internal point can be obtained readily from this equation. In fact, Eq. (8.29) is recast into

$$\frac{\partial\omega}{\partial n'}(P,t_k) = \int_\Gamma d\Gamma \int_{t_{k-1}}^{t_k} \omega(Q,t)\left\{-v\frac{\partial^2\omega^*}{\partial n'\,\partial n}(Q,t\colon P,t_k)\right\} dt$$

$$-\int_\Gamma d\Gamma \int_{t_{k-1}}^{t_k} \left\{-v\frac{\partial\omega}{\partial n}(Q,t)\right\}\frac{\partial\omega^*}{\partial n'}(Q,t\colon P,t_k)\,dt$$

$$+\int_\Omega \omega(Q,t_{k-1})\frac{\partial\omega^*}{\partial n'}(Q,t_{k-1}\colon P,t_k)\,d\Omega$$

$$-\int_\Omega d\Omega \int_{t_{k-1}}^{t_k} u_j(Q,t)\frac{\partial\omega}{\partial x_j}(Q,t)\frac{\partial\omega^*}{\partial n'}(Q,t\colon P,t_k)\,dt$$

$$+\int_\Omega d\Omega \int_{t_{k-1}}^{t_k} \alpha\frac{\partial T}{\partial x_1}(Q,t_{k-1})\frac{\partial\omega^*}{\partial n'}(Q,t\colon P,t_k)\,dt. \tag{8.30}$$

For example, if n' directs toward the positive x_1 direction, this equation gives the x_1-derivative of the vorticity at the internal point P at the time t_k in terms of boundary vorticities and boundary fluxes.

8.3.3 BIE of the Temperature

Let us consider the equation of heat conduction (8.6) in the conventional form:

$$\frac{\partial T}{\partial t} - \varkappa\nabla^2 T = -u_j\frac{\partial T}{\partial x_j} \tag{8.31}$$

in which $\varkappa = 1/Re\cdot Pr$. We regard the right-hand side as a pseudo heat source. The fundamental solution is given by the expression:

$$T^*(Q,t\colon P,t_k) = \begin{cases} 0 & (t_k < t) \\[2mm] \dfrac{1}{4\varkappa\pi\,(t_k - t)}\exp\left[-\dfrac{PQ^2}{4\varkappa\,(t_k - t)}\right] & (t < t_k). \end{cases} \tag{8.32}$$

Owing to the formal similarity between Eqs. (8.28) and (8.31), we can readily obtain the boundary integral equation for the temperature:

$$c(P) T(P, t_k) - \int_\Gamma d\Gamma \int_{t_{k-1}}^{t_k} T(Q, t) \left\{ -\varkappa \frac{\partial T^*}{\partial n} (Q, t: P, t_k) \right\} dt$$

$$= -\int_\Gamma d\Gamma \int_{t_{k-1}}^{t_k} \left\{ -\varkappa \frac{\partial T}{\partial n} (Q, t) \right\} T^* (Q, t: P, t_k) dt$$

$$+ \int_\Omega T(Q, t_{k-1}) T^* (Q, t_{k-1}: P, t_k) d\Omega$$

$$- \int_\Omega d\Omega \int_{t_{k-1}}^{t_k} u_j(Q, t_k) \frac{\partial T}{\partial x_j} (Q, t_{k-1}) T^* (Q, t: P, t_k) dt .$$ (8.33)

We can also derive the integral representation of heat fluxes at the internal point P from this equation. Differentiation in the direction n' yields

$$\frac{\partial T}{\partial n'} (P, t_k) = \int_\Gamma d\Gamma \int_{t_{k-1}}^{t_k} T(Q, t) \left\{ -\varkappa \frac{\partial^2 T^*}{\partial n' \partial n} (Q, t: P, t_k) \right\} dt$$

$$- \int_\Gamma d\Gamma \int_{t_{k-1}}^{t_k} \left\{ -\varkappa \frac{\partial T}{\partial n} (Q, t) \right\} \frac{\partial T^*}{\partial n'} (Q, t: P, t_k) dt$$

$$+ \int_\Omega T(Q, t_{k-1}) \frac{\partial T^*}{\partial n'} (Q, t_{k-1}: P, t_k) d\Omega$$

$$- \int_\Omega d\Omega \int_{t_{k-1}}^{t_k} u_j(Q, t_k) \frac{\partial T}{\partial x_j} (Q, t_{k-1}) \frac{\partial T^*}{\partial n'} (Q, t: P, t_k) dt .$$ (8.34)

For example, if n' directs toward the positive x_1 direction, this equation gives the x_1-derivative of the temperature at the internal point P at the time t_k in terms of boundary temperatures and boundary heat fluxes.

The system of Eqs. (8.24), (8.29), and (8.33) is the basic set of boundary integral equations for our present purpose.

8.4 Boundary Element Approximation

In order to obtain numerical solutions of the stream function, the vorticity, and the temperature by means of the boundary element method, we have to discretize not only the functions involved in the boundary integral equations, but also the surface bounding the space-time region.

We choose N nodes P_j ($j = 1, 2, \ldots, N$) on the boundary, which approximate the outline of the domain by a polygon. The boundary is subdivided into small boundary elements as

$$\Gamma = \bigcup_{j=1}^N \Gamma_j$$ (8.35)

in which Γ_j is the jth boundary element. If constant elements are used, the nodes are located at the centre of each Γ_j. If linear elements are used, then nodes are located at the ends of each Γ_j. The domain is subdivided into a series of internal cells.

8.4.1 Discretization of the Stream Function

The stream function $\psi(Q, t_k)$ on the boundary is approximated in terms of interpolating functions ϕ_j ($j = 1, 2, \ldots, N$) as

$$\psi^k(Q) = \sum_{j=1}^{N} \phi_j(Q)\, \psi_j^k \tag{8.36}$$

in which ψ_j^k is the approximated nodal value of the stream function to the exact $\psi(P_j, t_k)$. The interpolation functions are piecewise constant for the constant elements, and piecewise linear for the linear elements.

The boundary flux $\partial\psi(Q, t_k)/\partial n$ is approximated in the form:

$$\frac{\partial\psi^k}{\partial n}(Q) = \sum_{j=1}^{N} \phi_j(Q) \left(\frac{\partial\psi}{\partial n}\right)_j^k \tag{8.37}$$

in which $(\partial\psi/\partial n)_j^k$ is the approximation to the exact $\partial\psi(P_j, t_k)/\partial n$ on the boundary.

Boundary element discretization of the integral equation for the stream function can be readily obtained from Eq. (8.24) as follows:

$$c_i \psi_i^k + \sum_{j=1}^{N} \int_{\Gamma_j} \psi^k(Q)\, \frac{\partial\psi^*}{\partial n}(Q, P)\, d\Gamma$$

$$= \sum_{j=1}^{N} \int_{\Gamma_j} \frac{\partial\psi^k}{\partial n}(Q)\, \psi^*(Q, P)\, d\Gamma - \int_{\Omega} \omega^{k-1}(Q)\, \psi^*(Q, P)\, d\Omega \tag{8.38}$$

in which c_i is the value of $c(P_i)$ for $i = 1, 2, \ldots, N$. The integration on the boundary can be carried out using the standard Gaussian quadrature. The domain integral involved in this equation can be evaluated approximately on each cell by some appropriate numerical quadratures.

Let us denote by $\{\psi\}^k$ the column vector with components ψ_j^k ($j = 1, 2, \ldots, N$), $\{\partial\psi/\partial n\}^k$ the column vector with components $(\partial\psi/\partial n)_j^k$. We can express the set of Eq. (8.38) in the matrix form as follows:

$$[H^\psi]\{\psi\}^k = [G^\psi]\left\{\frac{\partial\psi}{\partial n}\right\}^k - \{b^\psi(\omega^{k-1})\} \tag{8.39}$$

in which $[H^\psi]$ and $[G^\psi]$ are square coefficient matrices of the order N independent on the time level, and the last term is the column vector depending on the approximate vorticity at the $(k-1)$th time level.

Velocity components given by Eqs. (8.1) and (8.2) can be approximated in the same way. Let us denote by $(\partial\psi/\partial n)_i^k$ the approximate derivative to the exact $\partial\psi(P_i, t_k)/\partial n'$ at the internal point P_i in the direction n'. The approximation is obtained from Eq. (8.25) by the expression:

$$\left(\frac{\partial \psi}{\partial n}\right)_i^k = -\sum_{j=1}^{N} \int_{\Gamma_i} \psi^k(Q) \frac{\partial^2 \psi^*}{\partial n' \partial n}(Q, P_i) \, d\Gamma$$

$$+ \sum_{j=1}^{N} \int_{\Gamma_i} \frac{\partial \psi^k}{\partial n}(Q) \frac{\partial \psi^*}{\partial n'}(Q, P_i) \, d\Gamma$$

$$+ \int_{\Omega} \omega^{k-1}(Q) \frac{\partial \psi^*}{\partial n'}(Q, P_i) \, d\Omega. \tag{8.40}$$

The velocity components are approximately obtained from $u_1^k(P_i) = (\partial \psi / \partial x_2)_i^k$ and $u_2^k(P_i) = -(\partial \psi / \partial x_1)_i^k$.

8.4.2 Discretization on Vorticity and Temperature

As far as the approximates of vorticity and temperature are concerned, we assume here that they are constant over the small time interval $t_{k-1} < t < t_k$. This seems reasonable in most practical problems in which the time variations in ω and T are smaller than those in the fundamental solutions ω^* and T^*.

Under this assumption, the vorticity $\omega(Q, t)$ and the temperature $T(Q, t)$ on the boundary between the small time interval are approximated respectively as

$$\omega^k(Q) = \sum_{j=1}^{N} \phi_j(Q) \, \omega_j^k \tag{8.41}$$

and

$$T^k(Q) = \sum_{j=1}^{N} \phi_j(Q) \, T_j^k \tag{8.42}$$

in which ω_j^k and T_j^k are approximations to the exact $\omega(Q_j, t_k)$ and $T(Q_j, t_k)$.

Corresponding to these approximations, the boundary fluxes are also approximated as follows:

$$-v \frac{\partial \omega^k}{\partial n}(Q) = \sum_{j=1}^{N} \phi_j(Q) \left(-v \frac{\partial \omega}{\partial n}\right)_j^k \tag{8.43}$$

and

$$-\varkappa \frac{\partial T^k}{\partial n}(Q) = \sum_{j=1}^{N} \phi_j(Q) \left(-\varkappa \frac{\partial T}{\partial n}\right)_j^k. \tag{8.44}$$

The application of a collocation method to Eq. (8.33) leads to the following discretized set of equations:

$$c_i \omega_i^k - \sum_{j=1}^{N} \int_{\Gamma_i} \omega^k(Q) \, d\Gamma \int_{t_{k-1}}^{t_k} \left\{-v \frac{\partial \omega^*}{\partial n}(Q, t: P, t_k)\right\} dt$$

$$= -\sum_{j=1}^{N} \int_{\Gamma_i} \left\{-v \frac{\partial \omega^k}{\partial n}(Q)\right\} d\Gamma \int_{t_{k-1}}^{t_k} \omega^*(Q, t: P, t_k) \, dt$$

$$+ \int_{\Omega} \omega^{k-1}(Q) \, \omega^*(Q, t_{k-1}: P, t_k) \, d\Omega$$

$$- \int_{\Omega} u_j^k(Q) \frac{\partial \omega^{k-1}}{\partial x_j}(Q) \, d\Omega \int_{t_{k-1}}^{t_k} \omega^*(Q, t: P, t_k) \, dt$$

$$+ \int_{\Omega} \alpha \frac{\partial T^{k-1}}{\partial x_1}(Q) \, d\Omega \int_{t_{k-1}}^{t_k} \omega^*(Q, t: P, t_k) \, dt \tag{8.45}$$

in which $u_j^k(Q)$ is the interpolated velocity component at t_k. The time integrals can be evaluated analytically.

Let us denote by $\{\omega\}^k$ the column vector with components ω_j^k $(j = 1, 2, \ldots, N)$, and $\{-\nu\, \partial\omega/\partial n\}^k$ with the components $(-\nu\, \partial\omega/\partial n)_j^k$. We can express Eq. (8.45) in the matrix form:

$$[H^\omega]\{\omega\}^k = [G^\omega]\left\{-\nu\frac{\partial\omega}{\partial n}\right\}^k + \{b^\omega(\omega^{k-1}, u_j^k, T^{k-1})\} \qquad (8.46)$$

in which $[H^\omega]$ and $[G^\omega]$ are square coefficient matrices of the order N, and $\{b^\omega\}$ is the column vector depending on the variables specified within the parentheses. If the time slice $\Delta t = t_k - t_{k-1}$ is kept constant, the coefficient matrices are independent on the time level.

The approximate flux to the exact $\partial\omega(P_i, t_k)/\partial n'$ at the internal point P_i can be obtained from Eq. (8.30) as follows:

$$\begin{aligned}
\left(\frac{\partial\omega}{\partial n}\right)_i^k = {} & \sum_{j=1}^{N} \int_{\Gamma_j} \omega^k(Q)\, d\Gamma \int_{t_{k-1}}^{t_k} \left\{-\nu\frac{\partial^2\omega^*}{\partial n'\, \partial n}(Q, t: P_i, t_k)\right\} dt \\
& - \sum_{j=1}^{N} \int_{\Gamma_j} \left\{-\nu\frac{\partial\omega^k}{\partial n}(Q)\right\} d\Gamma \int_{t_{k-1}}^{t_k} \frac{\partial\omega^*}{\partial n'}(Q, t: P_i, t_k)\, dt \\
& + \int_\Omega \omega^{k-1}(Q)\frac{\partial\omega^*}{\partial n'}(Q, t_{k-1}: P_i, t_k)\, d\Omega \\
& - \int_\Omega u_j^k(Q)\frac{\partial\omega^{k-1}}{\partial x_j}(Q)\, d\Omega \int_{t_{k-1}}^{t_k} \frac{\partial\omega^*}{\partial n'}(Q, t: P_i, t_k)\, dt \\
& + \int_\Omega \alpha\frac{\partial T^{k-1}}{\partial x_1}(Q)\, d\Omega \int_{t_{k-1}}^{t_k} \frac{\partial\omega^*}{\partial n'}(Q, t: P_i, t_k)\, dt .
\end{aligned} \qquad (8.47)$$

Under the assumption for the time variation in the temperature, the integral Eq. (8.33) can immediately be discretized into

$$\begin{aligned}
c_i T_i^k - \sum_{j=1}^{N} \int_{\Gamma_j} T^k(Q)\, d\Gamma \int_{t_{k-1}}^{t_k} \left\{-\varkappa\frac{\partial T^*}{\partial n}(Q, t: P, t_k)\right\} dt \\
= {} & -\sum_{j=1}^{N} \int_{\Gamma_j} \left\{-\varkappa\frac{\partial T^k}{\partial n}(Q)\right\} d\Gamma \int_{t_{k-1}}^{t_k} T^*(Q, t: P, t_k)\, dt \\
& + \int_\Omega T^{k-1}(Q)\, T^*(Q, t_{k-1}: P, t_k)\, d\Omega \\
& - \int_\Omega u_j^k(Q)\frac{\partial T^{k-1}}{\partial x_j}(Q)\, d\Omega \int_{t_{k-1}}^{t_k} T^*(Q, t: P, t_k)\, dt .
\end{aligned} \qquad (8.48)$$

Let us denote by $\{T\}^k$ the column vector with the components T_j^k $(j = 1, 2, \ldots, N)$, $\{-\varkappa\, \partial T/\partial n\}^k$ with the components $(-\varkappa\, \partial T/\partial n)_j^k$. We can express Eq. (8.48) in the form:

$$[H^T]\{T\}^k = [G^T]\left\{-\varkappa\frac{\partial T}{\partial n}\right\}^k + \{b^T(T^{k-1}, u_j^k)\} \qquad (8.49)$$

in which $[H^T]$ and $[G^T]$ are square coefficient matrices of the order N, and $\{b^T\}$ a column vector. Note that the suffix T does not denote the transpose of a matrix.

The approximate thermal flux to the exact $\partial T(P_i, t_k)/\partial n'$ at the internal point P_i can be obtained from Eq. (8.34) as follows:

$$\left(\frac{\partial T}{\partial n}\right)_i^k = \sum_{j=1}^{N} \int_{\Gamma_j} T^k(Q) \, d\Gamma \int_{t_{k-1}}^{t_k} \left\{ -\varkappa \frac{\partial^2 T^*}{\partial n' \, \partial n} (Q, t: P_i, t_k) \right\} dt$$

$$- \sum_{j=1}^{N} \int_{\Gamma_j} \left\{ -\varkappa \frac{\partial T^k}{\partial n}(Q) \right\} d\Gamma \int_{t_{k-1}}^{t_k} \frac{\partial T^*}{\partial n'}(Q, t: P_i, t_k) \, dt$$

$$+ \int_{\Omega} T^{k-1}(Q) \frac{\partial T^*}{\partial n'}(Q, t_{k-1}: P_i, t_k) \, d\Omega$$

$$- \int_{\Omega} u_j^k(Q) \frac{\partial T^{k-1}}{\partial x_j}(Q) \, d\Omega \int_{t_{k-1}}^{t_k} \frac{\partial T^*}{\partial n'}(Q, t: P_i, t_k) \, dt . \qquad (8.50)$$

8.5 Computational Scheme

Stream function, vorticity, and temperature are coupled in the set of the governing equations as discussed in Sect. 8.2. However, an iterative solution was presented in Sect. 8.4 for uncoupled problems. We shall try to use in turn such an iterative scheme for the solutions of nodal stream functions, vorticities, and temperatures. The diffusion equation is not coupled with viscous flow equations, and hence a direct use of the boundary element method leads to its solution.

Suppose that we know an initial value of the stream function, and an initial temperature distribution. We then calculate the initial vorticity, and the gradient of initial vorticity and temperature.

Using these initial values together with suitable boundary conditions, we can start computations with calculating stream functions at the first time step. Velocity components at this time step are evaluated from the calculated result. We then move to calculate vorticities, and then calculate temperatures. Using these results at the first time step, we can calculate unknowns at the second time step. This procedure is iterated until a quasi steady-state is attained.

The above algorithm can be summarized by using ALGOL-like statements as follows:

Given initial conditions ψ^0 and T^0.
Calculate ω^0, $\nabla\omega^0$, and ∇T^0.
For $k = 1, 2, 3, \ldots$, until satisfied, do:
    Calculate the stream function:
        Compute boundary values ψ^k and $\partial\psi^k/\partial n$ as the solution to Eq. (8.39).
        Compute the internal values ψ^k from Eq. (8.38).
        Compute $\nabla\psi^k$ using Eq. (8.40).
        Calculate the internal values u_j^k ($j = 1, 2$).
    Specify the boundary condition of Eq. (8.17).
    Calculate the vorticity:
        Compute the boundary values ω^k and $- \nu \, \partial\omega^k/\partial n$ as the solution of
        Eq. (8.46).

       Compute the internal values ω^k from Eq. (8.45).
       Compute $\nabla\omega^k$ using Eq. (8.47).
       └─Calculate the boundary values $\partial\omega^k/\partial x_j$.
    Calculate the temperature (or the concentration):
       Compute the boundary values T^k and $-\varkappa\,\partial T^k/\partial n$ as the solution of Eq. (8.49).
       Compute the internal values T^k from Eq. (8.48).
       Compute ∇T^k using Eq. (8.50).
       └─Calculate the boundary values $\partial T^k/\partial x_j$.
└──Next k.

The iteration is continued until a convergence criterion is satisfied. Such a criterion often employed is given by the inequality:

$$\max_{1\leq j\leq N}\left|\frac{\omega_j^k - \omega_j^{k-1}}{\omega_j^k}\right| < \varepsilon \tag{8.51}$$

for some small positive number ε.

In order to save c.p.u. time, the unknown derivatives such as $\nabla\psi^k$, $\nabla\omega^k$, ∇T^k, at internal nodes can be calculated directly by interpolation using nodal values, ψ_i^k, ω_i^k, T_i^k, over each cell in such a way as in the finite element method. For example, suppose that the stream function within each cell is interpolated in terms of nodal values associated to the cell as:

$$\psi^k(Q) = \sum_i N_i(Q)\,\psi_i^k \tag{8.52}$$

using some set of shape functions N_i, in which the summation index runs through the number of all nodes of the cell. The derivatives are obtained immediately from:

$$\nabla\psi^k(Q) = \sum_i \nabla N_i(Q)\cdot\psi_i^k. \tag{8.53}$$

From the practical point of view, the above method can be recommended, since the numerical implementation becomes easier in comparison with that of Eq. (8.40).

Integrations involved in the formation of boundary element equations are carried out using numerical quadratures. Line integrations over each boundary element are calculated by the standard Gaussian quadrature. Area integrations over each cell are calculated by a Hammer's quadrature on triangles, and by the product form of the Gaussian formula on quadrilaterals.

The accuracy depends not only on the integration formulas used, but also on the element subdivision of the boundary as well as the internal cell subdivision. From the computational experience, it can be suggested that the dimensionless time increment Δt is so chosen as to approximately satisfy the empirical relation

$$\Delta\Gamma = \sqrt{2\lambda\,\Delta t} \tag{8.54}$$

for representative dimensionless spatial mesh size $\Delta\Gamma$, in which λ denote the dimensionless parameters v and $\varkappa$. If the suggested Δt's are much different, then the larger value of Δt is recommended. Usually, the time increment in the boundary element method is allowed to be much larger than that in the finite element and finite difference methods, in order to obtain the same order of accuracy.

The above-mentioned scheme becomes unstable for increasingly large Reynolds numbers. In order to gain the stability, the following technique which is similar to that used in upwinding in domain-type methods could be useful.

Let (x_1^P, x_2^P) be the coordinates of the point P, and (u_1^P, u_2^P) the velocity vector at that point. We denote the point P simply by its ordinate x_i^P. The volume integral in Eq. (8.48) for the pseudo heat source is now replaced by the upwind weighted integral:

$$-\int_\Omega \beta u_i^k(x_i^Q)\frac{\partial T^{k-1}}{\partial x_j}(x_i^Q)\,d\Omega \int_{t_{k-1}}^{t_k} T^*(x_i^Q, t: x_i^P - \Delta t\, u_i^P, t_k)\,dt \qquad (8.55)$$

where β is the minus cosine of the vector PQ and the flow vector defined as follows:

$$(x_i^Q - x_i^P)\,u_i^Q = \|PQ\| \cdot \|(u_1^Q, u_2^Q)\| \cos A \qquad (8.56)$$

$$\beta = -\cos A, \quad \text{or} = 1\ . \qquad (8.57)$$

Therefore, not only the weight β is applied, but the charged point P is translated upstream by the distance $\Delta t\, u_i^P$.

8.6 Numerical Examples

8.6.1 Isothermal Channel Flow

As a simple example of isothermal viscous flow problem, we consider a channel flow between two infinite parallel plates. The fluid was initially motionless. A parabolic flow profile was suddenly applied on the artificial boundary upstream. Figure 8.1a shows boundary conditions on the stream function and Fig. 8.1b shows the condition on the vorticity. The material constant was set as $v = 1.0$. Uniform initial temperature was assumed, and all boundaries were thermally insulated.

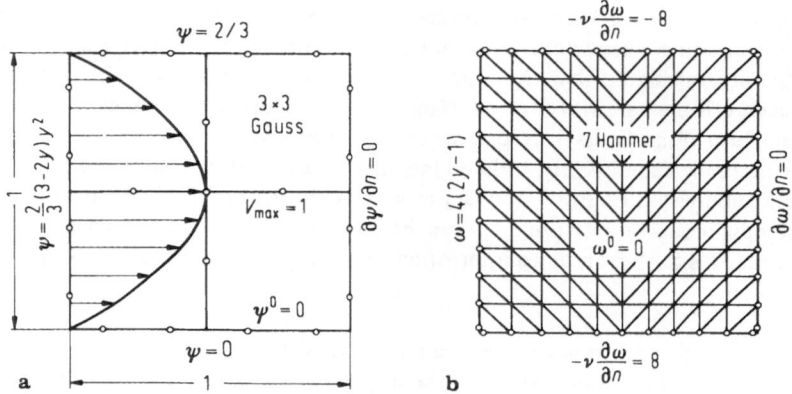

Fig. 8.1. Channel flow: **a** constant BE mesh, **b** linear BE mesh

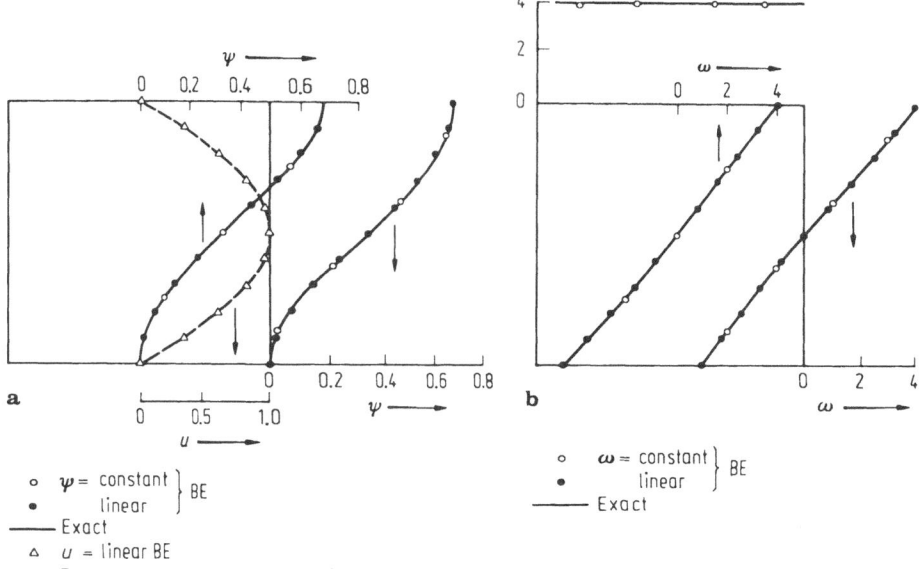

$\psi \longrightarrow$

0   0.2   0.4   0.6   0.8

0   0.2   0.4   0.6   0.8

$\psi \longrightarrow$

0   0.5   1.0

$u \longrightarrow$

a

○ $\psi = $ constant $\Big\}$ BE
●   linear
—— Exact
△ $u = $ linear BE
——— Exact

$\omega \longrightarrow$

0   2   4

0   2   4

$\omega \longrightarrow$

b

○ $\omega = $ constant $\Big\}$ BE
●   linear
—— Exact

Fig. 8.2.   Calculated values of stream function (**a**) and vorticity (**b**)

Two types of mesh were used as shown in Fig. 8.1. In the coarse mesh, we have sixteen constant boundary elements and only four eight-noded isoparametric cells. The 3 by 3 Gauss formula is used for cell integration. In the fine mesh, we have linear boundary elements and three-noded triangular cells. Double nodes are undertaken at corners. The seven-point Hammer integration is used in each cell. The time step size was taken as $\Delta t = 0.1$.

Calculated results are shown in Fig. 8.2. After eight iterations, calculated results converged to the quasi steady-state solution. The boundary solutions were accurate as compared to the exact values.

The temperature is not essentially involved in this example. However the temperature distribution was calculated as a part of the problem. We confirmed that the thermal energy is conserved in our computer codes.

8.6.2 Isothermal Flow Past a Cylinder

As an example of two-dimensional laminar flow problems, we consider the channel flow past a cylinder. The problem was described in Taylor and Hughes (1981). The fluid was initially motionless. A uniform flow profile was suddenly applied on the artificial boundary upstream. Figure 8.3 shows boundary conditions on stream function and the vorticity. Skin-deep internal nodes were taken on the dotted lines for the vorticity boundary condition. The material constant was set as $\nu = 1.0$. Uniform initial temperature was assumed, and all boundaries were thermally insulated.

The flow domain and the boundary were divided into constant boundary elements and eight-noded isoparametric cells as shown in Fig. 8.4. The time step size was taken as $\Delta t = 1.0$.

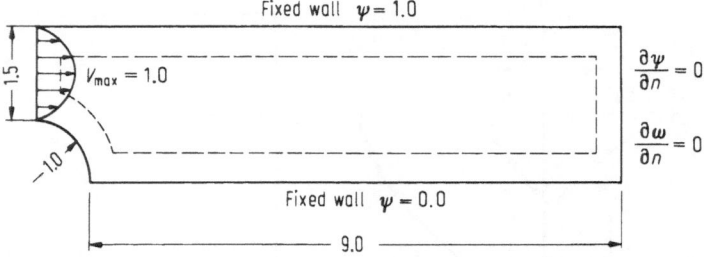

Fig. 8.3.   Boundary conditions for laminar flow past a cylinder

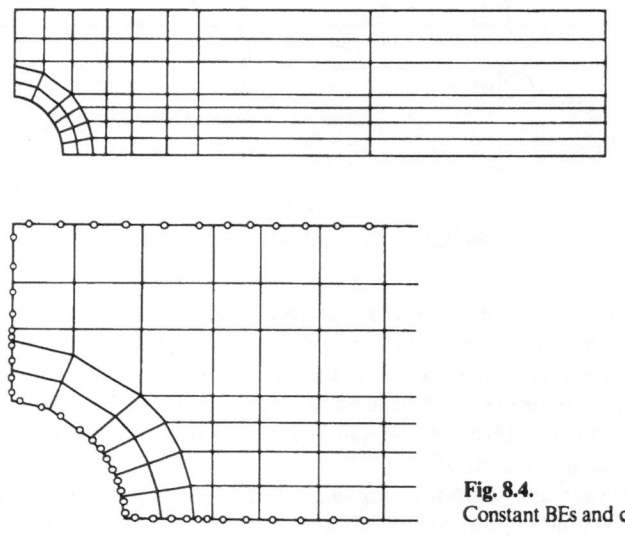

Fig. 8.4.
Constant BEs and cells

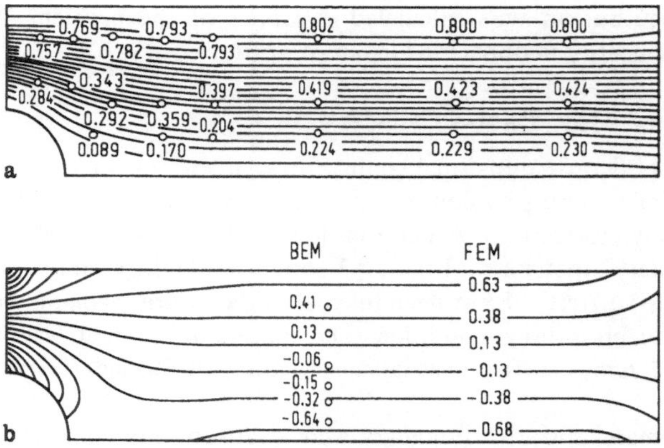

Fig. 8.5.   Calculated values of stream function (**a**) and vorticity (**b**)compared to FE solutions shown by solid curves

Calculated results are compared to the finite element solutions in Fig. 8.5. After twenty iterations, calculated results converged to the quasi steady-state solution. The boundary solutions were in good agreement with the finite element solutions.

8.6.3 Isothermal Driven-Cavity Flow

We consider a transient flow in a T-shirt cavity as illustrated in Fig. 8.6. Boundary conditions are given together with the dotted lines on which skin-deep nodes are taken. Initially, the fluid was assumed motionless. The uniform initial temperature was assumed, and the boundary was thermally insulated. The sliding lid on the top was suddenly started uniformly to the left at the unit velocity.

Figure 8.7 shows boundary elements and cell mesh: constant boundary elements and eight-noded isoparametric cells. The time step was taken as $\Delta t = 1.0$.

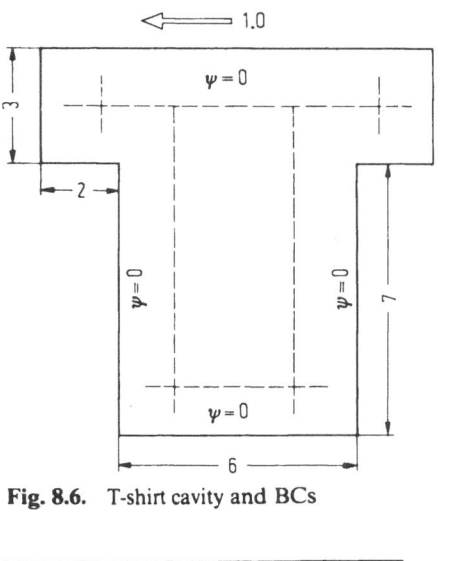

Fig. 8.6.   T-shirt cavity and BCs

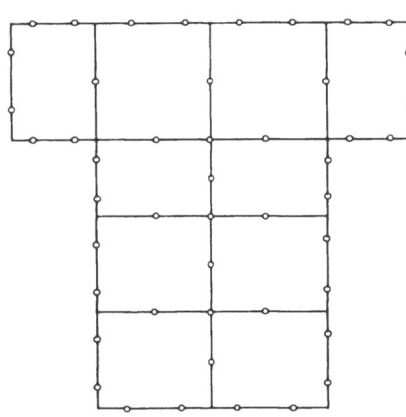

Fig. 8.7.   BE discretization and cells for the driven-cavity problem

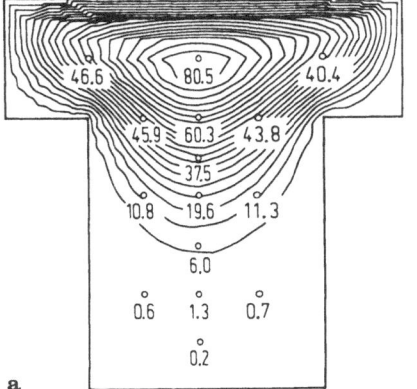

a

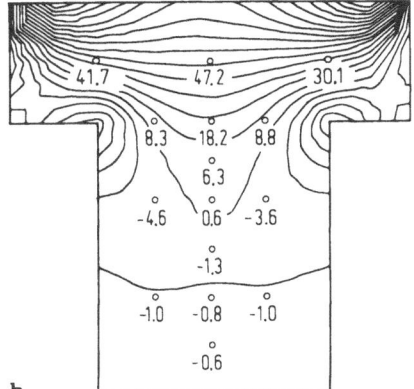

b

Fig. 8.8.   Calculated values ($\times$ 100) of stream function (**a**) and vorticity (**b**) compared to solid contours of FE solutions

Figure 8.8 shows convergent values of the stream function and vorticities. FE contours are shown for comparison. Locations of internal nodes used in BEM and FEM were not the same. Boundary solution was in good agreement with the FE solution.

8.6.4 Convection-Diffusion in Wind-Driven Flow

We consider a transient convection-diffusion of an aqueous solution with transient field velocity in a square cavity as illustrated in Fig. 8.9. Initially, zero concentration was assumed, and the fluid was assumed motionless. The wind along the top surface was at the velocities: $u_1 = -10.0$, $u_2 = 0.0$. The concentration on the top surface was maintained as $C = 1.0$. The reflexive condition was undertaken on the cavity wall.

Figure 8.10 also shows boundary elements and cell mesh: linear boundary elements and three-noded triangular cells. The time step was taken as $\Delta t = 0.5$.

Figure 8.11 shows equi-concentration curves after twenty iterations. The finite element solution is not available. Qualitatively good boundary solution was obtained.

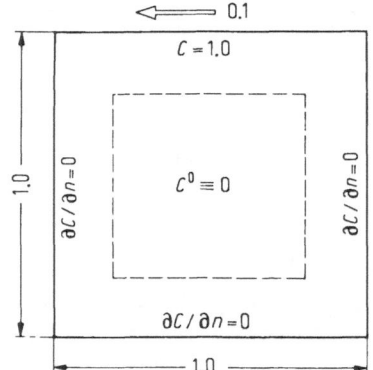

Fig. 8.9. Convection-diffusion problem in square cavity, $Re = 20$, $Pe = 100$

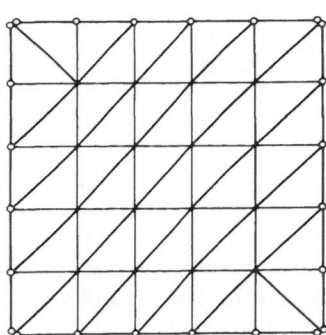

Fig. 8.10. Linear BEs and cells for convection-diffusion

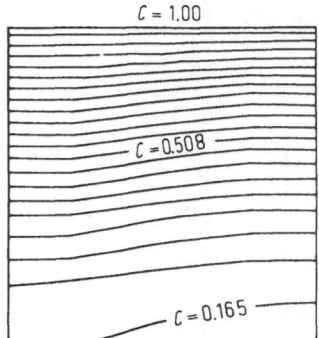

Fig. 8.11. Calculated equi-concentration

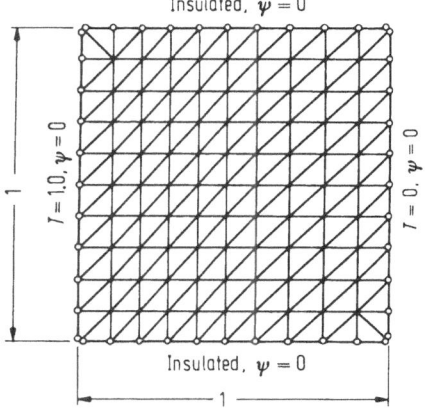

Fig. 8.12. Natural convection and BCs in a compartment, $Re = 100$, $Gr = 10^7$, $Pr = 0.1$

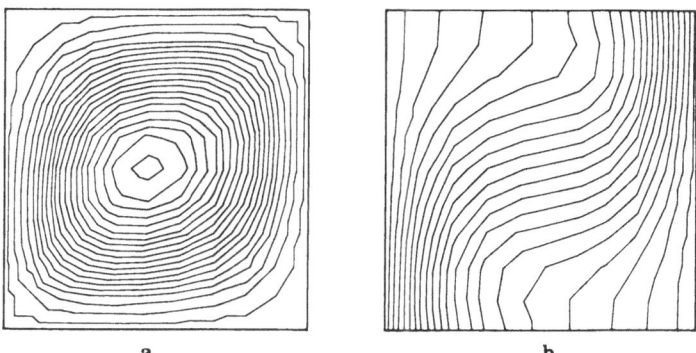

Fig. 8.13. Calculated streak lines (**a**) and isotherms (**b**)

8.6.5 Natural Convection in a Compartment

We consider buoyancy-driven natural convection in a compartment as illustrated in Fig. 8.12. The problem was described in Baker (1983). Initially, the fluid was assumed motionless. The uniform zero initial temperature was assumed. The right wall was maintained with zero temperature, while the left wall was heated with unit temperature. The rest of the boundary was adiabatic. Linear elements were used with the time step $\Delta t = 0.1$.

Figure 8.13 shows calculated values of the stream function and temperatures after forty iterations.

8.6.6 Natural Convection around a Heated Cylinder

We consider buoyancy-driven natural convection in a compartment as illustrated in Fig. 8.14. Initially, the fluid was assumed motionless. The uniform zero initial temperature was assumed. The wall was thermally insulated. The cylinder was

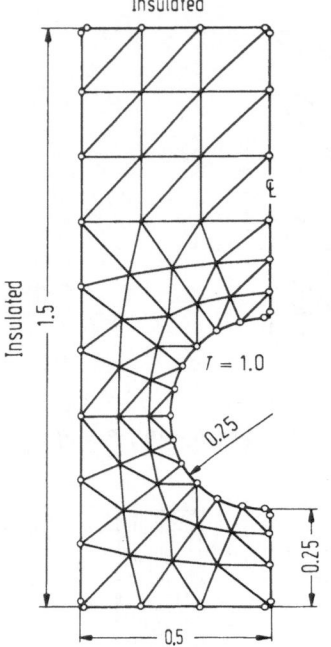

Fig. 8.14. Closed cavity with heated cylinder, $Re = 100$, $Gr = 10^7$, $Pr = 1.0$

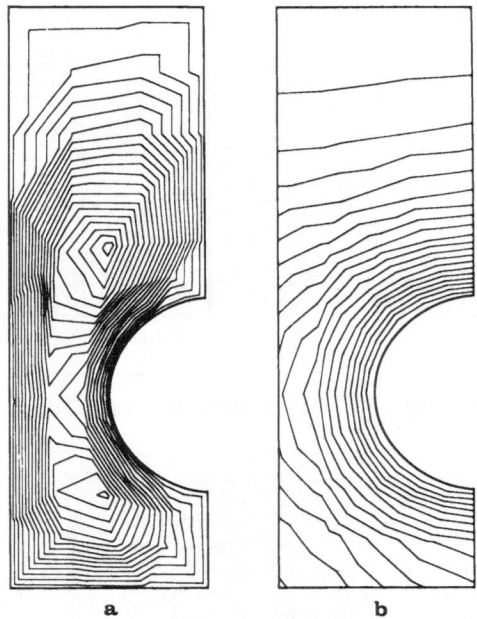

Fig. 8.15. Calculated streak lines (**a**) and isotherms (**b**) around a cylinder

heated with unit temperature. Linear elements were used with the time step $\Delta t = 0.1$.

Figure 8.15 shows calculated values of the stream function and temperatures after twenty iterations.

8.7 Conclusion

Boundary element method was applied to the numerical solution of two-dimensional Navier-Stokes equation and convective-diffusion equation. The laminar viscous flow was modelled in terms of stream function and vorticity.

Boundary solutions for problems of viscous flows between infinite plates, flows past a cylinder, cavity flows, and the flow of natural convection were obtained. The boundary solution of convective diffusion problems for unsteady viscous flow was also discussed. It is concluded that the laminar flows can be modelled by the direct boundary element method with much less unknowns than the number of unknowns necessary for the finite element method.

References

1 Baker, A.J., Finite Element Computational Fluid Mechanics. Series in Computational Methods in Mechanics and Thermal Sciences. Hemisphere Publishing Co., Washington, 1983

2 Brebbia, C.A. and Wrobel, L.C., Applications of Boundary Elements in Fluid Flow. Proceedings of the Second International Conference on Finite Elements in Water Resources, Imperial College, London, July, 4.67–4.85, Pentech Press, 1978

3 Farooq, M.U. and Kuwabara, S., Analysis for the Heat Convection Problems by Integral Equation Technique. RIMS Kokyuroku 477, 145–166, Research Institute for Mathematical Sciences, Kyoto University, 1983

4 Ikeuchi, M. and Onishi, K., Boundary Element Solutions to Steady Convective Diffusion Equations. Appl. Math. Modelling 7, 115–118, 1983

5 Khader, M.S., A Surface Integral Numerical Solution for Laminar Developed Duct Flow. Journal of Applied Mechanics, Transactions of the ASME 48, 695–700, 1983

6 Matsunashi, J., Applications of Boundary Element Method to Convection-Diffusion Problems (in Japanese). Proceedings of the 4th Symposium on Finite Element Methods in Flow Problems, JUSE, October, Tokyo, 145–152, 1983

7 Taylor, C. and Hughes, T.G., Finite Element Programming of the Navier-Stokes Equations. Pineridge Press Limited, Swansea, 1981

8 Wu, J.C. and Wahbah, M., Numerical Solution of Viscous Flow Equations Using Integral Representations. Lecture Notes in Physics 59, Springer-Verlag, 1976

Chapter 9

Asymptotic Accuracy and Convergence for Point Collocation Methods

by W. L. Wendland

9.1 Introduction

Here we continue the asymptotic error analysis of Wendland [118] where we have considered mainly the Galerkin methods. In most of the computer programs, however, the point collocation has been implemented as one of the weighted residual techniques for boundary element methods. In spite of the popularity of point collocation, its convergence properties are more delicate than those of the Galerkin procedures. Here we give a survey on the asymptotic convergence of point collocation boundary element methods for two- and three-dimensional stationary or time harmonic boundary value problems which lead to strongly elliptic boundary integral equations on the boundary curve or boundary surface Γ. As was pointed out in [118], these equations include some standard types of equations as Fredholm integral equations of the second kind with smooth or weakly singular kernels, singular integral equations with principal value integrals and strongly singular Cauchy or Giraud-Mikhlin kernels, respectively, Fredholm integral equations of the first kind with special weakly singular kernels or integro-differential operators on Γ which appear as "part finie" principal value integral operators with nonintegrable kernels. All these equations on Γ we shall write in short as

$$A u (x) + B (x) \omega = f (x), \quad \Lambda u = b \quad \text{for } x \in \Gamma \tag{9.1}$$

where $f(x)$, b denote a given vector-valued function with p components and a given constant vector in $\mathbb{R}^q$, respectively. x denotes the point on the boundary Γ, A denotes the $p \times p$ matrix of linear operators of our boundary integral equation, $B(x)$ denotes a given $p \times q$ matrix-valued function, Λ a vector of q given linear functionals presenting side conditions (like equilibrium conditions). u, ω denote the p-vector-valued unknown boundary charge and the unknown constant vector in $\mathbb{R}^q$, respectively. (In Sect. 9.2 we shall present some examples.) The linear operator A will be defined on some linear, normed complete function space X on Γ containing the solution u. In all boundary element methods, the space X will be approximated by boundary element functions spanning a finite, N-dimensional space $\tilde{H}_h$ where $h > 0$ denotes a parameter of maximal meshsize. If $\{\mu_j\}_{j=1}^N$ denotes a basis of the space $\tilde{H}_h$ then the approximate boundary charge has the form

$$u_\Delta (x) = \sum_{\ell=1}^N \gamma \; \mu_\ell (x)$$

with the constants $\gamma_1, \ldots, \gamma_N$ to be computed. For the point collocation we choose a set Ξ of N points $x_i \in \Gamma$, $i = 1, \ldots, N$ and compute $\gamma_1, \ldots, \gamma_N, \omega_\Delta$ by solving the quadratic system of linear equations

$$\sum_{\ell=1}^{N} a_{\ell i} \gamma_\ell + B(x_i) \omega_\Delta := \sum_{\ell=1}^{N} (A \mu_\ell)(x_i) + B(x_i) \omega_\Delta = f(x_i), \quad i = 1, \ldots, N,$$

$$\sum_{\ell=1}^{N} A_\ell \gamma_\ell := \sum_{\ell=1}^{N} (A \mu_\ell) \gamma_\ell = b. \tag{9.2}$$

The concept of asymptotic error estimates is based on the idea to consider not only one space $\tilde{H}_h$ with associated Ξ but a whole *family* with $N \to \infty$, $h \to 0$ corresponding to mesh refinements. In order to have a sensible approximation available. the least requirement will be that the spaces $\tilde{H}_h$ approximate X and that in addition the interpolations of the point values $(A u)(x_i)$, $x_i \in \Xi$ converge to the function $(A u)(x)$ in a reasonable sense; i.e. *consistency* of the approximation scheme. These two requirements are well accepted and usually are fulfilled. However, it is well known that for linear problems the convergence $u_\Delta, \omega_\Delta \to u, \omega$ takes place *if and only if* the discrete Eqs. (9.2) are *stable* (see Stummel's theory [107]). The stability is by no means a trivial property but is often neglected in practice. For assuring stability of the point collocation one needs to relate the choice of the collocation points Ξ with rather strong properties of the operator A. If Ξ is a unisolvent set defining intgerpolations by $\tilde{H}_h$-functions which converge for $h \to 0$ then for two-dimensional problems where Γ is a curve the operator A must necessarily be *strongly elliptic* [92, 93, 103, 104]. Unique solvability of the boundary integral Eq. (9.1) alone is by no means sufficient for stability and convergence (on the contrary to pp. 64−65 in [123]). In Ivanov's book [58] one finds on pp. 200−205 an example with piecewise linear splines and collocation at the nodal points Ξ where the discrete Eqs. (9.2) have vanishing determinants for $h \to 0$. There A is just the Hilbert transformation which can be used in plane elasticity. The Hilbert transform on the unit circle is unitary and, hence, invertible. Nevertheless, standard collocation fails (as well as the least squares method with one point Gaussian integration at Ξ for the outer integration would fail on the contrary to common belief, e.g. p. 34 in [45]).

G. Schmidt showed in [105] for the equations on curves Γ that in general Ξ must be chosen specifically depending on the properties of the principal part of A. For the three-dimensional case where Γ is a surface these relations yet are not clear at all.

Here we consider only standard point collocation. In numerical computations the final errors have mainly two reasons, the systematical errors due to the method (9.2) and additional consistency errors due to numerical quadrature. If the systematical errors can be estimated asymptotically then the influence of numerical quadrature can also be estimated by using stability and the second Strang lemma (similarly to [118]). These results allow to choose the numerical quadratures such that on one hand the asymptotic convergence corresponds to the chosen order and that on the other hand the least nodal function values are to be computed to reduce the computing time (see [7]).

Our results are different for the two- and three-dimensional problems. For the two-dimensional problems where Γ is a curve we can use the sharp results

obtained in [8, 9, 101, 104, 105]. These provide asymptotic error estimates in
Sobolev spaces as well as pointwise estimates. The error analysis is based on
coercivity of the variational formulation of the boundary integral equations in the
form of *strong ellipticity*. This concept is easily understood for differential
operators. For our more general boundary integral operators, however, one needs
for its explanation the mathematical tool of Fourier transformation and the theory
of pseudodifferential operators providing the appropriate framework. Unfortunate-
ly this part of mathematical analysis is rather new and therefore not common
knowledge yet. (There are several new books as e.g. [28, 31, 89, 110, 111].)
Nevertheless we shall try to explain the important properties elementarily leaving
out technical details.

The strongly elliptic equations already describe a large class and therefore our
two-dimensional asymptotic error results are valid for many boundary element
methods. However, there remain a number of important collocation methods to
which our results do not apply. In particular we cannot yet handle systems of
equations of differing orders. We also assume that Γ is smooth enough excluding
boundary element methods arising from boundary value problems on corner
domains or mixed boundary value problems.

For the three-dimensional problems where Γ is a surface we give only a state of
the art survey for Fredholm integral equations of the second kind. Here a large
literature is available [10, 14, 88] for integral operators with smooth kernels. We
present results based on the concept of collectively compact operator families
[4, 17, 20, 107] which provide uniform pointwise asymptotic error bounds also
for the weakly singular integral operators which appear in many applications. It
should be noted that this approach can be extended to some problems where Γ
possesses corners and edges in two- as well as in three-dimensional problems (see
[11, 20, 64, 100, 115, 116, 119]).

9.2 Examples of Boundary Integral Equations

In this section we first recall the definitions of Sobolev spaces which will play the
role of X and then present four typical examples which resemble the properties of
many of the boundary integral equations underlying boundary element procedures
in applications.

For two-dimensional problems, Γ is a curve. For simplicity, let us consider first
the case that Γ is just one simple closed boundary curve. Let Γ be given by a
regular parameter representation

$$\Gamma: x = x(t), \quad t \in \mathbb{R} \tag{9.3}$$

with $x(t)$ a 1-periodic sufficiently smooth 2-vector valued function satisfying

$$\left| \frac{dx}{dt} \right| = \varrho(t) \geqq \varrho_0 > 0 \quad \text{for all } t$$

where ϱ denotes the Jacobian. Via (9.3), we have a one-to-one correspondence
between functions on Γ and 1-periodic functions of t. More general, for a system of

mutually disjoint closed Jordan curves $\Gamma = \bigcup\limits_{j=1}^{L} \Gamma_j$ we may parametrize each Γ_j and then identify functions on Γ with L-vector-valued 1-periodic functions.

For the three-dimensional problems, the boundary Γ is a surface. For simplicity, we suppose that Γ is a closed Ljapounov surface in C^r with $r \geq 2$.

As usual, we introduce the Sobolev spaces $H^\sigma(\Gamma)$ by

$$H^\sigma(\Gamma) := \left\{ f \mid \|f\|_{H^\sigma}^2 := \sum_{|\alpha| \leq [\sigma]} \int_\Gamma |D^\alpha f|^2 \, ds \right.$$

$$\left. + \sum_{|\alpha| = [\sigma]} \int_\Gamma \int_\Gamma \frac{|D^\alpha f(x) - D^\alpha f(y)|^2}{|x - y|^{n-1+2(\sigma - [\sigma])}} \, ds_x \, ds_y < \infty \right\} \qquad (9.4)$$

for real $\sigma \geq 0$ where $[\sigma]$ denotes the integer part of σ. ds_x denotes the arc length or surface element, respectively, at $x \in \Gamma$.

$$D^\alpha = \frac{\partial^{\alpha_1}}{\partial t_1^{\alpha_1}} \cdots \frac{\partial^{\alpha_{n-1}}}{\partial t_{n-1}^{\alpha_{n-1}}}$$

denotes covariant differentiations of order α within Γ. For $\sigma < 0$, the spaces $H^\sigma(\Gamma)$ are defined by duality with respect to the L_2-scalar product

$$(f, g) = \int_\Gamma f \cdot g \, ds$$

(see p. 292 in [118]). For $n = 2$, one can also use simpler definitions based on Fourier series in t and Parseval's equality (see e.g. in [8]). For our operators A we assume that they define continuous linear mappings $A: H^\sigma \to H^{\sigma - 2\alpha}$ for a whole scale of real σ (which depends on r). We further assume that A admits a decomposition

$$(A u)(X(\tau)) = \int_{\mathbb{R}^{n-1}} \int_{\mathbb{R}^{n-1}} \exp(2\pi i(\tau - t) \cdot \xi) \, a_0(\tau, \xi) \, \Psi(|\tau - t|) \, u(X(t)) \, dt \, d\xi + C_1 u$$

$$= A_0 u + C_1 u$$

where $C_1: H^\sigma \to H^{\sigma - 2\alpha}$ is compact for suitable σ near α. $x = X(\tau)$ and $y = X(t)$ are regular local parameter representations of Γ; $t, \tau \in \mathbb{R}^{n-1}$, $X \in C^r$. $\Psi(\varrho)$ is an even C^∞ cut-off function with compact support and $\Psi(\varrho) \equiv 1$ for $|\varrho| \leq \delta$ with suitable $\delta > 0$. The *principal symbol* $a_0(\tau, \xi)$ is C^∞ with respect to ξ and sufficiently smooth with respect to τ satisfying

$$a_0(\tau, \lambda \xi) = \lambda^{2\alpha} a_0(\tau, \xi) \quad \text{for all } \tau, \text{ all } \lambda \geq 1 \text{ and } |\xi| \geq 1.$$

For a differential operator, a_0 is just the highest order characteristic polynomial. In our more general case a_0 is the leading term of highest order in ξ in the asymptotic expansion of the symbol given on p. 44 ff. in [110]; the symbol a can be given by

$$a(\tau, \xi) = e^{-2\pi i \tau \cdot \xi} A(\Psi(|\tau - X^{-1}(\cdot)|) \exp(2\pi i \xi \cdot X^{-1}(\cdot))).$$

Now A is called *strongly elliptic of order* 2α if A admits the above representation and if *there exist a positive constant* γ_0 and a complex valued C^∞ matrix function

$\theta(x)$ such that

$$\text{Re}\,\zeta^T \theta(X(\tau))\,a_0(\tau, \xi)\bar\zeta \geq \gamma_0\,|\zeta|^2 \quad \text{for all } x \in \Gamma, |\xi| = 1, \zeta \in \mathbb{C}^P.$$

The strongly elliptic operators provide coercivity in form of Gårding's inequality for θA (see Inequality (9.8) in [118]) which can sometimes also be proved directly, in particular for $n = 2$ with Fourier series. Besides the Sobolev spaces in Eq. (9.4) we will also need the supremum norm Sobolev spaces

$$W^\sigma_\infty(\Gamma) := \left\{ f \mid \|f\|_{W^\sigma_\infty} := \sum_{|\alpha| \leq \sigma} \operatorname{ess\,sup}_{x \in \Gamma} |D^\alpha f(x)| < \infty \right\}$$

for integers $\sigma \in \mathbb{N}_0$. By C^σ we denote the subspace of σ times continuously differentiable functions f where

$$\|f\|_{C^\sigma} := \|f\|_{W^\sigma_\infty} = \sum_{|\alpha| \leq \sigma} \max_{x \in \Gamma} |D^\alpha f(x)|.$$

As the first example of boundary integral equations we present the classical scattering problem with an obstacle whose boundary is Γ (see [24] or [119]). Let the velocity potential of the total field be U and let an incident plane wave move in the x_1-direction. Then $U(x)$ has the representation

$$U(x) = e^{ikx_1} + \int_\Gamma \left\{ U(y)\left(\frac{\partial}{\partial v_y} F_k(x, y)\right) - \frac{\partial U}{\partial v}(y)\,F_k(x, y) \right\} ds_y \quad \text{for } x \notin \Gamma \quad (9.5)$$

in the exterior. Here $\partial/\partial v_y = v(y) \cdot \text{grad}$ denotes the derivative in the direction of the exterior normal $v(y)$ at $y \in \Gamma$. $F_k(x, y)$ is the fundamental solution of the Helmholtz equation given by

$$F_k(x, y) = \begin{cases} \dfrac{i}{4}\,H_0^{(1)}(k\,|x - y|) & \text{for } n = 2, \\[2ex] \dfrac{1}{4\pi}\,\dfrac{1}{|x - y|}\,\exp(i\,k\,|x - y|) & \text{for } n = 3. \end{cases}$$

If we consider a *hard scatterer* then $(\partial U/\partial v)_\Gamma = 0$ and (9.5) yields with the jump relation the

Fredholm integral equation of the second kind for $u = U|_\Gamma$ on Γ:

$$Au := u(x) - 2\int_\Gamma \left(\frac{\partial}{\partial v_y} F_k(x, y)\right) u(y)\, ds_y = 2\,e^{ikx_1} \quad \text{for } x \in \Gamma. \quad (9.6)$$

The kernel $\partial F_k(x, y)/\partial v_y$ in Eq. (9.6) is continuous for $n = 2$ and weakly singular for $n = 3$ (see also [24, 64]).

For a *soft scatterer* one has $U|_\Gamma = 0$. Then (9.5) yields an integral equation of the first kind. However, the potential

$$U(x) = \int_\Gamma \left(\frac{\partial}{\partial v_y} F_k(x, y)\right) u(y)\, ds_y + \omega\, F_k(x, 0) + e^{ikx_1}$$

with unknown density u and unknown ω yields on Γ the modified Fredholm integral equation of the second kind for u, ω,

$$u(x) + 2 \int_\Gamma \left(\frac{\partial}{\partial \nu_y} F_k(x, y) \right) u(y)\, ds_y + 2 F_k(x, 0)\, \omega = -2 e^{ikx_1} \quad \text{for } x \in \Gamma,$$

$$\int_\Gamma u\, ds_y = 0. \tag{9.7}$$

Except for special frequencies, the Eqs. (9.6) and (9.7) are uniquely solvable. (For corresponding modifications see e.g. Chap. 3.4 in [24].) The kernel of the integral operator is in C^{r-2} for $n = 2$ and weakly singular for $n = 3$.

Although these equations seem to model rather special problems they appear in many and also more general applications in acoustics (see [2, 18, 33, 34, 40, 41, 43, 49, 63, 66, 78, 91, 102]).

In the case $k = 0$ (here set $F_0 = -\frac{1}{2} \pi \log |x - y|$ for $n = 2$) such equations appear in the direct method applied to ideal flow problems (see [115, 116] and also Chaps. 4.11, 4.12, 22.2 in [42]) and for Stokes flows [124].

Equations of the same type but with the adjoint kernel $\partial F_k(x, y)/\partial \nu_x$ have also been used in ideal flow computations [50, 73] and for electrostatic problems [45, 60]. Systems of the integral equations with smooth or weakly singular kernels, respectively, appear also for interior and exterior Maxwell problems and for electric and magnetic boundary value problems of electromagnetic fields (see e.g. [2, 78] and Eqs. (4.29), (4.30), (4.32), (4.33), (4.35), (4.36) in Colton and Kress [24]). In the framework of pseudodifferential operators, the type of Eqs. (9.6), (9.7) is trivial. Here the principal part of A is the identity, hence $a_0 = 1$ and the order $2\alpha = 0$. The additional integral operators have either smooth kernels for $n = 2$, then on a C^∞ curve Γ they have order $-\infty$, or they have weakly singular kernels for $n = 3$ defining pseudodifferential operators of order -1 (see pp. 177f. in [119]).

The vibration problems in elasticity and the elstostatic problems, however, yield boundary integral equations with *strong* singularities. As an example let us consider the exterior boundary value problems in elasticity [57]. Let $U(x)$ denote the displacement field. Solutions of the Navier equations with finite energy admit the modified Betti formula,

$$U(x) = \int_\Gamma \{ (T_y E(y, x))^T U(y) - E(y, x)(T_y U)(y) \}\, ds_y + M(x)\, \omega. \tag{9.8}$$

Here T_y denotes the traction operator on Γ with respect to y,

$$T_y = \lambda v(x) \operatorname{div} \cdot + 2\mu \frac{\partial \cdot}{\partial \nu_y} + \mu\, v(y) \wedge \operatorname{curl} \cdot,$$

$E(y, x)$ is the Kelvin matrix

$$E(y, x) = \frac{\lambda + 3\mu}{4(n-1)\,\pi(\lambda + 2\mu)} \left\{ \gamma(y, x)\, I + \frac{\lambda + \mu}{\lambda + 3\mu} \frac{(x-y)(x-y)^T}{|x-y|^n} \right\}$$

with

$$\gamma(y, x) = \begin{cases} \log \dfrac{1}{|x-y|} & \text{for } n = 2, \\[2mm] \dfrac{1}{|x-y|} & \text{for } n = 3. \end{cases}$$

λ and μ are the Lamé constants and $M(x)\,\omega$ describes the rigid motion at infinity. The rigid motions are given by

$$M(x) = \begin{pmatrix} 1 & 0 & -x_2 & 0 & 0 & x_3 \\ 0 & 1 & x_1 & 0 & -x_3 & 0 \\ 0 & 0 & 0 & 1 & x_2 & -x_1 \end{pmatrix} \quad \text{for } n = 3.$$

In case $n = 2$ skip the last row and the last three columns.

In the *traction problem*, $T_y\,U|_\Gamma = \psi$ is given on the boundary Γ. Then the jump relations yield from $x \to \Gamma$ in Eq. (9.8) the *singular integral equation* for $u = U|_\Gamma$ and $\omega \in \mathbb{R}^{3(n-1)}$

$$u(x) - 2 \int_\Gamma (T_y\,E(y, x))^T\,u(y)\,ds_y - 2M(x)\,\omega = -2 \int_\Gamma E(y, x)\,\psi(y)\,ds_y \quad \text{for } x \in \Gamma,$$
$$\tag{9.9}$$
$$\int_\Gamma M^T(y)\,u(y)ds_y = b.$$

The integral operator in these equations contains strong singularities and defines a pseudodifferential operator of order $2\alpha = 0$ having the principal symbol

$$a_0 = \begin{pmatrix} 1 & i\gamma\dfrac{\xi}{|\xi|} \\ -i\gamma\dfrac{\xi}{|\xi|} & 1 \end{pmatrix} \quad \text{for } n = 2$$

and

$$a_0 = M^{-1}(x) \begin{pmatrix} 1 & 0 & i\gamma\dfrac{\xi_1}{|\xi|} \\ 0 & 1 & i\gamma\dfrac{\xi_2}{|\xi|} \\ -i\gamma\dfrac{\xi_1}{|\xi|} & -i\gamma\dfrac{\xi_2}{|\xi|} & 1 \end{pmatrix} M(x) \quad \text{for } n = 3$$

with $M(x)^T = (X_{t_1}, X_{t_2}, v(x))$ provided X_{t_i} are orthonormal at x. Here $0 < \gamma = \dfrac{\mu}{\lambda + 2\mu} < 1$ guarantees strong ellipticity. Eqs. (9.9) are uniquely solvable.

Again, the Eqs. (9.9) are just one special case of a large variety of singular integral equations and can be used for many elasticity problems [27, 46, 61, 62, 67, 68, 75, 77, 79, 80, 87, 96, 97, 113]. Our results on the collocation method, however, apply only to the case of two-dimensional problems, $n = 2$. Here our estimates are sharp. Antes gives in [5, 6] a-posteriori error estimates but cannot prove stability. Even in light of our results in Sect. 3 and our sharp stability estimates his convergence orders are weaker than ours by at least one order.

For the next type of boundary integral equations we choose the

displacement problem where $U|_\Gamma = \phi$ is given and the boundary traction $u = TU|_\Gamma$ is unknown. Here the Betti formula Eq. (9.8) yields on the boundary a system of

Fredholm integral equations of the first kind whose kernel is $E(y, x)$:

$$\int_\Gamma E(y, x)\, u(y)\, ds_y - M(x)\, \omega = -\tfrac{1}{2}\phi(x) + \int_\Gamma (T_y\, E(y, x))^T \phi(y)\, ds_y \quad \text{for } x \in \Gamma,$$

$$(9.10)$$

$$\int_\Gamma M^T(y)\, u(y)\, ds_y = b.$$

Here the additional conditions with $b \in \mathbb{R}^{3(n-1)}$ corresponds to equilibrium conditions. The operator is a pseudodifferential operator of order $2\alpha = -1$ having the principal symbol

$$a_0 = \frac{\lambda + 3\mu}{4(\lambda + 2\mu)} \frac{1}{|\xi|} \begin{pmatrix} 1 & 0 \\ 0 & 1 \end{pmatrix} \quad \text{for } n = 2$$

and

$$a_0 = \frac{\lambda + 3\mu}{8(\lambda + 2\mu)} M^{-1}(x) \frac{1}{|\xi|^3} \begin{pmatrix} |\xi|^2 + \varkappa\,\xi_2^2 & -\varkappa\,\xi_1\,\xi_1 & 0 \\ -\varkappa\,\xi_1\,\xi_2 & |\xi|^2 + \varkappa\,\xi_1^2 & 0 \\ 0 & 0 & |\xi|^2 \end{pmatrix} M(x) \quad \text{for } n = 3$$

where $\varkappa = \dfrac{\lambda + \mu}{\lambda + 3\mu}$. Eqs. (9.10) are uniquely solvable and the operators are always strongly elliptic if $\lambda + 3\mu \neq 0$. Systems of this form arise in a huge variety of applications. For $n = 2$ the method goes back to Fichera [32] and has been extended to rather general situations in [25]. Examples are Symm's integral equation of conformal mapping [22, 37, 48, 51, 53, 108, 117]; viscous flow problems and Stokes flows [35, 36, 52, 53, 54, 55]; electrostatics and electromagnetic problems [26, 72, 84, 86, 98, 99]; acoustics [2, 33, 34, 106, 119]; elasticity [16, 19, 23, 27, 46, 53, 57, 59, 60, 84, 87]; plate bending [46, 53, 54]; and torsion problems [60, 76]; see also p. 246 ff. in [109].

Our asymptotic error results for the two-dimensional case and collocation of these equations are new. Preliminary results for special cases and with lower order convergence have already been obtained in [1, 3, 112]. An extension of our convergence to less regular Γ can be found in [51].

(Differentiation of (9.10) yields a system with the Hilbert transform for which standard collocation fails, see also [103].)

Our last example is a *second formulation* of the *traction problem* which is obtained from applying the traction operator to Eq. (9.78). This yields the

hypersingular boundary equation

$$-\int_\Gamma T_x(T_y\, E(y, x))^T (u(y) - u(x))\, ds_y + M(x)\, \alpha = -\tfrac{1}{2}\psi(x) - \int_\Gamma T_x\, E(y, x)\, \psi(y)\, ds_y,$$

$$(9.11)$$

$$\int_\Gamma M^T(y)\, u(y)\, ds_y = b.$$

The operator A in Eq. (9.11) is a pseudodifferential operator of order $2\alpha = 1$ and has the principal symbol

$$a_0 = \frac{\mu(\lambda + \mu)}{\lambda + 2\mu} \begin{pmatrix} |\xi| & 0 \\ 0 & |\xi| \end{pmatrix} \quad \text{for } n = 2$$

and

$$a_0 = \frac{\mu^2}{2} M(x)^{-1} \frac{1}{|\xi|} \begin{pmatrix} |\xi|^2 + \varepsilon\xi_1^2 & \varepsilon\xi_1\xi_2 & 0 \\ \varepsilon\xi_1\xi_2 & |\xi|^2 + \varepsilon\xi_2^2 & 0 \\ 0 & 0 & \frac{2}{\gamma}|\xi|^2 \end{pmatrix} M(x)$$

for $n = 3$ with $\gamma = \mu/(\lambda + 2\mu)$, $\varepsilon = \lambda/(\lambda + 2\mu)$. Hence, A is strongly elliptic if $\mu \neq 0$, $\lambda + 2\mu \neq 0$ and $\lambda + \mu > 0$ for $n = 3$, $\lambda + \mu \neq 0$ for $n = 2$. Eqs. (9.11) are in these cases uniquely solvable. Equations of this type arise in acoustics and electrostatics [2, 12, 26, 29, 34, 38, 39, 44, 83, 106, 119, 122] and in elasticity [15, 47, 57, 59, 81, 85]. For a general approach see Costabel [25]. Our convergence results for equations of the type (9.11) in case $n = 2$ apply to the methods in [65, 113, 114].

9.3 Standard Collocation for Two-Dimensional Problems

In the two-dimensional problems we consider all functions on Γ as 1-periodic functions depending on the parameter t.

For the collocation we select an increasing sequence of mesh points $\Delta = \{t_i\}_{i=-\infty}^{+\infty}$ satisfying $t_{i+N} = t_i + 1$, for a fixed natural number N and for all integers $i \in \mathbf{Z}$. Let $h_\Delta = \max(t_{i+1} - t_i)$. By $S_d(\Delta)$ we denote the space of all 1-periodic, $(d-1)$ times continuously differentiable splines of degree d subordinate to the partition Δ. Sometimes we shall consider only equidistant partitions for which $t_{i+1} - t_i = h = 1/N$. Note that, since $\chi^{(d)}$ (the superscript denotes differentiation) is a step function for $\chi \in S_d(\Delta)$, $S_d(\Delta) \subset H^s(\Gamma)$ if and only if $s < d + \frac{1}{2}$.

We recall the *approximation properties* of $S_d(\Delta)$. *Let* $-\infty < \sigma \leq s \leq d + 1$ *and* $\sigma < d + \frac{1}{2}$. *Then to any* $u \in H^s(\Gamma)$ *and* $S_d(\Delta)$ *there exists* $\psi \in S_d(\Delta)$ *and a constant* $c > 0$ *such that*

$$\| u - \psi \|_\sigma \leq c h_\Delta^{s-\sigma} \| u \|_s \tag{9.12}$$

where ψ *is independent of* σ. *The constant* c *is independent of* u, ψ, h_Δ.

(Throughout, c denotes a generic constant independent of Δ, not the same at each occurrence. $\| \cdot \|_\sigma$ denotes the H^s-norm.)

For *families of equidistant partitions*, the spline functions also provide the *inverse assumptions; i.e. for any* σ *and* s *with* $\sigma \leq s < d + \frac{1}{2}$ *there exists a constant* c *such that the inequality*

$$\| \chi \|_s \leq c h_\Delta^{\sigma-s} \| \chi \|_\sigma \quad \text{holds for all } \chi \in S_d(\Delta). \tag{9.13}$$

For *odd* splines, i.e. d is odd, we choose for the set of collocation points $\Xi = \{x_i = X(t_i)\}$ whereas for *even* d we choose the points corresponding to midpoints, $\Xi = \{x_i = X(\frac{1}{2}(t_{i+1} + t_i))\}$. (These choices correspond to Definition (9.50) in [118] with $x_i = z_{i-[(d+1)/2]}$.)

The standard collocation equations (9.2) now read as follows.

Find

$$u_\Delta \in \frac{1}{\varrho} S_d(\Delta) := \left\{ \psi = \frac{1}{\varrho} \chi \,|\, \chi \in S_d(\Delta) \right\} \quad \text{and} \quad \omega_\Delta \in \mathbf{R}^q$$

such that

$$Au_\Delta(x_i) + B(x_i)\,\omega_\Delta = f(x_i), \quad i = 1, \ldots, N; \quad x_i \in \Xi,$$

$$\Lambda u = b. \tag{9.14}$$

For these two-dimensional problems we have the following asymptotic convergence result.

Theorem 3.1: *Let A be strongly elliptic and Eq. (9.1) be uniquely solvable. Then the standard collocation method viz. Eqs. (9.14) converges asymptotically with optimal orders; i.e. there exist positive constants c, h_0 such that Eqs. (9.14) are uniquely solvable for every h with $0 < h_\Delta \le h_0$, and for the corresponding approximate solutions u_Δ there holds the estimate*

$$|\omega_\Delta - \omega| + \|u_\Delta - u\|_\sigma \le ch_\Delta^{\tau - \sigma}\|u\|_\tau \tag{9.15}$$

where c is independent of h_Δ, u, ω, u_Δ, ω_Δ, provided the following assumptions (i) or (ii) are satisfied:

(i) *Δ is a uniform family,*

$$2\alpha \le \sigma < d + \tfrac{1}{2}, \quad \max\{\sigma, \alpha + \tfrac{1}{2}(d+1)\} \le \tau \le d+1 \tag{9.16}$$

and either $2\alpha < d$ or d is even and $d \le 2\alpha < d + \tfrac{1}{2}$ and, in addition, $\sigma < \tau$ or $2\alpha + \tfrac{1}{2} < \sigma = \tau$.

(ii) *Δ is an arbitrary family but d is odd, $2\alpha < d$ and*

$$2\alpha \le \sigma \le \alpha + \tfrac{1}{2}(d+1) \le \tau \le d+1. \tag{9.17}$$

The proof for the first case in (i) and even d and equations with a purely convolutional principal part (as for Eqs. (9.37) in [118]) is given by Saranen and Wendland [101]. The proof for the case (ii) is given by Arnold and Wendland [8]. The remaining cases in (i) are proved by Arnold and Wendland [9]. For the case (ii), the proof rests on the formulation of the collocation equations (9.14) as modified Galerkin equations which are obtained by $\tfrac{1}{2}(d+1)$-fold integrations by parts [8]. In the remaining cases the proofs are based on explicit Fourier expansions and recurrence relations between the Fourier coefficients for spline functions [9, 101].

Note that our convergence results and asymptotic error estimates in particular apply to piecewise constant trial functions for equations with $\alpha \le 0$ and, hence, for point collocation of the direct method as used in [16, 22, 23, 45, 48, 60, 76, 80, 97, 98, 108] and to more general splines on Γ as in [3, 5, 6, 29, 33, 34, 51, 65, 113, 114].

The restrictions on the Sobolev index inequalities (9.16), (9.17), respectively, show that – for the *same* splines – the collocation method converges at most with the order $h^{d+1-2\alpha}$ whereas the Galerkin method converges at most with the order $h^{2d+2-2\alpha}$ (see [8, 56] and Theorem 4.1 in [118]). This situation is summarized in Fig. 9.1.

Hence, to obtain the *same* order of convergence as in the standard Galerkin method, the collocation method requires splines whose degrees exceed those employed for the Galerkin method. More precisely, for

$$d_c + 1 - 2\alpha = 2d_G + 2 - 2\alpha$$

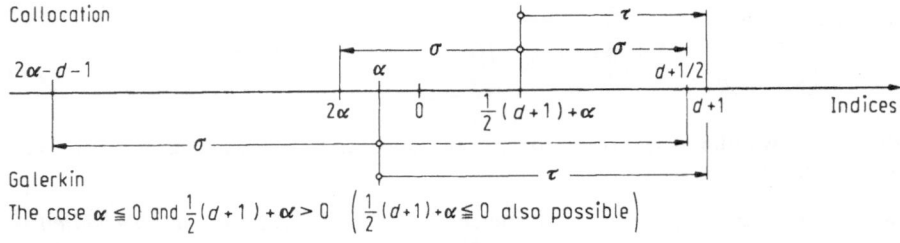

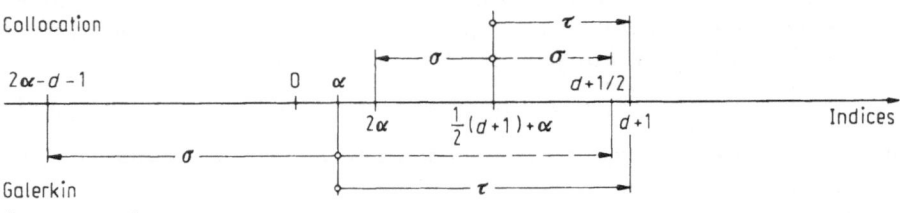

Fig. 9.1. The indices $\sigma \leq \tau$ for which $\|u - v\|_\sigma \leq c h^{\tau - \sigma} \|u\|_\tau$. Dashed lines indicate estimates requiring a quasiuniform mesh family

we must choose

$$d_c = 2 d_G + 1 \tag{9.18}$$

where d_c denotes the spline degree for collocation and d_G that for Galerkin's procedure, respectively. On the other hand, for Galerkin's method one needs double integration to compute the weights of the influence matrix (see Eq. (9.33) in [118]) in contrary to the collocation method where we need only one integration. This effect is reflected in the necessary accuracy of the corresponding numerical integrations so that — after all — the computing times and corresponding costs are essentially the same in both methods provided both converge asymptotically with the same highest orders, i.e. if Eq. (9.18) holds. (See also [7].)

For the necessary accuracy of the numerical integrations one needs stability estimates for the collocation equations (9.14) e.g. in $L_2(\Gamma)$. For their formulation let us introduce the interpolation operator I_Δ mapping continuous functions onto $S_d(\Delta)$ via

$$I_\Delta f = \sum_{j=1}^{N} \alpha_j \mu_j \quad \text{with} \quad \sum_{j=1}^{N} \alpha_j \mu_j(x_i) = f(x_i) \quad \text{for } x_i \in \Xi. \tag{9.19}$$

For the interpolation I_Δ in Eq. (9.19) we have the approximation properties

$$\| I_\Delta f - f \|_\sigma \leq c h_\Delta^{\tau - \sigma} \| f \|_\tau$$

provided $0 \leq \sigma \leq \tau \leq d + 1$, $\frac{1}{2} < \tau$, $\sigma < d + \frac{1}{2}$. (See Chapter III in [120] and p. 377 in [8].)

Let us further restrict to the case of equidistant partitions. Then we have the following stability result.

Theorem 3.2 (see Theorem 2.1.8 in [8]): *Let A be strongly elliptic and Eqs. (9.1) be uniquely solvable. Let $S_d(\varDelta)$ provide the inverse assumptions (9.13) and let $u_\varDelta$, $\omega_\varDelta$ solve (9.14) with any given $f \in S_d(\varDelta)$. Then there hold the conditioning estimates*

$$|\omega_\varDelta| + \|u_\varDelta\|_0 \le ch^{2\alpha'}(\|f\|_0 + |b|),$$

$$\|f\|_0 + |b| \le ch^{-2\alpha + 2\alpha'}(|\omega_\varDelta| + \|u_\varDelta\|_0)$$

(9.20)

where the constants c are independent of $h, f, b; u_\varDelta$, $\omega_\varDelta$ and $\alpha' = \min\{0, \alpha\}$.

(Note that Theorem 3.2 is the proposed Theorem 7.1 in [118]. But also note that our estimates with inequalities (9.16), (9.17) in Theorem 3.1 are now valid for less restrictions on d and α than in Theorem 7.2 in [118].)

The error estimates in Theorem 3.1 and conditioning estimates in Theorem 3.2 allow to give estimates for the *numerical integration* which is needed for the implementation of the point collocation. That means that the weights $a_{\ell i}$ and $\varLambda_\ell$ in Eq. (9.2) are to be computed by numerical quadrature. The corresponding numerical values let us denote by $\tilde{a}_{\ell i}$, $\tilde{\varLambda}_\ell$, respectively. Then let us require the accuracy

$$|a_{\ell i} - \tilde{a}_i| + |\varLambda_\ell - \tilde{\varLambda}_\ell| \le ch^{\varrho+1}.$$

(9.21)

Since the operator A is usually defined by integral operators having singular kernels as in Sect. 9.2 the requirement (9.21) with $\varrho > 0$ has strong consequences for the numerical quadrature formulas used. Namely, the principal parts of the singular integrals are either to be integrated analytically (see e.g. p. 201 in [119]) or by corresponding highly accurate numerical integrations. For Cauchy principal value integrals as in Eqs. (9.9), Piessens [90] presents Gauss-Legendre quadrature formulas which are exact for polynomial weights as with our splines. The remainders with smooth kernels can be integrated numerically either by using special quadrature formulas assigned to the grid $\varDelta$ and the splines $S_d(\varDelta)$ (as e.g. in formula (4.24) in [119]), or by using Gaussian quadrature on each subinterval $[t_i, t_{i+1}]$. (See also [7].) In the latter case, however, the order of accuracy in the numerical integration is reduced by the degree d of our splines since they appear in $a_{\ell i}$ and $\varLambda_\ell$ as special weights (see Lemma 3.3 in [69]). Therefore even for smooth integrands one needs rather excessively high order Gaussian quadratures. It should be noted that the development of economical and accurate procedures for the evaluation of the influence matrix $a_{\ell i}$, $\varLambda_\ell$ is still an important task for the boundary element methods (see also [7]).

If $w = \dfrac{1}{\varrho} \sum\limits_{\ell=1}^{N} \alpha_\ell \mu_\ell$ is an arbitrary spline function in $\dfrac{1}{\varrho} S_d(\varDelta)$ then the spline $f \in S_d(\varDelta)$ is uniquely determined by the equations

$$\sum_{\ell=1}^{N} \alpha_\ell \tilde{a}_{\ell i} = f(x_i) \quad \text{for} \quad x_i \in \varXi$$

(9.22)

and defines a linear mapping $\tilde{A}_\varDelta : \dfrac{1}{\varrho} S_d(\varDelta) \to S_d(\varDelta)$ via

$$\tilde{A}_\varDelta w = f.$$

(9.23)

Correspondingly we define $\varLambda_\varDelta : \dfrac{1}{\varrho} S_d(\varDelta) \to \mathbb{R}^q.$

Lemma 3.3: *If the coefficients $\tilde{a}_{\ell i}$, $\tilde{\Lambda}_\ell$ provide accuracy inequality (9.21) then there hold the consistency estimates*

$$\| (I_\Delta A - \tilde{A}_\Delta) \, w \|_{L_2} + |(\Lambda - \tilde{\Lambda}_\Delta) \, w| \leq ch^\varrho \| w \|_{L_2}. \tag{9.24}$$

The constant c is independent of h and $w \in S_d(\Delta)$.

Since the proof of Lemma 3.3 is almost exactly the same as of Theorem 6.1 in [118] we omit here the presentation.

The computational solution $\bar{u}_\Delta = \sum_{\ell=1}^{N} \bar{\gamma}_\ell \mu_\ell$, $\bar{\omega}_\Delta$ is obtained by solving the linear equations

$$\sum_{\ell=1}^{N} \tilde{a}_{\ell i} \, \bar{\gamma}_\ell + B(x_i) \, \bar{\omega}_\Delta = f(x_i), \quad x_i \in \Xi,$$
$$\sum_{\ell=1}^{N} \tilde{\Lambda}_\ell \, \bar{\gamma}_\ell = b \tag{9.25}$$

or, in short,

$$\tilde{A}_\Delta \, \bar{u}_\Delta + I_\Delta B \omega_\Delta = I_\Delta f,$$
$$\tilde{\Lambda}_\Delta \, \bar{u}_\Delta = b. \tag{9.26}$$

Hence, $\bar{u}_\Delta$, $\bar{\omega}_\Delta$ is the solution of the system

$$I_\Delta A \bar{u}_\Delta + I_\Delta B \bar{\omega}_\Delta = (I_\Delta A - \tilde{A}_\Delta) \, \bar{u}_\Delta + I_\Delta f,$$
$$\Lambda \bar{u}_\Delta = (\Lambda - \tilde{\Lambda}_\Delta) \, \bar{u}_\Delta + b \tag{9.27}$$

to which we may apply estimate (9.20) of Theorem 3.2 yielding with the consistency estimate (9.24)

$$|\bar{\omega}_\Delta| + \| \bar{u}_\Delta \|_0 \leq ch^{2\alpha'} \{ch^\varrho \| \bar{u}_\Delta \|_0 + \| I_\Delta f \|_0 + |b|\}.$$

Therefore, if $\varrho > -2\alpha'$, we find for Eqs. (9.26) also the stability estimate

$$|\bar{\omega}_\Delta| + \| \bar{u}_\Delta \|_0 \leq ch^{2\alpha'}\{\| I_\Delta f \|_0 + |b|\} \tag{9.28}$$

(provided $h \leq h_0$ with some $h_0 > 0$ small enough). Now we are in the position to obtain error estimates for the actually computed approximate solutions which contain the influence of numerical integration.

Theorem 3.4: *Let A be strongly elliptic, Eqs. (9.1) be uniquely solvable, $S_d(\Delta)$ provide the inverse assumptions and the numerical integrations be accurate with $\varrho > -2\alpha'$. Then there exists $h_0 > 0$ such that the numerically integrated (boundary element) Eqs. (9.25) are uniquely solvable for every h with $0 < h \leq h_0$ provided the assumptions of Theorem 3.1 are satisfied. The corresponding solutions satisfy the asymptotic error estimates*

$$|\omega - \bar{\omega}_\Delta| + \| u - \bar{u}_\Delta \|_\sigma \leq ch^{\tau-\sigma} \| u \|_\tau + ch^{2\alpha'+\varrho-\max\{0,\sigma\}} \| u \|_\tau. \tag{9.29}$$

Remarks: One needs the accuracy

$$\varrho \geq \tau - 2\alpha' + \max\{0, -\sigma\}$$

to recover the systematical order of convergence as in Theorem 3.1 according to estimate (9.16). Hence, the highest possible order of convergence, i.e. $\sigma = 2\alpha$, $\tau = d + 1$ requires an accuracy

$$\varrho \geq d + 1 - 4\alpha'.$$

Proof: The difference $u_\Delta - \bar{u}_\Delta$, $\omega_\Delta - \bar{\omega}_\Delta$ satisfies modified Eqs. (9.26) in the form

$$\tilde{A}_\Delta (u_\Delta - \bar{u}_\Delta) + I_\Delta B(\omega_\Delta - \bar{\omega}_\Delta) = (\tilde{A}_\Delta - I_\Delta A) u_\Delta,$$

$$\tilde{A}_\Delta (\omega_\Delta - \bar{\omega}_\Delta) = (\tilde{A}_\Delta - A) u_\Delta$$

which yield with estimate (9.24) and estimate (9.28),

$$|\omega_\Delta - \bar{\omega}_\Delta| + \|u_\Delta - \bar{u}_\Delta\|_0 \leq \mathrm{ch}^{2\alpha' + \varrho} \|u_\Delta\|_0 \leq \mathrm{ch}^{2\alpha' + \varrho} \|u_\Delta\|_{\alpha + 1/2(d+1)}. \qquad (9.30)$$

From estimate (9.15) we have with assumption (9.16) and triangle inequality

$$\|u_\Delta\|_{\alpha + 1/2(d+1)} \leq c \|u\|_\tau.$$

The inverse assumption, triangle inequality and estimate (9.15) eventually yield

$$|\omega - \bar{\omega}_\Delta| + \|u - \bar{u}_\Delta\|_\sigma \leq \mathrm{ch}^{\tau - \sigma} \|u\|_\tau + |\omega_\Delta - \bar{\omega}_\Delta| + \mathrm{ch}^{-\max\{0,\sigma\}} \|u_\Delta - \bar{u}_\Delta\|_0$$

$$\leq \mathrm{ch}^{\tau - \sigma} \|u\|_\tau + \mathrm{ch}^{2\alpha' + \varrho - \max\{0,\sigma\}} \|u\|_\tau,$$

the proposed estimate (9.29), which completes the proof.

Sobolev norms measure the errors not uniformly but in average. Often one is also interested in *uniform estimates*. Since for splines on uniform partitions there holds the inverse assumption

$$\max_{x \in \Gamma} |\chi(x)| \leq \mathrm{ch}^{-1/2} \|\chi\|_{L_2} \qquad (9.31)$$

(see e.g. Lemma 2 in [55] in connection with estimate (9.13)) we have the pointwise estimates

$$\max_t \sum_{j=0}^{\ell} \left| \frac{d^j}{dt^j} (u - \bar{u}_\Delta)(t) \right| \leq \mathrm{ch}^{-1/2-\ell} \|u - \bar{u}_\Delta\|_{L_2} + \mathrm{ch}^{\tau - \ell - 1/2} \|u\|_{H^\tau} \qquad (9.32)$$

from the Sobolev imbedding theorem provided $\ell \leq d$ and $\ell + \frac{1}{2} < \tau \leq d + 1$. Together with Theorem 3.4 this provides the highest order $h^{d+1/2-\ell}$ for $t = d + 1$ which is $\frac{1}{2}$ order less than optimal. This loss has been removed in [94] (see also [95]).

Theorem 3.5 [94]: *Let all the assumptions of Theorems 3.1 and 3.4 be satisfied and, in addition, let d be odd, $-1 - 2\alpha < d$, $2\alpha + 1 \leq d$, $2\alpha + \frac{1}{2} \leq \ell \leq d$ and $u \in C^{d+1}(\Gamma)$. Then the collocation method converges pointwise asymptotically as*

$$\sum_{0 \leq j \leq \ell} \left| \frac{d^j}{dt^j} (u - u_\Delta)(t) \right| \leq \mathrm{ch}^{d+1-\ell} \|u\|_{C^{d+1}} \qquad (9.33)$$

where c is independent of h and u, u_Δ.

Corollary 3.6: *If, under the assumption of Theorem 3.5, the accuracy of the numerical integration satisfies $\varrho \geq d + \frac{3}{2} - 2\alpha'$ then*

$$\sum_{0 \leq j \leq \ell} \left| \frac{d^j}{dt^j} (u - \bar{u}_\Delta)(t) \right| \leq \mathrm{ch}^{d+1-\ell} (\|u\|_{C^{d+1}} + \|u\|_{\alpha + 1/2(d+1)}).$$

The Corollary 3.6 follows with inequality (9.31) in connection with estimate (9.30) from Theorem 3.5:

$$\| u - \bar{u}_\Delta \|_{C^\ell} \leqq \| u - u_\Delta \|_{C^\ell} + \mathrm{ch}^{-1/2-\ell} \| u_\Delta - \bar{u}_\Delta \|_0$$

$$\leqq \mathrm{ch}^{d+1-\ell} \| u \|_{C^{d+1}} + \mathrm{ch}^{2\alpha'+\varrho-1/2-\ell} \| u_\Delta \|_{\alpha+1/2(d+1)} \cdot$$

Now $\tau = \alpha + \frac{1}{2}(d+1)$ in estimate (9.15) gives the desired pointwise estimate in Corollary 3.6.

Clearly, our error estimates for the densities also imply corresponding error estimates for the generated fields in the form of potentials. These are given e.g. by Eqs. (9.5) or (9.8) in the form

$$U(x) = \int_\Gamma k(x, y) \, u(y) \, ds_y + B(x) \, \omega + F(x) =: V(u, \omega)(x) + F(x) \quad \text{for } x \notin \Gamma$$

where $k(x, y)$ is a C^∞-kernel for $x \neq y$ (but is singular at $x = y$) and the approximation

$$\bar{U}_\Delta(x) = V(\bar{u}_\Delta, \bar{\omega}_\Delta) + F(x) = \int_\Gamma k(x, y) \, \bar{u}_\Delta(y) \, ds_y + B(x) \, \bar{\omega}_\Delta + F(x).$$

Usually V is a continuous linear mapping from $H^\sigma(\Gamma) \times \mathbb{R}^q$ into $H^{\sigma+\beta+1/2}(\Omega_R)$ and has boundary values from Ω in $H^{\sigma+\beta}(\Gamma)$ where

$$\Omega_R = \begin{cases} \Omega & \text{for interior problems} \\ \Omega \cap \{x \,|\, |x| \leqq R\} & \text{for exterior problems}. \end{cases}$$

(See e.g. Chap. 8 by Eskin [31]. In our scattering problem and the displacement problems there holds $\beta = 1$ and in the traction problem $\beta = 0$.)

As a simple consequence we have the following lemma.

Lemma 3.6: *Let all assumptions of Theorems* 3.1 *and* 3.4 *be satisfied. Then the generated potential fields satisfy asymptotic error estimates up to the boundary*

$$\| U - \bar{U}_\Delta \|_{H^{\sigma+\beta+1/2}(\Omega_R)} \leqq c \{ \| u - \bar{u}_\Delta \|_{H^\sigma(\Gamma)} + | \omega - \bar{\omega}_\Delta | \}$$

$$\leqq (c_1 h^{\tau-\sigma} + c_2 h^{2\alpha'+\varrho-\max\{0, \sigma\}}) \| u \|_{H^\tau(\Gamma)} \tag{9.34}$$

and away from the boundary

$$\| U - \bar{U}_\Delta \|_{C^\nu(\tilde{\Omega})} \leqq c \{ \| u - \bar{u}_\Delta \|_{H^{2\alpha}(\Gamma)} + | \omega - \bar{\omega}_\Delta | \}$$

$$\leqq (c_1 h^{\tau-2\alpha} + c_2 h^{4\alpha'+\varrho}) \| u \|_{H^\tau(\Gamma)} \tag{9.35}$$

with any $\nu \in \mathbb{N}_0 = \{0, 1, 2, \ldots\}$ *where* $\tilde{\Omega}$ *denotes any compact subset of* Ω_R *with* $\text{dist}(\Gamma, \tilde{\Omega}) > 0$ *and where the constants* c, c_1, c_2 *depend on* ν *and* $\text{dist}(\Gamma, \tilde{\Omega})$ *but do not depend on* $h, u, \bar{u}_\Delta, \omega, \bar{\omega}_\Delta$.

Obviously, the highest rates in Lemma 3.6 are attained for $\tau = d + 1$.

9.4 Standard Collocation for Three-Dimensional Problems with Fredholm Boundary Integral Equations of the Second Kind

For the three-dimensional problems we restrict our presentation to smooth closed Ljapounov boundaries Γ. But it should be noted that most of the following analysis can be extended to surfaces having corners and edges (see Wendland [119] and the references given there). Γ can be partitioned into finitely many pieces S,
$\Gamma = \bigcup_{\ell=1}^{L} S_{\ell}$ and each S_{ℓ} can be considered to be the image of a C^{r}-application $x = X(t_1, t_2) = X(t)$ defined on a polygonal parameter domain $U_{\ell} \subset \mathbb{R}^2$, $S = X(U_{\ell})$. Then a *regular* family of triangular partitions of U_{ℓ} with triangles T_j of maximal side length h defines a corresponding family of partitions of each S_{ℓ} and, eventually, on Γ, $\Gamma = \bigcup_{j=1}^{N} F_j$, $F_j = X(T_j)$. The approximating finite element spaces can be obtained by lifting appropriate finite elements $\phi_h(t)$ from the parameter domain onto Γ by $w_h(p) := \phi_h(X^{-1}(p))$. Then a regular (m, d)-system of finite elements on the U_{ℓ} defines a corresponding finite boundary element family $\tilde{H}_h$ on Γ — provided that across the boundaries of S_{ℓ} the functions w_h satisfy the same transition conditions as across the boundaries of F_j within S_{ℓ}—. These transplanted finite elements provide the approximation property (9.12) for $\sigma \leq s \leq m$, $\sigma \leq d \leq m-1$ and $r \geq \max\{m+1, |\sigma|\}$ in the Sobolev spaces on Γ and (because of the *regular* family) also the inverse assumption (9.13) for $\sigma \leq s \leq d$ (see Babuška and Aziz [13]).

For the collocation method we need to choose the set Ξ of collocation points such that the collocation equations (9.2) are uniquely solvable for all h small enough and such that, in addition, the sequence of boundary element solutions $u_{\Delta}, \omega_{\Delta}$ converges to u, ω. In view of Sect. 9.3 it is desirable to use for Ξ *the set of unisolvent points* of $\tilde{H}_h$ defining interpolation operators I_{Δ} which converge to the identity I, and to apply the corresponding collocation method *without* modifications to Eqs. (9.1) with strongly elliptic operators A. The question of convergence is in general for such A still open. For the special case $A_0 = I$, however, that is for Fredholm integral equations of the second kind we do have asymptotic convergence.

For the basis $\{\mu_j\}_{j=1}^{N}$ of $\tilde{H}_h$ we now require that the *Lagrangian interpolation problem*, i.e. find $\alpha_1, \ldots, \alpha_N$ satisfying

$$\sum_{j=1}^{N} \alpha_j \mu_j(x_i) = f(x_i) \quad \text{for } x_i \in \Xi \tag{9.36}$$

is uniquely solvable for any given values $f(x_i)$ and every $h > 0$. The corresponding interpolation operator I_{Δ} is then defined by

$$I_{\Delta} f := \sum_{j=1}^{N} \alpha_j \mu_j$$

for every function f with finite values $f(x_i)$.

We now require an approximation property for the family of interpolation operators I_{Δ}:

Let $0 \leq \sigma \leq s \leq m$, $\sigma \leq d \leq m - 1$. For any $f \in C^s(\Gamma)$ there holds

$$\|I_{\Delta} f - f\|_{W_2^{\sigma}(\Gamma)} \leq c h^{s - \sigma} \|f\|_{W_2^{s}(\Gamma)} \qquad (9.37)$$

where the constant c is independent of h and f.

Since inequality (9.37) also holds for $s = \sigma$ and $C^1(\Gamma)$ is dense in $C^0(\Gamma)$, this implies

$$\lim_{h \to 0} \|I_{\Delta} f - f\|_{W_2^{0}(\Gamma)} = 0 \quad \text{for every } f \in C^0(\Gamma)$$

together with the uniform boundedness

$$\|I_{\Delta} f\|_{W_2^{0}(\Gamma)} \leq c \|f\|_{W_2^{0}(\Gamma)} . \qquad (9.38)$$

The simplest boundary elements $\tilde{H}_h$ are piecewise constant functions which form a regular $(1, 0)$-system of boundary elements. Here $x_i \in F_i$ are chosen to be the images of the centers of gravity in the corresponding triangles T_i of the parameter domain.

Then there holds estimate (9.37) with $\sigma = 0$ and $s = 1$ and estimate (9.38) with $c = 1$. For second order approximation introduce the piecewise linear Courant elements on the partitions $\{T_j\}$ of U_ℓ and lift these functions by $X(t): U_\ell \to S_\ell$ on Γ. On joint boundary curves $S_\ell \cap S_m \neq 0$ we require the traces to coincide. This again defines a boundary element space $\tilde{H}_h$. Now I_{Δ} can be chosen to be the collocation operator with the *corner points* of F_j as collocation points x_i. From these boundary elements we have estimate (9.37) with $m = 2$ and $d = 1$.

Now the weights $a_{\ell i} = (A \mu_\ell)(x_i)$ can be expressed in terms of integrals having integration domains in the parameter domains in $\mathbb{R}^2$. But the boundary integral equations on the surface Γ mostly contain singular integrals with weak or even strong singularities. Here already the approximation with piecewise constant trial functions confronts us with a large amount of computational difficulties, even in the case of Fredholm integral equations of the second kind as in Eqs. (9.6), (9.7) (see 64, [119], Sect. 3 in connection with exact integration of the principal part and one point Gaussian quadrature). For higher order methods the computational expense will grow tremendously due to higher degree Gaussian quadratures. Here formulas using the grid points of the triangulations only, should also be developed. Therefore most numerical computations in [27, 38, 39, 43, 44, 49, 50, 76, 82, 83, 85, 86, 113, 114] are based on simultaneous approximations of the geometry as well as of the desired densities similarly to the finite element treatment of shell problems by Ciarlet [21], Chap. 8. The corresponding error analysis for this approximation is based on the fundamental paper by Nedelec [82]. The boundary elements are constructed as follows. To the triangles T_ℓ of the *parameter triangulations* we associate a C^0 finite element $(\varkappa + 1, 1)$-system S_h of Lagrangian type, $\varkappa \geq 1$, containing piecewise polynomials of degree $\varkappa$ and also associated with a unisolvent set of grid points $\{p_{\ell i}\}$ such that the interpolation problem $\Phi_h(p_{\ell i}) = \Phi(p_i)$ for $\Phi_h \in S_h$ is uniquely solvable. The corresponding interpolation operator denote by I_h. Let $X_j(t)$ be the parameter representation of Γ over the parameter domain U_j. Then the approximate surface is defined by $\Gamma_h: X_{hj} = I_h X_j(t), t \in U_j$. It is further assumed that along the curves $\bar{F}_j \cap \bar{F}_i \subset \Gamma$ adjacent to F_j and F_i we have the coincidence $X_{hj} = X_{hi}$. For the construction of $\tilde{H}_h$ we choose a regular (m, d)-

system $\tilde{S}_h$, $m \leqq \varkappa + 1$, $d \leqq 1$ of finite elements ϕ_h associated with the regular triangulations in the parameter polygonal domains and a corresponding interpolation operator I_A. On the boundary of U_j we require via $\bar{F}_j \cap \bar{F}_i$ identical interpolations. The finite elements ϕ_h are lifted with X_h onto Γ_h by

$$\chi_h(x) = \phi_h(X_h^{-1}(x)) \quad \text{for } x \in \Gamma_h \tag{9.39}$$

defining H_h on Γ_h. Now let $\Gamma \in C^{\varkappa+2}$.

For $x \in \Gamma$ let the straight line through x in the $\pm$ directions of the normal vector $\nu(x)$ hit Γ_h at $\psi^{-1}(x)$ being the nearest hit. For h small enough this mapping and its inverse ψ exist. Then $\tilde{H}_h$ on Γ is defined by the functions

$$w_h(x) := \phi_h(X_h^{-1}(\psi^{-1})(x))) \quad \text{where } \phi_h \in \tilde{S}_h. \tag{9.40}$$

For any of the operators in Eqs. (9.6), (9.7), (9.9), (9.10) of (9.11) replace Γ by Γ_h and x, y by corresponding points on Γ_h as well as $r = |x - y|$, $\nu(x)$, $\nu(y)$ and the surface elements. With the basis of $\tilde{H}_h$, respectively H_h on Γ_h this gives rise to coefficients $\tilde{a}_{\ell i}$, Λ_ℓ which are defined by integrals on the parameter domains. A particularly simple choice of approximation is $\varkappa = 1$ and $m = 1$, $d = 0$ i.e. Γ_h is a polyhedron and w_h is piecewise constant.

The corresponding collocation method is often called the "panel method" and has been used in [49, 50, 91, 124[1]]. The numerical evaluation of the above collocation weights with appropriate Gaussian quadratures has been developed in [19, 113, 114]. Extrapolation methods for the singular integrals one finds in [70, 71].[1]

Since the asymptotic error analysis is yet available only for Fredholm integral equations of the second kind let us now restrict ourselves to equations

$$u + Cu + B\omega = f,$$
$$\Lambda u = b. \tag{9.41}$$

Our examples Eqs (9.6) and (9.7) are of this form and correspond to the classical boundary integral equations. (In [119] we consider a slightly larger class corresponding to strongly elliptic equations of order zero.) There the operator C is always given by a *weakly singular integral operator* or by a matrix of such operators having the form

$$Cu(x) = \int_\Gamma \frac{k(x, y)}{|x - y|} u(y) \, ds_y.$$

If we fix the observation point $x \in \Gamma$ and introduce (locally) about x by $\varrho = |y - x|$ and the trajectories $\phi = \text{const}$ orthogonal to $\varrho = \text{const}$ Martensen's *surface polar coordinates* on Γ (see Sect. 2.15 in [74]) then C admits an asymptotic expansion as

$$Cu(x) = \int_{\varrho=0}^{R_0} \int_{\phi=0}^{2\pi} \left\{ \sum_{\ell=0}^{N} b_\ell(x, \cos\phi, \sin\phi) \varrho^\ell \right\} u(x + \bar{y}(x, \varrho, \phi)) \, d\varrho \, d\phi$$
$$+ \int_{\Gamma \cap |y-x| \leqq R_0} R_K(x, y) u(y) \, ds_y + \int_{\Gamma \cap |y-x| > R_0} \frac{k(x, y)}{|x - y|} u(y) \, ds_y \tag{9.43}$$

---

[1] See note and references added in proof, p. 257.

provided that $\Gamma \in C^{K+2}$. Here $R_0 > 0$ is a fixed number, the functions $b_\ell(x, \xi_1, \xi_2)$ are homogeneous polynomials of the form

$$b_\ell(x, \xi_1, \xi_2) = \sum_{j=0}^{\ell+2} b_{\ell j}(x)\, \xi_1^j\, \xi_2^{\ell+2-j}$$

with coefficients $b_{\ell j} \in C^K$ and $R_K(x, y)$ is a remainder with $|x - y|^{-K} R_K \in C^0$.

A similar expansion holds for $\tilde{y}(x, \varrho, \phi) = y - x$ (see Formula (2.63) in [74]). Note that corresponding expansions are valid for all our boundary integral operators in Sect. 9.2. See also Chap. X in [77].

Equations (9.41) and their approximation will be considered with respect to uniform convergence. Accordingly we use the space $X_0 = C^0(\Gamma)$ and $X = W_\infty^0(\Gamma)$ if $d = 0$, $X = X_0$ if $d \geq 1$, respectively, equipped with the $W_\infty^0(\Gamma)$-norm. Operators of the form (9.43) provide several welcome properties (see p. 178 in [118]). In particular, C is for $K \geq 0$ a compact mapping from $W_\infty^0(\Gamma)$ into the Hölder space $C^\alpha(\Gamma)$ with any $0 < \alpha < \alpha' < 1$ since $C: W_\infty^0(\Gamma) \to C^{\alpha'}(\Gamma)$ for $0 < \alpha < \alpha' < 1$ is continuous (see (2.23) in [24]) and $C^{\alpha'}(\Gamma)$ is compactly imbedded into $C^\alpha(\Gamma)$ (see § 28 in [121]). The collocation method for Eqs. (9.41) reads as

$$u_\Delta + I_\Delta\, C u_\Delta + I_\Delta\, B \omega_\Delta = I_\Delta f,$$
$$\Lambda u_\Delta = b. \tag{9.44}$$

For the operators in these equations we have the following properties:

$$C: X \to X_0 \text{ is a compact linear mapping},$$

$$\| C_h\, \psi \|_{W_\infty^0} \leq c \, \| \psi \|_{W_\infty^0} \quad \text{for all } \psi \in X,$$

$$\lim_{h \to 0} (\| C_h v - C v \|_{W_\infty^0} + \| B_h \omega - B \omega \|_{W_\infty^0}) = 0 \quad \text{for all } (v, \omega) \in X_0 \times \mathbb{R}^q,$$

the set

$$\bigcup_{0 < h} \{ w = (C_h - C)\, \psi + (B_h - B)\, \omega \mid (\psi, \omega) \in X \times \mathbb{R}^q \text{ and } \| \psi \|_{W_\infty^0} + |\omega| \leq 1 \}$$

is relatively compact in X_0.

Here

$$C_h = I_\Delta\, C \quad \text{and} \quad B_h = I_\Delta\, B. \tag{9.46}$$

Under these assumptions one can prove the following theorem which can be found in [4, 18, 20, 107].

Theorem 4.1: Let Eqs. (9.41) be uniquely solvable and let $f \in X_0$. Suppose (9.45)–(9.46). Then there exist positive constants h_0 and c_1, c_2 such that the equation

$$u_h + C_h\, u_h + B_h\, \omega_h = f, \quad \Lambda_h = b \tag{9.47}$$

a unique solution (u_h, ω_h) for any $0 < h \leq h_0$.

We also have the asymptotic estimates

$$\| u_h \|_{W_\infty^0} + |\omega_h| \leq c_1 \{ \| u \|_{C^0} + |\omega| \} \leq c_2 \{ \| f \|_{C^0} + |b| \} \tag{9.48}$$

and

$$\| u - u_h \|_{W_\infty^0} + |\omega - \omega_h| \leq c_2 \{ \| (C_h - C)\, u \|_{C^0} + \| (B_h - B)\, \omega \|_{C^0} \}. \tag{9.49}$$

Proof (Stummel [107]): The classical Fredholm alternative implies that (9.47) for any fixed $h > 0$ is either uniquely solvable or has an eigensolution v_h, η_h with $\| v_h \| + | \eta_h | = 1$ and

$$v_h + C_h v_h + B_h \eta_h = 0,$$
$$\Lambda v_h = 0. \tag{9.50}$$

If there were infinitely many $h > 0$ with $h \to 0$ admitting the above eigensolutions then there would exist a subsequence $\eta_{h'} \to \eta$ and

$$C_{h'} v_{h'} = (C_{h'} - C) v_{h'} + (B_{h'} - B) \eta_{h'} + C v_{h'} + B \eta_{h'} \to \chi \in X_0$$

converging due to Eqs. (9.45). Then from (9.50) also followed the convergence of $v_{h'} \to \psi \in X_0$. Hence, we would find for the limit

$$\psi + C\psi + B\eta = 0,$$
$$\Lambda \psi = 0$$

with $\| \psi \| + | \eta | = 1$ in contrary to the *unique* solvability of Eqs. (9.41) and the classical Fredholm alternative.

Consequently, there exists $h_0 > 0$ such that Eqs. (9.47) are uniquely solvable with solutions u_h, ω_h for every h, $0 < h \leq h_0$.

(ii) Now the Eqs. (9.41), (9.47) yield

$$u_h + C u_h + B \omega_h = u + Cu + B\omega + ((C - C_h) u_h + (B - B_h) \omega_h , \tag{9.51}$$
$$\Lambda u_h = \Lambda u$$

for every $0 < h \leq h_0$. If estimate (9.48) were not true then there would exist a sequence $(\tilde{u}_h, \omega_h) \in X_0 \times \mathbb{R}^q$ with $\| \tilde{u}_h \| + | \tilde{\omega}_h | = 1$ and corresponding solutions $(u_h, \omega_h) \in \tilde{H}_h \times \mathbb{R}^q$ of the collocation Equations (9.47) with

$$f = \tilde{u}_h + C \tilde{u}_h + B \tilde{\omega}_h, \quad b = \Lambda \tilde{u}_h$$

such that $\| u_h \| + | \omega_h | \to \infty$. Deviding Eq. (9.51) by $\| u_h \| + | \omega_h |$ and introducing

$$\psi_h = u_h / (\| u_h \| + | \omega_h |), \quad \alpha_h = \omega_h / (\| u_h \| + | \omega_h |)$$

would yield

$$\psi_h + C \psi_h + B \alpha_h = \chi_h + (C - C_h) \psi_h + (B - B_h) \alpha_h .$$
$$\Lambda_h \psi_h = \Lambda u / (\| u_h \| + | \omega_h |)$$

where $\chi_h \to 0$. Now we could choose a subsequence with converging $\alpha_h \to \alpha$ and $(C - C_h) \psi_j \to \lambda \in X_0$. But this would imply the convergence of a subsequence of

$$\psi_h = \{ \chi_h + (C - C_h) \psi_h + (B - B_h) \alpha_h - C \psi_h - B \alpha_h \} \to \psi \in X_0$$

with $\| \psi \| + | \alpha | = 1$ and satisfying

$$\psi + C\psi + B\alpha = 0, \quad \Lambda \psi = 0$$

implying $\psi = 0$, $\alpha = 0$, a contradition.

Inequality (9.49) now follows from

$$(u - u_h) + C_h (u - u_h) + B_h (\omega - \omega_h) = (C_h - C) u + (B_h - B) u,$$
$$\Lambda (u - u_h) = 0$$

with estimate (9.48) since for $u \in X_0$ the right hand side also belongs to X_0. This completes the proof of Theorem 4.1.

Theorem 4.2: *Let Eq. (9.41) be uniquely solvable with $f \in C^s$ and $u \in C^s$, $0 \leq s \leq m$. Then the collocation method with Eqs. (9.44) provides a unique boundary element solution z_Δ, ω_Δ for every h, $0 < h \leq h_0$ and we have the asymptotic error estimate*

$$\| u - u_\Delta \|_{W^0_\infty} + | \omega - \omega_\Delta | \leq c_2 \{ \| (I_\Delta - I) u \|_{W^0_\infty} + 2 \| (I_\Delta - I) Bw \|_{W^0_\infty} + \| (I_\Delta - I)f \|_{W^0_\infty} \}$$

$$\leq ch^s \{ \| u \|_{W^s_\infty} + \| f \|_{W^s_\infty} + | \omega | \}. \tag{9.52}$$

Remark: If f is given in the Höder space $C^{s+\varepsilon}$, $s \in \mathbb{N}_0$, $0 < \varepsilon < 1$ and $K \geq s$ then the solution u of Eq. (9.41) belongs to $C^{s+\varepsilon} \subset C^s$. This follows from the continuity of $C: C^{t+\varepsilon} \to C^{t+1+\varepsilon}$ for $t = 0, 1, ..., s-1$ which can be obtained from differentiating (9.43), integrating by parts (p. 249ff. in [77]) and the Giraud theorem (p. 239 in [77]).

Proof of Theorem 4.2: With Theorem 4.1 and Eqs. (9.46) we have

$$\| u - u_\Delta \|_{W^0_\infty} + | \omega - \omega_\Delta | \leq c_2 \{ \| (I_\Delta C - C) u \|_{W^0_\infty} + \| (I_\Delta B - B) \omega \|_{W^0_\infty} \}. \tag{9.53}$$

Now we replace Cu by Eq. (9.41) and obtain

$$(I_\Delta - I) Cu = (I_\Delta - I)(f - u - B\omega).$$

Inserting this relation into estimate (9.53) yields with property (9.37) the desired estimate (9.52).

For the actual computations, Eqs. (9.44) will be implemented via the quadratic system of linear equations,

$$\sum_{\ell=1}^{N} \gamma_\ell \left(\mu_\ell(x_i) + C\mu_\ell(x_i) \right) + B(x_i) \, \omega = f(x_i),$$

$$\sum_{\ell=1}^{N} \Lambda_\ell \gamma_\ell = b,$$

where the weights are computed with numerical quadrature giving

$$\tilde{\Lambda}_\ell \cong \Lambda \mu_\ell = \Lambda_\ell, \quad \tilde{c}_{\ell i} \cong \int_\Gamma \frac{k(x_i, y)}{|x_i - y|} \mu_\ell(y) \, ds_y \quad \text{and where} \quad u_\Delta = \sum_{\ell=1}^{N} \gamma_\ell \mu_\ell.$$

The numerical weights $\tilde{c}_{\ell i}$ now define a linear mapping $\tilde{C}_\Delta: \tilde{H}_h \to \tilde{H}_h$ and the $\tilde{\lambda}_\ell$ a mapping $\tilde{\Lambda}_\Delta: \tilde{H}_h \to \mathbb{R}^q$ as in Eqs. (9.22), (9.23). Corresponding to Lemma 3.3 we now have

Lemma 4.3: *If the numerical coefficients satisfy the accuracy conditions*

$$| \tilde{c}_{\ell i} - C\mu_\ell(x_i) | + | \tilde{\Lambda}_\ell - \Lambda\mu_\ell | \leq ch^{\varrho+2} \tag{9.54}$$

then the corresponding operators satisfy the asymptotic consistency estimates

$$\| (I_\Delta C - \tilde{C}_\Delta) w \|_{W^0_\infty} + | (\Lambda - \tilde{\Lambda}_\Delta) w | \leq ch^\varrho \| w \|_{W^0_\infty}$$

for all $w \in \tilde{H}_h$.

The proof follows elementarily with triangle inequality from

$$N \le ch^{-2} \quad \text{and} \quad \max_{1 \le j \le N} |\alpha_j| \le c_1 \left\| \sum_{j=1}^{N} \alpha_j \mu_j \right\|_{W_x^0} \le c_2 \max_{1 \le j \le N} |\alpha_j|.$$

These inequalities are both simple consequences of the *regularity assumptions* for the boundary element family $\tilde{H}_h$.

Now the combination of Theorem 4.2 and Lemma 4.3 again allows to estimate asymptotically the errors between the computed solution

$$\tilde{u}_\Delta = \sum_{l=1}^{N} \tilde{\gamma}_l \mu_l, \ \tilde{\omega}_\Delta$$

obtained from

$$\sum_{l=1}^{N} \tilde{\gamma}_l \{\mu_l(x_i) + \tilde{c}_{li}\} + B(x_i) \tilde{\omega}_\Delta = f(x_i) \quad \text{for } x_i \in \Xi,$$

$$\sum_{l=1}^{N} \tilde{\gamma}_l \tilde{A}_l = b,$$

(9.55)

and the actual solution u, ω.

Theorem 4.4: *If $\varrho > 0$ and if all the foregoing assumptions for Eq. (9.41) and its approximation are satisfied then there exists $h_0 > 0$ such that for all h with $0 < h \le h_0$ the numerical Eqs. (9.55) are uniquely solvable. The solutions $\tilde{u}_\Delta$, $\tilde{\omega}_\Delta$ satisfy the asymptotic error estimate*

$$|\omega - \tilde{\omega}_\Delta| + \|u - \tilde{u}_\Delta\|_{W_\infty^0} \le ch^s \{\|u\|_{W_\infty^s} + \|f\|_{W_\infty^s} + |\omega|\} + ch^\varrho \{\|f\|_{W_\infty^0} + |b|\}$$

where $0 \le s \le m$.

For the proof observe that one may proceed in the same manner as for estimate (9.29) since here $A = I + C$, estimate (9.52) corresponds to estimate (9.20) with the W_x^0-norm instead of the L_2-norm and $\alpha' = 0$ whereas Eqs. (9.25)−(9.27) are the same. With these correspondences the remaining steps of the proof are the same as in the proof of Theorem 3.4.

In order to assure an accuracy of the numerical integrations as in estimates (9.54) one needs appropriate techniques for the numerical quadrature of the weakly singular integrals as e.g. in Eqs. (9.43) with $u = \mu_l$. For *fixed* $R_0 > 0$, i.e. not depending on h, and $|y - x| \ge R_0$ the kernels in C are real analytic functions and two-dimensional Gaussian quadrature (see Engels [30]) will provide sufficient accuracy. The integrands simplify significantly if the integrals are taken on Γ_h according to Eq. (9.39). In the triangles T_j in the parameter domain U_l these are eventually compositions of polynomials in t_1, t_2 and the expressions in Eq. (9.43). Here the extrapolation method by Lyness [70, 71] provides an efficient scheme[2]. However, it is too costly[3].

For $\varkappa = 1$, $d = 0$, $m = 1$ and the principal part of the kernel in Eq. (9.6) the integrals are solid angles which can be computed explicitly. (For a numerically stable procedure see [115].)

---

[2] I owe this remark to Mr. Schwab who also made experiments with this method applied to the operator in Eq. (9.6).
[3] See note and references added in proof, p. 257.

Hence, for our error estimates we may now assume that estimate (9.54) is either already satisfied ($|y - x| \geq R_0$) or the integrals on Γ_h are evaluated accurately up to machine accuracy. Then the error can be estimated as follows.

Theorem 4.5: *If we use the isoparametric boundary elements* (9.39) *for $\bar{H}_h$ and the integrations on Γ_h are accurate then for the corresponding elements $\bar{c}_{\ell i}$ and $\bar{\Lambda}_\ell$ there hold the estimates* (4.54) *with $\varrho = \varkappa + 1$.*

The proof of this theorem is based on Nedelec's work in [82] and has been performed by Giroire in [38] for the special kernels in Eq. (9.6). The proof is rather tedious but carries over to our more general kernels of the form in Eq. (9.43) without difficulties. We omit here the presentation.

Remarks: An even simpler numerical method based on one-point Gaussian integration with $m = 1$, $d = 0$ and without the approximation of Γ by Γ_h has been analyzed in [64, 116, 119].

In case of the approximation with polyhedral Γ_h, $\varkappa = 1$ and piecewise constant μ_ℓ, i.e. $m = 1$, $d = 0$, Theorems 4.5 and 4.4 assure the convergence of the panel method by Hess and Smith [50, 51] with order h^1. Further methods of the same kind can be found in [91]. But for $\varkappa = 1$ and collocation, the choice $m = 1$ is not optimal. Here the polyhedral approximation Γ_h and the choice of the *Courant elements* is optimally combined, i.e. $\varkappa = 1$, $m = 2$, $d = 1$, and collocation at the corner points gives for the Fredholm integral equations of the second kind with the weakly singular kernels already an order h^2 i.e. quadratic asymptotic convergence.

References

1   Abou El-Seoud, M.S., Numerische Behandlung von schwach singulären Integralgleichun-
    gen erster Art. Doctoral Dissertation, D 17, Technische Hochschule Darmstadt, Ger-
    many, 1979
2   Agranovich, M.S., Spectral properties of diffraction problems. In *The General Method of
    Natural Vibrations in Diffraction Theory* by N.N. Voitovic, B.Z. Katzenellenbaum and
    A.N. Sivov (Russian). Izdat. Nauka, Moscow, 1977
3   Aleksidze, M.A., *The Solution of Boundary Value Problems with the Method of the
    Expansion with Respect to Nonorthonormal Functions.* Nauka, Moscow (Russian), 1978
4   Anselone, P.M., *Collectively Compact Operator Approximation Theory.* Prentice Hall,
    Englewood Cliffs, N.J., 1971
5   Antes, H., Die Splineinterpolation zur Lösung von Integralgleichungen und ihre An-
    wendung bei der Berechnung von Spannungen in krummlinig berandeten Scheiben.
    Doctoral Dissertation R-W-TH Aachen, Germany, 1970
6   Antes, H., Splinefunktionen bei der Lösung von Integralgleichungen. Numer. Math. **19**,
    116–126, 1972
7   Arnold, D.N. and Wendland, W.L., Collocation versus Galerkin procedures for boundary
    integral methods. In *Boundary Element Methods in Engineering* (C.A. Brebbia, ed.).
    Springer, Berlin, Heidelberg, New York, 18–33, 1982
8   Arnold, D.N. and Wendland, W.L., On the asymptotic convergence of collocation
    methods. Math. Comp. **41**, 349–381, 1983
9   Arnold, D.N. and Wendland, W.L., The convergence of spline collocation for strongly
    elliptic equations on curves. Numer. Math. (to appear)
10 Atkinson, K.E., A Survey of Numerical Methods for the Solution of Fredholm Integral
    Equations of the Second Kind. Soc. Ind. Appl. Math. Philadelphia, 1976

11 Atkinson, K.E. and de Hoog, F., Collocation methods for a boundary integral equation on a wedge. In *Treatment of Integral Equations by Numerical Methods*. C.T. Baker and G.F. Miller (ed.). Academic Press, London, 253–260, 1982

12 Aziz, A.K., Dorr, M.R., and Kellog, R.B., Calculation of electromagnetic scattering by a perfect conductor. *Naval Surface Weapons Center Report* TR 80-245, Silver Spring, Maryland 20910, 1980

13 Babuška, I. and Aziz, A.K., Survey lectures on the mathematical foundations of the finite element method. In *The Mathematical Foundation of the Finite Element Method with Applications to Partial Differential Equations*. A.K. Aziz (ed.). Academic Press, New York, 3–359, 1972

14 Baker, C., *The Numerical Treatment of Integral Equations*. Clarendon Press, Oxford, 1977

15 Bamberger, A., Approximation de la diffraction d'ondes elastiques – une nouvelle approche (I). Rapport Int. **91**, Centre Math. Appl. Ecole Polytechnique Palaiseau, France, 1983

16 Bolteus, L. and Tullberg, O., BEMSTAT – A new type of boundary element program for two-dimensional elasticity problems. In *Boundary Element Methods* (C.A. Brebbia, ed.). Springer, Berlin, Heidelberg, New York, 518–537, 1981

17 Brakhage, H., Über die numerische Behandlung von Integralgleichungen nach der Quadraturformelmethode. Num. Math. **2**, 183–196, 1960

18 Brakhage, H. and Werner, P., Über das Dirichletsche Außenraumproblem für die Helmholtzsche Schwingungsgleichung. Arch. Math. **16**, 325–329, 1965

19 Brebbia, C.A., *The Boundary Element Method for Engineers*. Pentech Press, London, 1978

20 Bruhn, G. and Wendland, W.L., Über die näherungsweise Lösung von linearen Funktionalgleichungen. In *Funktionalanalysis, Approximationstheorie, Numerische Mathematik*. L. Collatz and H. Ehrmann (ed.). Intern. Ser. Numer. Math. **7**. Birkhäuser Basel, 136–164, 1967

21 Ciarlet, P.G., *The Finite Element Method for Elliptic Problems*. North Holland, Amsterdam, New York, Oxford, 1978

22 Christiansen, S., Condition number of matrices derived from two classes of integral equations. Math. Methods Appl. Sci. **3**, 364–392, 1981

23 Christiansen, S. and Hansen, E., A direct integral equation method for computing the hoop stress in plane isotropic sheets. J. Elasticity **5**, 1–14, 1975

24 Colton, D. and Kress, R., *Integral Equation Methods in Scattering Theory*. John Wiley & Sons, New York, 1983

25 Costabel, M., Starke Elliptizität von Randintegraloperatoren erster Art. Habilitationsschrift, Technische Hochschule Darmstadt, Germany, 1984

26 Costabel, M. and Stephan, E., A direct boundary integral equation method for transmission problems. J. Math. Anal. Appl. (in print)

27 Cruse, T.A., Application of the boundary integral equation method to three-dimensional stress analysis. Comp. Struct. **3**, 309–369, 1973

28 Dieudonné, *Eléments d'Analyse. Tome VII*. Gauthier-Villars, Paris, 1980

29 Durand, M., Layer potentials and boundary value problems for the Helmholtz equation in the complement of a thin obstacle. Math. Meth. Appl. Sci. **5**, 389–421, 1983

30 Engels, H., *Numerical Quadrature and Cubature*. Academic Press, London, New York, 1980

31 Eskin, G.I., Boundary Value Problems for Elliptic Pseudodifferential Equations. AMS, Transl. Math. Mon. 52, Providence, Rhode Island, 1980

32 Fichera, G., Linear elliptic equations of higher order in two independent variables and singular integral equations. In Proc. Conf. Partial Differential Equations and Cont. Mechanics. Univ. of Wisconsin Press, 55–80, 1961

33 Filippi, P., Potentiels de couche pour les ondes mécaniques scalaires. Révue de Cethedec **51**, 121–175, 1977

34 Filippi, P., Layer potentials and acoustic diffraction. J. Sound and Vibration **54**, 473–500, 1977

35 Fischer, T., Ein Verfahren zur Berechnung schleichender Umströmungen beliebiger, dreidimensionaler Hindernisse mit Hilfe singulärer Störungsrechnung, Integralgleichungen erster Art und Randelementmethode. Diplom Thesis, Technische Hochschule Darmstadt, 1980

36 Fischer, T., An integral equation procedure for the exterior three-dimensional viscous flow. Integral Equations and Operator Theory **5**, 490–505, 1982

37 Gaier, D., Integralgleichungen erster Art und konforme Abbildung. Math. Z. **147**, 113–129, 1976

38 Giroire, J. Integral equation methods for exterior problems for the Helmholtz equation. Rapport Int. **40**, Centre Math. Appl., Ecole Polytechnique, Palaiseau, France, 1978

39 Giroire, J. and Nedelec, J.C., Numerical solution of an exterior Neumann problem using a double layer potential. Math. Comp. **32**, 973–990, 1978

40 Goldstein, C., Numerical methods for Helmholtz type equations in unbounded domains. BNL-26543, Brookhaven Lab., Brookhaven, N.Y., 1979

41 Gregoire, J.P., Nedelec, J.C., and Planchard, J., Problèmes relatifs à l'équation d'Helmholtz. Serv. Inf. et Math. Appl. Bull. Direction des Etudes et Recherches, Ser. C, **2**, 15–32, 1974

42 Haack, W. and Wendland, W.L., *Lectures on Partial and Pfaffian Differential Equations.* Pergamon Press, Oxford, 1971

43 Ha Duong, T., La méthode de Schenck pour le résolution numérique du problème de radiation acoustique. E.D.F., Bull. Dir. Etudes Recherches, Ser. C, Math., Informatique, Service Informatique et Mathématiques Appl. **2**, 15–50, 1979

44 Ha Duong, T., A finite element method for the double-layer potential solutions of the Neumann exterior problem. Math. Meth. Appl. Sci. **2**, 191–208, 1980

45 Harrington, R.F. and Sarkar, T.K., Boundary elements and the method of moments. In *Boundary Elements.* C.A. Brebbia, T. Futagami, M. Tanaka (eds.). Springer, Berlin, Heidelberg, New York, Tokyo, 31–40, 1983

46 Hartmann, F., Elastostatics. In *Progress in Boundary Element Methods*, Vol. 1, C.A. Brebbia (ed.). Pentech Press, London, Plymouth, 84–167, 1981

47 Hartmann, F., Kompatibilität auf dem Rand (in preparation)

48 Hayes, J.K., Kahaner, D.K., and Keller, R.G., An improved method for numerical conformal mapping. Math. Comp. **26**, 327–334, 1972

49 Hess, J.L., Calculation of acoustic fields about arbitrary three-dimensional bodies by a method of surface source distributions based on certain wave number expansions. Report DAC 66901, McDonnell Douglas, 1968

50 Hess, J.L. and Smith, A.M.O., Cacluation of potential flow about arbitrary bodies. In *Progress in Aeronautical Sciences.* D. Kuchemann (ed.). Pergamon, Oxford, **8**, 1–138, 1967

51 Hoidn, H.-P., Die Kollokationsmethode angewandt auf die Symmsche Integralgleichung. Doctoral Dissertation, ETH Zürich, Switzerland, 1983

52 Hsiao, G.C., Kopp, P., and Wendland, W.L., A Galerkin collocation method for some integral equations of the first kind. Computing **25**, 89–130, 1980

53 Hsiao, G.C., Kopp, P., and Wendland, W.L., Some applications of a Galerkin collocation method for integral equations of the first kind. Math. Meth. Appl. Sci. **6**, 280–325, 1984

54 Hsiao, G.C. and Mac Camy, R.C., Solution of boundary value problems by integral equations of the first kind. SIAM Rev. **15**, 687–705, 1973

55 Hsiao, G.C. and Wendland, W.L., A finite element method for some integral equations of the first kind. J. Math. Anal. Appl. **58**, 449–481, 1977

56 Hsiao, G.C. and Wendland, W.L., The Aubin-Nitsche lemma for integral equations. J. Integral Equations **3**, 299–315, 1981

57 Hsiao, G.C. and Wendland, W.L., On a boundary integral method for some exterior problems in elasticity. In *Dokl. Akad. Nauk SSSR.* Special issue ded. Academician V.D. Kupradze 80th birthday (in print)

58 Ivanov, V.V., *The theory of Approximate methods and their Application to the Numerical Solution of Singular Integral Equations.* Noordhoff Int. Publ., Leyden, The Netherlands, 1976

59 Jaswon, M.A., Some theoretical aspects of boundary integral equations. In *Boundary Element Methods.* C.A. Brebbia (ed.). Springer, Berlin, Heidelberg, New York, 399–411, 1981

60 Jaswon, M.A. and Symm, G.T., *Integral Equation Methods in Potential Theory and Elastostatics.* Academic Press, London, 1977

61 Jentsch, L., Über stationäre thermoelastische Schwingungen in inhomogenen Körpern. Math. Nachr. **64**, 171–231, 1974

62 Jentsch, L., Stationäre thermoelastische Schwingungen in stückweise homogenen Körpern infolge zeitlich periodischer Außentemperatur. Math. Nachr. **69**, 15–37, 1975

63 Kleinman, R.E. and Roach, G.F., Boundary integral equations for the three-dimensional Helmholtz equation. SIAM Review **16**, 214–236, 1974

64 Kleinman, R. and Wendland, W.L., On Neumann's method for the exterior Neumann problem for the Helmholtz equation. J. Math. Anal. Appl. **57**, 170–202, 1977

65 Krawietz, A., Energetische Behandlung des Singularitätenverfahrens. Doctoral Dissertation, Technical University, Berlin, D 83, Germany, 1972

66 Kupradze, W.D., *Randwertaufgaben der Schwingungstheorie und Integralgleichungen*. Dt. Verlag d. Wissenschaften, Berlin, 1956

67 Kupradze, V.D., *Potential Methods in the Theory of Elasticity*. Israel Program Scientific Transl., Jerusalem, 1965

68 Kupradze, V.D., Gegelia, T.G., Basheleishvili, M.O., Burchuladze, T.V., *Three-Dimensional Problems of the Mathematical Theory of Elasticity and Thermoelasticity*. North Holland, Amsterdam, 1979

69 Lamp, U., Schleicher, T., Stephan, E., and Wendland, W.L., Galerkin collocation for an improved boundary element method for a plane mixed boundary value problem. Computing **33**, 269–296, 1984

70 Lyness, J.N., Applications of extrapolation techniques to multidimensional quadrature of some integrand functions with a singularity. J. Comp. Physics **20**, 346–364, 1976

71 Lyness, J.N., An error functional expansion for N-dimensional quadrature with an integrand function singular at a point. Math. Comp. **30**, 1–23, 1976

72 MacCamy, R.C. and Stephan, E., A boundary element method for an exterior problem for three-dimensional Maxwell's equations. Appl. Analysis **16**, 141–163, 1983

73 Martensen, E., Berechnung der Druckverteilung an Gitterprofilen in ebener Potentialströmung mit einer Fredholmschen Integralgleichung. Arch. Rational Mech. Anal. **3**, 235–270, 1959

75 Maul, J., Eine einheitliche Methode zur Lösung der ebenen Aufgaben der linearen Elastostatik. Schriftenreihe ZIMM Akad. Wiss. DDR **24**, Berlin, 1976

76 Mehlhorn, G., Ein Beitrag zum Kipp-Problem bei Stahlbeton- und Spannbetonträgern. Doctoral Dissertation, D 17, Technische Hochschule Darmstadt, Germany, 1970

77 Michlin, S.G. and Prössdorf, S., *Singuläre Integraloperatoren*. Akademie-Verlag, Berlin, 1980

78 Müller, C., *Foundations of the Mathematical Theory of Electromagnetic Waves*. Springer, Berlin, Heidelberg, New York, 1969

79 Muskhelishvili, N.I., *Some Basic Problems of the Mathematical Theory of Elasticity*. Noordhoff, Groningen, 1963

80 Mustoe, G.G. and Mathews, I.C., Direct boundary integral methods, point collocation and variational procedures (to appear)

81 Mustoe, G.W., Volait, F., and Zienkiewicz, O.C., A symmetric direct boundary integral equation method for two-dimensional elastostatics. Res. Mechanica **4**, 57–82, 1982

82 Nedelec, J.C., Curved finite element methods for the solution of singular integral equations on surfaces in $\mathbb{R}^3$. Comp. Math. Appl. Mech. Eng. **8**, 61–80, 1976

83 Nedelec, J.C., Approximation par potentiel de double couche du problème de Neumann extérieur. C.R. Acad. Sci. Paris, Ser. A **286**, 616–619, 1977

84 Nedelec, J.C., Approximation des Equations Intégrales en Mécanique et en Physique. Lectures Notes, Centre de Mathématiques Appliquées, Ecole Polytechnique, Palaiseau, France, 1977

85 Nedelec, J.C., Formulations variationelles de quelques équations intégrales faisant intervenir des parties finies. In *Innovative Numerical Analysis for the Engineering Sciences*. R. Shaw (ed.). Univ. Press of Virginia, Charlottesville, 517–524, 1980

86 Nedelec, J.C. and Planchard, J., Une méthode variationelle d'éléments finis pour la résolution numérique d'un problème extérieur das $\mathbb{R}^3$. R.A.I.R.O. **7**, R 3, 105–129, 1973

87 Niwa, Y., Kaboyashi, S., and Kitahara, M., Applications of the boundary integral equation method to eigenvalue problems of elastodynamics. In *Boundary Element*

Methods in Engineering. C.A. Brebbia (ed.). Springer, Berlin, Heidelberg, New York, 297−311, 1982

88 Noble, Ben, Error analysis of collocation methods for solving Fredholm integral equations. In *Topics in Numerical Analysis*, J.H. Miller (ed.). Academic Press, London, 3−359, 1972

89 Petersen, B.E., *Introduction to the Fourier Transform and Pseudo-Differential Operators*, Pitman, London, 1983

90 Piessens, R., Numerical evaluation of Cauchy principal values of integrals. BIT **10**, 476−480, 1970

91 Poggio, A.J. and Miller, E.K., Integral equation solutions of three-dimensional scattering problems. In *Computer Techniques for Electromagnetics*. R. Mittra (ed.). Pergamon, Oxford, 1973

92 Prössdorf, S. and Schmidt, G., A finite element collocation method for singular integral equations. Math. Nachr. **100**, 33−66, 1981

93 Prössdorf, S. and Schmidt, G., A finite element collocation method for systems of singular equations. Preprint P-MATH-26/81, Akademie d. Wissenschaften DDR, Inst. Math., DDR-1080 Berlin, Mohrenstr. 39, 1981

94 Rannacher, R. and Wendland, W.L., The order of pointwise convergence of some boundary element methods (Part II: Operators of positive order) (in preparation)

95 Rannacher, R. and Wendland, W.L., On pointwise error bounds for boundary element methods. Engineering Analysis (to appear)

96 Rieder, G., Iterationsverfahren und Operatorgleichungen in der Elastizitätstheorie. Abh. d. Braunschweigischen Wiss. Ges. **14**, 109−443, 1962

97 Rizzo, F.J., An integral equation approach to boundary value problems of classical elastostatics. Quart. Appl. Math. **25**, 83−95, 1967

98 Le Roux, M.N., Résolution Numérique du Problème du Potential dans le Plan par une Méthode Variationelle d'Elements Finis. Doctoral Dissertation, L'Université de Rennes, Ser. A, No. d'ordre 347, No. Ser. 38, 1974

99 Le Roux, M.N., Equations intégrales pour le problème électrique das le plan. C.R. Acad. Sci. Paris, Ser. A **278**, 1974

100 Ruland, C., Ein Verfahren zur Lösung von $(\Delta + k^2)u = 0$ in Außengebieten mit Ecken. Appl. Analysis **7**, 69−79, 1978

101 Saranen, J. and Wendland, W.L., On the asymptotic convergence of collocation methods with spline functions of even degree. Math., Com., to appear. Preprint Nr. 690, Technical Univ. Darmstadt, Dept. Mathematics, D-6100 Darmstadt, Fed. Rep. Germany, 1982

102 Schenck, H.A., Improved integral formulation for acoustic radiation problems. J. Acoustics Soc. Amer. **44**, 41−58, 1968

103 Schmidt, G., On spline collocation for singular integral equations. Math. Nachr. **111**, 177−196, 1983

104 Schmidt, G., The convergence of Galerkin and collocation methods with splines for pseudodifferential equations on closed curves. Zeitschrift Analysis u.i. Anwendungen **3**, 371−384, 1984

105 Schmidt, G., On spline collocation methods for boundary integral equations in the plane. Math. Meth. Appl. Sci., to appear

106 Stephan, E., Solution procedure for interface problems in acoustics and electromagnetics. In *Theoretical Acoustics and Numerical Treatments*. D. Filippi (ed.). CISM Courses and Lectures No. 277, Springer-Verlag, Wien, New York, 291−348, 1983

107 Stummel, F., Diskrete Konvergenz linearer Operatoren I und II. Math. Zeitschr. **120**, 231−264, 1971

108 Symm, G.T., Numerical mapping of exterior domains. Numer. Math. **10**, 437−445, 1967

109 Szabo, I., *Höhere Technische Mechanik*. Springer, Berlin, Göttingen, Heidelberg, 1956

110 Taylor, M.E., *Pseudodifferential Operators*. Princeton Univ. Press, Princeton, N.J., 1981

111 Treves, F., *Introduction to Pseudodifferential and Fourier Integral Operators I*. Plenum Press, New York, London, 1980

112 Voronin, V.V. and Cecoho, V.A., An interpolation method for solving an integral equation of the first kind with a logarithmic singularity. Soviet. Math. Dokl. **15**, 949−952, 1974

113 Watson, J.O., Advanced implementation of the boundary element method for two- and three-dimensional elastostatics. In *Developments in Boundary Element Methods – 1*. P.K. Banerjee and R. Butterfield (ed.). Appl. Science Publ. Ltd., London, 31–63, 1979

114 Watson, J.O., Hermitian cubic boundary elements for plane problems of fracture mechanics. Res. Mechanica **4**, 23–43, 1982

115 Wendland, W.L., Lösung der ersten und zweiten Randwertaufgaben des Innen- und Außengebietes für die Potentialgleichung im $\mathbb{R}_3$ durch Randbelegung. Doctoral Dissertation, Technical University, Berlin, D 83, Germany, 1965

116 Wendland, W.L., Die Behandlung von Randwertaufgaben im $\mathbb{R}_3$ mit Hilfe von Einfach- und Doppelschichtpotentialen. Num. Math. **11**, 380–404, 1968

117 Wendland, W.L., On Galerkin collocation methods for integral equations of elliptic boundary value problems. In *Numerical Treatment of Integral Equations*. J. Albrecht and L. Collatz (ed.). Intern. Ser. Num. Math. **53**, Birkhäuser, Basel, 244–275, 1980

118 Wendland, W.L., Asymptotic accuracy and convergence. In *Progress in Boundary Element Methods*, Vol. 1. C.A. Brebbia (ed.). Pentech Press, London, Plymouth, 289–313, 1981

119 Wendland, W.L., Boundary element methods and their asymptotic convergence. In *Theoretical Acoustics and Numerical Treatments*. P. Filippi (ed.). CISM Courses and Lectures No. 277, Springer-Verlag, Wien, New York, 135–216, 1983

120 Werner, H. and Schaback, R., *Praktische Mathematik*, Vol. II. Springer-Verlag, Berlin, New York, 1972

121 Wloka, J., *Funktionalanalysis und Anwendungen*. Walter de Gruyter, Berlin, New York, 1971

122 Wolfe, P., An integral operator connected with the Helmholtz equation. *J. Functional Anal.* **36**, 105–113, 1980

123 Zienkiewicz, O.S., *The Finite Element Method*. McGraw-Hill, London, 1977

Added in Proof

124 Hebeker, F.-K., A boundary element method for Stokes' equations in 3-D exterior domains. In *The Mathematics of Finite Elements and Applications* V. J. Whiteman (ed.). Academic Press, London, New York, 1985

For one singular integral the Lyness method required several thousand evaluations of the integrand. For special kernels one finds a method combining analytic integration and triangular coordinates in

125 Hong-Bao Li, Guo-Ming Han, and Mang, H.A., A new method for evaluating singular integrals in stress analysis of solids by the direct BEM. Comp. Meth. Appl. Mech. Eng. (to appear)

For general operators as in Eq. (9.34) appropriate fast cubature formulas have been developed in

126 Schwab, C. and Wendland, W.L., On numerical quadrature in boundary element methods. Numer. Meth. Partial Diff. Equations (to appear)

even for strong singularities and have been applied in

127 Stock, B., Über die Anwendung der Randelementmethode zur Lösung des linearen Molodenskiischen und verallgemeinerten Neumannschen geodätischen Randwertproblems. Doctoral Dissertation, D 17, Technische Hochschule Darmstadt, Germany, 1985.

Subject Index

Topics in Boundary Element Research

Editor: C.A.Brebbia

Volume 1

Basic Principles and Applications

Editor: **C.A.Brebbia**

1984. 144 figures, 11 tables. XIII, 256 pages.
ISBN 3-540-13097-7

Contents: Boundary Integral Formulations. – A Review of the Theory. – Applications in Transient Heat Conduction. – Fracture Mechanics Application in Thermoelastic States. – Applications of Boundary Element Methods to Fluid Mechanics. – Water Waves Analysis. – Interelement Continuity in the Boundary Element Method. – Applications in Geomechanics. – Applications in Mining. – Finite Deflections of Plates. – Trefftz Method. – Subject Index.

This book deals with the B.E.M. solution of a number of non-linear and time-dependent problems which have only recently become amenable to solution using boundary elements.
The first chapter presents a new approach on weighted residual and error approximations which permits easy construction of the governing boundary integral equations. One chapter is dedicated to the interpretation of boundary integral methods using Trefftz' original idea. Another two chapters of the book deal with problems in geomechanics, including the basic formulation for non-tension and joint problems and the full viscoplastic analysis.
The book also includes a chapter describing the use of B.E. for solving problems in aerodynamics and hydro-dynamics, free body problems and water resources, while another is dedicated to the application of B.E. to find water wave forces on fixed free-floating or moored offshore structures.
Finally, the last chapter reviews the theory and in particular its application in potential and elastostatic problems, including the indirect formulation.

Springer-Verlag
Berlin
Heidelberg
New York
Tokyo